APPLIED DESCRIPTIVE GEOMETRY

Kathryn Holliday-Darr

Delmar Publishers

an *International Thomson Publishing* company

Albany • Bonn • Boston • Cincinnati • Detroit • London • Madrid
Melbourne • Mexico City • New York • Pacific Grove • Paris • San Francisco
Singapore • Tokyo • Toronto • Washington

NOTICE TO THE READER

Cover photo by Evan Lauber.

Delmar Staff
Acquisitions Editor: Sandy Clark
Senior Project Editor: Christopher Chien
Production Coordinator: Jennifer Gaines
Art and Design Coordinator: Mary Beth Voight
Editorial Assistant: Christopher Leonard

COPYRIGHT© 1998
By Delmar Publishers
a division of International Thomson Publishing Inc.

The ITP logo is a trademark under license.

Printed in the United States of America

Online Services

Delmar Online
To access a wide variety of Delmar products and services on the World Wide Web, point your browser to:
http://www.delmar.com
or email: info@delmar.com

thomson.com
To access International Thomson Publishing's home site for information on more than 34 publishers and 20,000 products, point your browser to:
http://www.thomson.com
or email: findit@kiosk.thomson.com

A service of I(T)P®

For more information, contact:
Delmar Publishers
3 Columbia Circle, Box 15015
Albany, New York 12212-5015

International Thomsom Publishing Europe
Berkshire House
168-173 High Holborn
London, WC1V 7AA
England

Thomas Nelson Australia
102 Dodds Street
South Melbourne, 3205
Victoria, Australia

Nelson Canada
1120 Birchmount Road
Scarborough, Ontario
Canada, M1K 5G4

International Thomson Editores
Campos Eliseos 385, Piso 7
Col Polanco
11560 Mexico D F Mexico

International Thomson Publishing GmbH
Konigswinterer Strasse 418
53227 Bonn
Germany

International Thomson Publishing Asia
221 Henderson Road
#05-10 Henderson Buiding
Singapore 0315

International Thomson Publishing—Japan
Hirakawacho Kyowa Building, 3F
2-2-1 Hirakawacho

Chiyoda-ku, Tokyo 102
Japan

1 2 3 4 5 6 7 8 9 10 XXX 04 03 02 01 00 99 98

Library of Congress Cataloging-in-Publication Data

Holliday –Darr, Kathryn.
    Applied descriptive geometry. - - 2nd ed. / Kathryn Holliday-Darr.
        p.   cm.
    Rev. ed. of: Applied descriptive geometry / Susan A. Stewart.
c1986
    ISBN 0-8273-7912-9
    1. Geometry, Descriptive.   I. Stewart, Susan A. (Susan Ann),
1947 – Applied descriptive geometry.   II. Title.
QA501.H7 1998
516- - dc21
                                        98-14712
                                    CIP

This text is dedicated to my family and friends, the greatest gift life has to give. To my parents and family, in-laws, and friends, I thank you for support and understanding. Most of all I would like to give a special thanks to my multi-talented husband, John, for without his love, support, technical advice, ideas, and the dinners and snacks that magically appeared at my computer, this project would not have been possible.

CONTENTS

PREFACE

PURPOSE

This book has been developed to teach college-level engineering and technology students the fundamental concepts of descriptive geometry through an emphasis on logical reasoning, visualization, and practical applications. Special emphasis has been made throughout the text on applications in the various engineering disciplines. (Examples have been taken from mechanical, plastics, industrial, piping, aerospace, marine, civil, structural, and architectural applications.) The value of descriptive geometry can be readily seen by students as they apply the tools and techniques to practical problems.

FORMAT

The book is constructed in a text/workbook format. Each chapter includes relevant instructional concepts and examples, practice problems, study questions, and a selection of applied problems. Helpful Hints sections have been added where beneficial. The information included in the Helpful Hints sections is designed to repeat, in detail, the concept previously covered and to help the student to organize where to start and how to proceed in solving the problem. The practice problems are designed to be solved directly on the page, with the solution provided in the figure following. Cut-out paper models have been provided throughout the text to aid in visualizing a concept. The Study Questions may be answered directly on the workbook pages, removed from the workbook, and submitted for evaluation. The Chapter Problems at the end of the chapters may be solved directly on the worksheets, two-dimensionally or three-dimensionally on a CAD system. The text is based on following a specific set of rules necessary to solve descriptive geometry problems. These Steps of Procedure are listed in Appendix B.

LENGTH OF COURSE

The course content is designed to be used in an eleven- to fifteen-week time frame, with variability resulting from the quantity of problem assignments, the depth of classroom discussion, and instructor's prerogative. Generally speaking, each chapter can be covered in a one-hour lecture.

COURSE PLAN

Most courses in descriptive geometry are based on the principles of orthographic projection. This text is based on the four fundamental views: (1) true length of a line, (2) point view of a line, (3) edge view of a plane, and (4) true shape of a plane. It progresses in small steps from the simple to the more complex. There are Helpful Hint sections that give extra attention to the areas where students traditionally need a little more detail in understanding a concept. To help the students test their understanding of the material, practice problems with answers, are included. Special emphasis has been made throughout the text on applications in the various engineering disciplines. Examples have been taken from mechanical, plastics, industrial, piping, aerospace, marine, civil, structural, and architectural applications. The value of descriptive geometry can be readily seen by students as they apply the tools and techniques to practical problems.

The review of orthographic projection, scales, terminology, and the visual orientation for descriptive geometry are presented in Chapters 1 and 2. The concepts of auxiliary views, both primary and secondary, are discussed in Chapter 3. This chapter demonstrates that once the line of sight has been determined, the rest of the steps necessary to solve the problem are always the same.

Chapter 4 introduces the concept of the first two fundamental views (or first two lines of sight). Line characteristics using the first two fundamental views are explained in Chapter 5. Chapter 6 introduces concepts of the third and fourth fundamental views (or last two lines of sight). Plane relationships using all four fundamental views are explained in Chapters 7 and 8. The concept of revolution as an alternative method of problem solution is presented in Chapter 9, and applied specifically to sheet metal developments in Chapter 10. Chapter 11, Mining and Civil Engineering Applications, provides the student with an additional and very different application for the tools of descriptive geometry. The instructor has some flexibility in choosing the order of the topics.

APPLICATIONS

Special emphasis has been made throughout the text on applications in the various engineering disciplines. Examples have been taken from mechanical, industrial, piping, aerospace, marine, civil, structural, and architectural applications. The value of descriptive geometry can be readily seen by students as they apply the tools and techniques to practical problems.

INSTRUCTOR'S MANUAL

The Instructor's Manual contains answers to all chapter problems and study questions. Also provided is a section on solving descriptive geometry problems on a 3D CAD system.

ACKNOWLEDGMENTS

I am deeply grateful to the professors, my husband, and students, especially my graphic's assistants Trevor Wilhelm, Daniel McCullough, and James Wolfe, who reviewed the manuscript and provided valuable suggestions for improvement. I would also like to thank the students who took the time and effort to provide me with a list of strengths and weaknesses for each chapter. Their enthusiasm and support have been invaluable. I would also like to thank the following companies and individuals for providing chapter opening photographs: Chapter 1—Steele Enterprises; Chapter 2—Brian Feeney; Chapter 3—David Roth; Chapter 4—Lowry Photography; Chapter 5—Lowry Photography; Chapter 6—ERIEZ Magnetics; Chapter 7—James Sharpe; Chapter 8—Cannondale; Chapter 9—Fanuc Robotics; Chapter 10—ERIEZ Magnetics; and Chapter 11—Bethlehem Steel Corporation.

A special thank you is extended to all the students at Pennsylvania State Erie, The Behrend College, who class tested this text. Their many suggestions helped refine and improve the text. In addition, I, along with Delmar Publishers, would like to thank those professional individuals who reviewed the manuscript and offered invaluable suggestions and feedback. Their assistance is greatly appreciated:

Bruce Bainbridge, Clinton Community College; Robert Chin, East Carolina University; David Freeland, Los Angeles Trade and Technical College; Larry Lamont, Moraine Park Technical College; Ed Loft, West Shore Community College; Ron Meloche, Lansing Community College; Nicholas G. Pocock, Texas State Technical College; Mark Schwendau, Kishwaukee College; Ken Williams, Navarro College.

A special acknowledgment for a thorough technical edit is extended to Morgan Stout, Austin Community College.

CHAPTER 1

Orthographic Projection and Descriptive Geometry

In order to manufacture three-dimensional objects, detailed information such as dimensions must be known.

This chapter is intended as a review of the engineer's scale, architect's scale, and the terminology used throughout the text. After completing this chapter, you will be able to:

▸ *Read an engineer's scale.*

▸ *Read an architect's scale.*

▸ *Describe the relationship between orthographic projection and descriptive geometry.*

▸ *Explain the fundamental orientation required in orthographic projection.*

▸ *Explain how the projection planes are unfolded.*

▸ *Define projection plane, projection line, reference line, line of sight, adjacent view, related view, elevation view, and origin.*

We live in a world of three-dimensional objects. The problem encountered in drawing three-dimensional objects is in accurately describing the three-dimensional object in two-dimensional views, complete with dimensions. A photograph of a pictorial drawing shows well enough how an object appears, but generally it is not adequate for transmitting the details of the idea from the designer or inventor to the manufacturer. Before an object can be made, an exact description must be conveyed to the manufacturer. This is the purpose of orthographic projection and descriptive geometry.

Descriptive geometry is the graphical solution to three-dimensional spatial problems. Spatial problems were originally solved by mathematics. Gaspard Monge

(1746–1818) spent many long hours proving to his military school headmaster that his graphical methods produced the same results in less time. Because of this, Monge's methods were kept a military secret for 15 years. Today Monge's methods teach the analysis of problems through visualization and reasoning.

Today orthographic projection and descriptive geometry are among the most valuable subjects in technical and engineering education. The fundamentals of descriptive geometry are based on the principles of orthographic projection. These types of drawings are practically indispensable to the engineer. They are the language of the engineering profession.

Some drawings use orthographic projection to transmit an exact description of an object to the manufacturer, while other drawings use descriptive geometry and orthographic projection to make a type of layout drawing that aids in the solution of a spatial problem. So, orthographic projection and descriptive geometry are not really different subjects, but rather they are different areas within the same subject that make use of the same tools and thought processes.

Descriptive geometry differs from orthographic projection in that it is not necessarily concerned with a finished drawing that communicates between the designer and the manufacturer. Instead it is concerned with a study of spatial relationships that occur during the design process. A spatial relationship could be the relationship of one part of a structure to another, of one part of a mechanism to another part of the mechanism, of one surface of a part to another surface on the same part, or of a highway surface to its adjacent surroundings. The examples are endless.

Specific examples of the areas listed above will be studied in the following chapters. Using the tools of descriptive geometry, the study of these relationships is a process of drawing the objects or situations in a series of successive auxiliary views until the required solution is found. The graphical solutions must be accurate. Therefore, line work, measuring, lettering, and notation must be of the *highest quality*.

▼ SCALE

Since accuracy is very important, take time to review the architect's and engineer's scales.

Architect's Scale

The architect's scale is used to make structural drawings. On a triangular-shaped architect's scale, one side is full scale (Figure 1–1). The rest of the scales are divided into lines that represent feet and inches. The architect's scale is represented by $X = 1'0''$ where X is the fractional part of an inch used on the drawing to represent each foot on the actual structure.

Some scales have more than one scale on a side. The scale at one end of the scale is either one-half or double the scale on the other end. For demonstration purposes, measure line A-B with the scales ¾" = 1'0" and ⅜" = 1'0" (Figure 1–2). Steps to reading the rest of the scales are listed below.

Figure 1–1 Full scale on the architect's scale (16 divisions per inch).

1. Identify the correct scale to be used (Figure 1–2).

   Example scales: ¾" = 1'0", and ⅜" = 1'0"

Figure 1–2 Appropriate architect's scale for examples listed in step 1.

2. Read the maximum number of feet first (Figure 1–3).

   2' on the ¾" = 1'0" scale, and 5' on the ⅜" = 1'0" scale

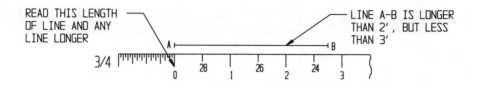

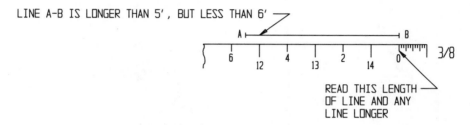

Figure 1–3 Read the maximum number of feet on the architect's scale first.

3. Read the inches next (Figure 1–4).

   9" on the ¾" = 1'0" scale, and 6" on the ⅜" = 1'0" scale

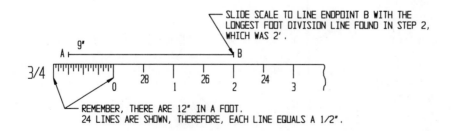

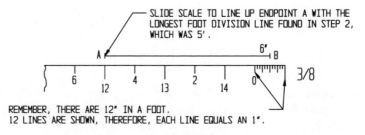

Figure 1–4 Read the inches on the architect's scale next.

Example

Figure 1–5 is an example using 1" = 1'0" scale.

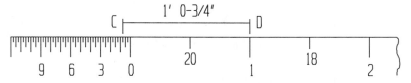

Figure 1–5 Architect's scale example

EXERCISE

Problem: Find the length of line E-F using a 1½" = 1'0"scale (Figure 1–6).

Figure 1–6 Architect's scale problem.

Answer

The answer is 1'7". Remember to look at the zero (O) line and count any foot division line that same length and longer.

Problem: Find the length of line G-H using the ⅛" = 1'0" scale (Figure 1–7).

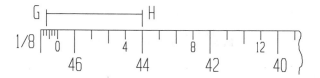

Figure 1–7 Architect's scale problem.

Answer

Remember to look at the zero (O) line and count any foot division line that same length and longer. In this example, it is 5 feet. Each line to the left of the zero (O) represents 2 inches because there are 12 inches in a foot and only six division lines showing. Therefore, the answer is 5'8".

Engineer's Scale

The engineer's scale, also known as the civil engineer's scale, can be used for outdoor projects, such as roadways and site plans. Each division on the scale is conveniently divided into 10 equal units, resulting in the measurement being in decimal form. The scale is represented by 1" = x or x = 1", where x is the number of units represented by each inch of the drawings. These may be written as 1:x or x:1. It is important to understand that the mathematics has been done and a real inch lines up with the number corresponding to the scale.

Refer to Figure 1–8. Notice on the scale marked 10 that the 1 lines up with a real inch. The 1 may represent .1, 1, 10, 100, 1000, and so on. The units may be inches, feet, miles, or other measurements.

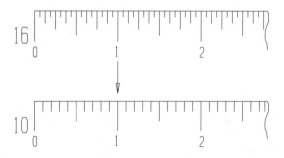

Figure 1–8   Engineer's scale. Notice the 1 on the 10 scale lines up with a real inch.

Notice on the scale marked 20 (Figure 1–9) that the 2 lines up with a real inch. The 2 may represent .2, 2, 20, 200, 2000, and so on.

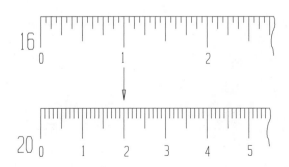

Figure 1–9   Engineer's scale. Notice the 2 on the 20 scale lines up with a real inch.

Notice on the scale marked 30 (Figure 1–10) that the 3 lines up with a real inch. The 3 may represent .3, 3, 30, 300, 3000, and so on.

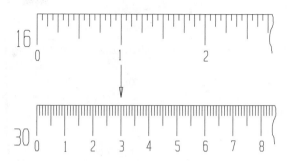

Figure 1–10   Engineer's scale. Notice the 3 on the 30 scale lines up with a real inch.

Notice on the scale marked 40 (Figure 1–11) that the 4 lines up with a real inch. The 4 may represent .4, 4, 40, 400, 4000, and so on.

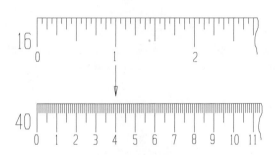

Figure 1–11   Engineer's scale. Notice the 4 on the 40 scale lines up with a real inch.

. . . and so on for the 50 and 60 scale.

Steps to reading the engineer's scale are listed below. Line I-J is used for demonstration purposes. Several illustrations are shown in each set.

Set 1

1. Identify the correct scale to be used.

   The 10 scale will be used for the following problems: 1" = 1", 1" = 1', 1" = 10", 1" = 10 miles, 1" = 100", 1" = 100', etc.

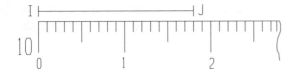

Figure 1–12 Engineer's scale.

2. Decide what the 1 division line on the 10 scale represents and calculate the number of units that line I-J measures. (The units may represent inches, feet, miles, etc. When initially reading the scale, it does not matter what the unit represents.)

Scale 1" = 1 (Figure 1-13)

For the 1" = 1 scale, the 1 represents 1 unit and each division line represents .1 unit; Therefore, line I-J is 1.8 units.

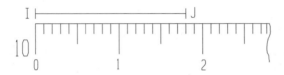

Figure 1–13 Engineer's scale. If 1" = 1 unit, then line I-J is 1.8 units long.

Scale 1" = 10 (Figure 1-14)

For 1" = 10, the 1 represents 10 units and each division line represents 1 unit; therefore, line I-J is 18 units.

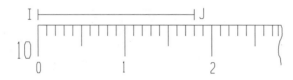

Figure 1–14 Engineer's scale. If 1" = 10 units, then line I-J is 18 units long.

Scale 1" = 100 (Figure 1-15)

For 1" = 100, the 1 represents 100 units and each division line represents 10 units; therefore, line I-J is 180 units.

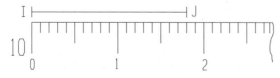

Figure 1–15 Engineer's scale. If 1" = 100 units, then line I-J is 180 units long.

Scale 1" = 1000 (Figure 1–16)

For 1" = 1000, the 1 represents 1000 units and each division line represents 100 units; therefore, line I-J is 1800 units.

Scale 1" = .1 (Figure 1–17)

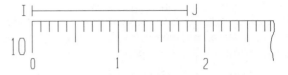

Figure 1–16 Engineer's scale. If 1" = 1000 units, then line I-J is 1800 units long.

For 1" = .1, the 1 represents .1 units and each division line represents .01 units; therefore, line I-J is 0.18 units.

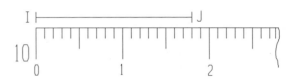

Figure 1–17 Engineer's scale. If 1" = .1 units, then line I-J is 0.18 units long.

Review

Scale	Answer
1" = 1	1.8 Note answers for line I-J. When read at 1" = 1 (Figure 1–13) line I-J =1.8. At 1" = 2, line I-J= 3.6. (The mathematics have already been done—do not double the answer.)
1" = 10	18
1" = 100	180
1" = 1000	1800
1" = .1	.18

3. Add the units.

For 1" = 10" line I-J reads 18".

For 1" = 10' line I-J reads 18'.

For 1" = 10 miles line I-J reads 18 miles.

Set 2

1. Identify the correct scale to be used.

   The 20 scale, will be used for the following problems: 1" = 2", 1" = 2', 1" = 20", 1" = 20 miles, 1" = 200", 1" = 200', etc.

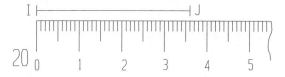

Figure 1–18 Engineer's scale.

2. Decide what the 2 division line on the 20 scale represents. Remember that the 2 lines up with a real inch; therefore, the scaling has been done. That means you do not have to double the answer.

Scale	Answer
1" = 2	3.6 Note answers for line I-J. When read at 1" = 1 (Figure 1–18), line I-J =1.8. At 1" = 2, line I-J= 3.6. (The mathematics have already been done—do not double the answer.)
1" = 20	36
1" = 200	360
1" = 2000	3600
1" = .2	.36

3. Add the units.

   For 1" = 20", line I-J reads 36".

   For 1" = 20', line I-J reads 36'.

   For 1" = 20 miles, line I-J reads 36 miles.

Set 3

1. Identify the correct scale to be used.

   The 30 scale will be used for the following problems: 1" = 3", 1" = 3', 1" = 30", 1" = 30 miles, 1" = 300", 1" = 300', etc.

2. Decide what the 3 division line on the 30 scale represents. Remember the 3 lines up with a real inch, therefore; all mathematics have been done.

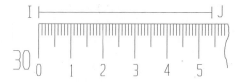

Figure 1–19 Engineer's scale.

Scale	Answer
1" = 3	5.4
1" = 30	54 Note answers for line I-J. When read at 1" = 10 (Figure 1–19) line I-J reads 18. Line I-J at 1" = 30, reads 54—the mathematics have already been done.
1" = 300	540
1" = 3000	5400
1" = .3	.54

3. Add the units.

   For 1" = 30", line I-J reads 54".

   For 1" = 30', line I-J reads 54'.

   For 1" = 30 miles, line I-J reads 54 miles.

Example

Figure 1–20 is an example using 1" = 2' scale.

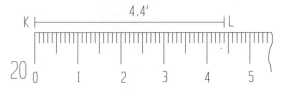

Figure 1–20 Engineer's scale example problem.

EXERCISE

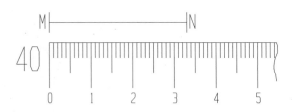

Figure 1–21 Engineer's scale problem.

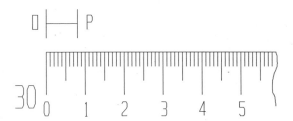

Figure 1–22 Engineer's scale problem.

Problem: Find the length of line M-N using 1" = 400' scale (Figure 1–21).

Answer

The answer is 330'.

Problem: Find the length of line O-P using 1" = 3" scale (Figure 1–22).

Answer

The answer is .8".

▼ DEFINITIONS

In an orthographic drawing, it is common practice to draw different views of an object. To do this, you, as the draftperson must first imagine the object is placed in a definite position—usually a normal or natural position. If you wish to see a different view, simply imagine that you have moved around the object to view it from another position in space. In this method, you the observer change position, not the object. In order to more fully understand and use this system of projection, several definitions must be reviewed or learned. Study the following illustrations and definitions carefully.

Isometric A pictorial view that shows three sides of an object, usually the top, front and right side, so that they are equally foreshortened (Figure 1–23).

Surfaces The exterior of an object (Figure 1–24).

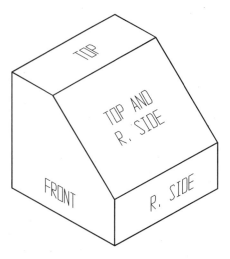

Figure 1–23 Isometric.

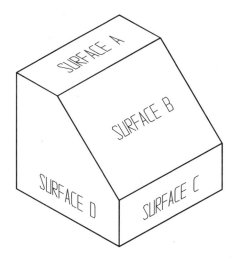

Figure 1–24 Surfaces.

Line of sight (LOS) An imaginary straight line from the eye of the observer to a point on the object or the direction the observer is looking. An arrow is used to show the direction of the line of sight. The line of sight is always perpendicular to a projection plane. All lines of sight for a particular view are assumed to be parallel (Figures 1–25 and 1–26).

Projection plane A flat surface that the view of the object is projected onto, such as paper, blackboard,

or the face of a plastic box. Your (the observer's) lines of sight are always perpendicular to the projection plane (Figure 1–26).

Projection lines Lines that are parallel to the lines of sight and perpendicular to the projection plane. They transfer the two-dimensional shape from the object to the projection plane, but are rarely shown on a finished drawing (Figure 1–26).

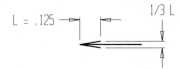

Figure 1–25 Line of sight.

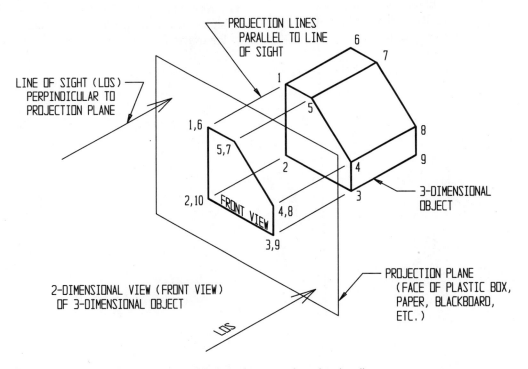

Figure 1–26 Line of sight, projection plane, and projection lines.

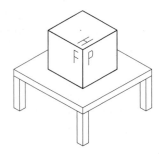

Figure 1-27 Plastic box.

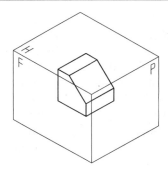

Figure 1-28 Object inside plastic box.

Plastic box A six-sided transparent box (Figure 1–27). The sides represent six possible projection planes which are at 90 degrees to each other. These six projection planes are known as the principal planes and the lines of sight must always be perpendicular to them. The object is set inside of the plastic box with the majority of the surfaces parallel to the box sides (Figure 1–28).

Horizontal planes (H) Level planes. The lines of sight for horizontal planes are vertical (perpendicular to the level planes). This rule holds true for three-dimensional as well as two-dimensional drawings (Figure 1–29).

Frontal planes (F) Vertical planes. The lines of sight for frontal planes are horizontal (perpendicular to the frontal planes) (Figure 1–30).

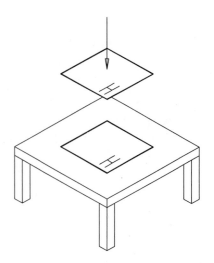

Figure 1-29 Horizontal planes.

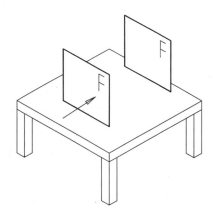

Figure 1-30 Frontal planes.

Profile planes (P)   Vertical planes perpendicular to frontal planes. The lines of sight for profile planes are also in a horizontal (perpendicular to the profile planes) position (Figure 1–31).

Dimensions   True **Height, Width,** and **Depth** are perpendicular to the planes (Figure 1–32). Height is used to indicate the dimension perpendicular to

the horizontal planes. Width is used to indicate the dimension perpendicular to the profile planes. Depth is used to indicate the dimension perpendicular to the frontal planes.

Orthographic projection   Two or more views projected at right angles to each other. It is a method of drawing that uses parallel lines of sight at right

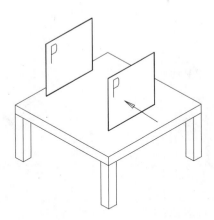

Figure 1–31   Profile planes.

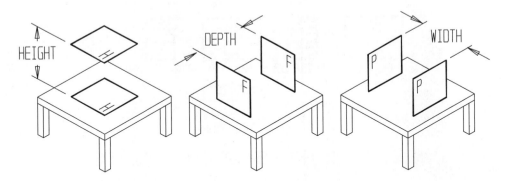

Figure 1–32   Dimensions.

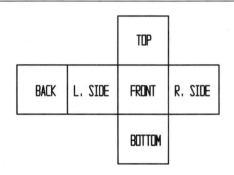

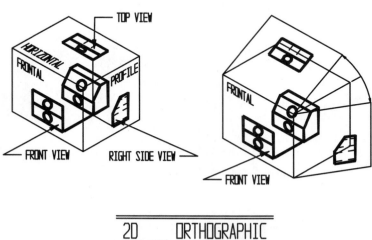

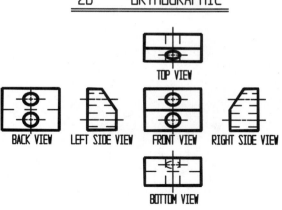

Figure 1–33 Orthographic projection—opening the box to show views in their proper positions.

angles (90°) to a projection plane. Visualize the plastic box opened up into what is called third-angle projection. All views *must be aligned* in their projectable position (Figure 1–33). The primary purpose of an orthographic drawing, or multiview drawing, is to present views of an object on which actual measurements can be shown. Some-

times additional projection planes, called auxiliary planes, are required to show all of the necessary details of the object. Auxiliary planes will be discussed in Chapter 3.

Reference line (fold line or hinged line) Edges of the plastic box or the intersection of the perpendicular planes. The reference line is only drawn when needed to aid in constructing additional views. It may be placed anywhere, but once it has been referred to, do not move it. All projection lines will be perpendicular to its corresponding reference line. The reference line is represented by a phantom line. Draw the short dashes ⅛", gaps 1⁄16", and long dashes ¾" to 1-½" (Figure 1–34).

Figure 1–34   Reference line.

The reference line must be labeled to show its association between the planes it is representing. The lettering should be ⅛" high and 1⁄16" away from the reference line. Figure 1–35 represents the fold line between the horizontal and frontal planes.

H
————————————————   —— —— ————————
F

Figure 1–35   Reference line labeled.

Origin   A starting point that may be located anywhere on the object or near it. Many computer-aided-drafting (CAD) packages use the lower left corner as their origin. Coordinates of the origin are 0,0,0 (0 width, 0 height, 0 depth).

Adjacent views   Views aligned next to each other to share a common dimension (Figure 1–36). The front and side views share the height projection lines; therefore, they are adjacent views.

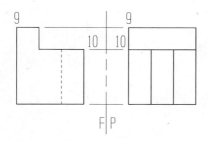

Figure 1–36   Adjacent views.

Related views   Views adjacent to the same view, sharing a common dimension which must be transferred (Figure 1–37). The dimension may be transferred using dividers or a scale. Students familiar with the mitered method will only be able to use it when transferring dimensions between related principal views.

The top and profile views share the same depth, but the information must be transferred; therefore, the views are called related.

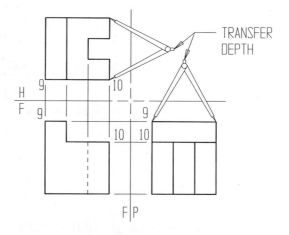

Figure 1–37   Related views.

Elevation view   A view in which the lines of sight are level. Elevation views, such as front, right side, and top-adjacent views, all have one feature in common—they show the perpendicular distance between horizontal planes referred to as elevation or true height.

Points   Indicate the coordinates of a location in space. Points are drawn with ⅛"-long intersecting perpendicular lines (Figure 1–38).

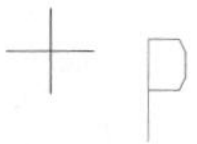

Figure 1–38   Representation of a point.

Do *not* show individual points at the end of each line (Figure 1–39).

Figure 1–39   Endpoints of a line.

When drawing the edge view of a plane, draw a perpendicular line locating the point only if it falls in between the endpoints (Figure 1–40). The perpendicular line should be drawn ⅛" long. Notice the placement of the line of sight to view the front view.

Review

Figure 1–41 shows a pictorial of orthographic projection.

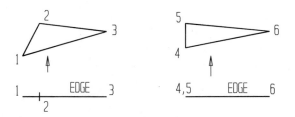

Figure 1–40 Endpoints of a plane.

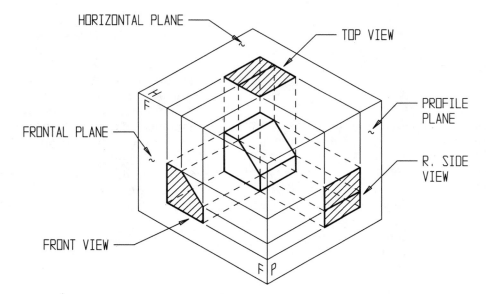

Figure 1–41 Orthographic projection using the plastic box method.

Figure 1–42 shows a pictorial of the orthographic box unfolded.

Projection planes are unfolded to show a three-dimensional object two-dimensionally in a single plane, such as a sheet of paper. This representation is called a multiview drawing (Figure 1–43).

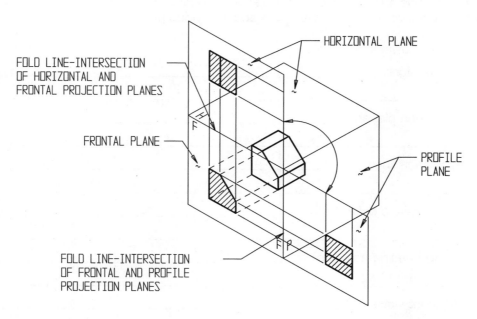

Figure 1–42 Plastic box unfolded.

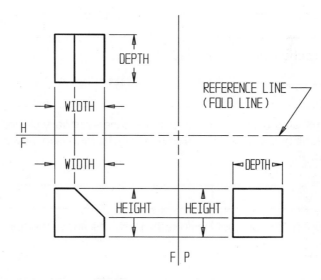

Figure 1–43 Multiview drawing.

CHAPTER 1 STUDY QUESTIONS

The Study Questions are intended to assess your comprehension of chapter material. Please write your answers to the questions in the space provided.

1. Define the following:

   a. Line of sight: _____

   _____

   b. Projection planes: _____

   _____

   c. Projection lines: _____

   _____

   d. Orthographic projection: _____

   _____

   e. Reference line (fold line): _____

   _____

   f. Adjacent views: _____

   _____

   g. Related views: _____

   _____

2. What is the primary purpose of a multiview drawing?

   _____

   _____

3. Why are the projection planes unfolded?

   _____

   _____

4. In a standard multiview drawing, including the three primary views, (a) which two views will show the height of the object viewed?, and (b) which two views will show the overall width of the object?

   a. _____

   b. _____

5. What relationship do the projection lines have to the lines of sight and reference lines?

   _____

   _____

6. Using the appropriate scale, measure line 1-2.

1 |————————————| 2

1" = 400' _____

¾" = 1'0" _____

CHAPTER 1 PROBLEMS

The problems contained in this chapter are meant to provide a review in orthographic projection and isometric drawing. They are designed to test your understanding of the basic fundamentals and of the theoretical methods used, to teach you to apply the principles while finding solutions to a large variety of engineering problems, and to practice your visualization skills. Many of the problems are derived from the broad spectrum of engineering settings.

The following problems may be sketched or drawn, using instruments, on the page provided, created two-dimensionally on CAD, or a combination of drawing board and CAD.

Chapter 1, Problem 1: GIVEN A PICTORIAL DRAWING OF A SLIDE BRACKET, DRAW THE TOP, FRONT, AND RIGHT SIDE VIEWS IN THEIR CONVENTIONAL RELATIONSHIP, AS INDICATED BY THE REFERENCE LINES.

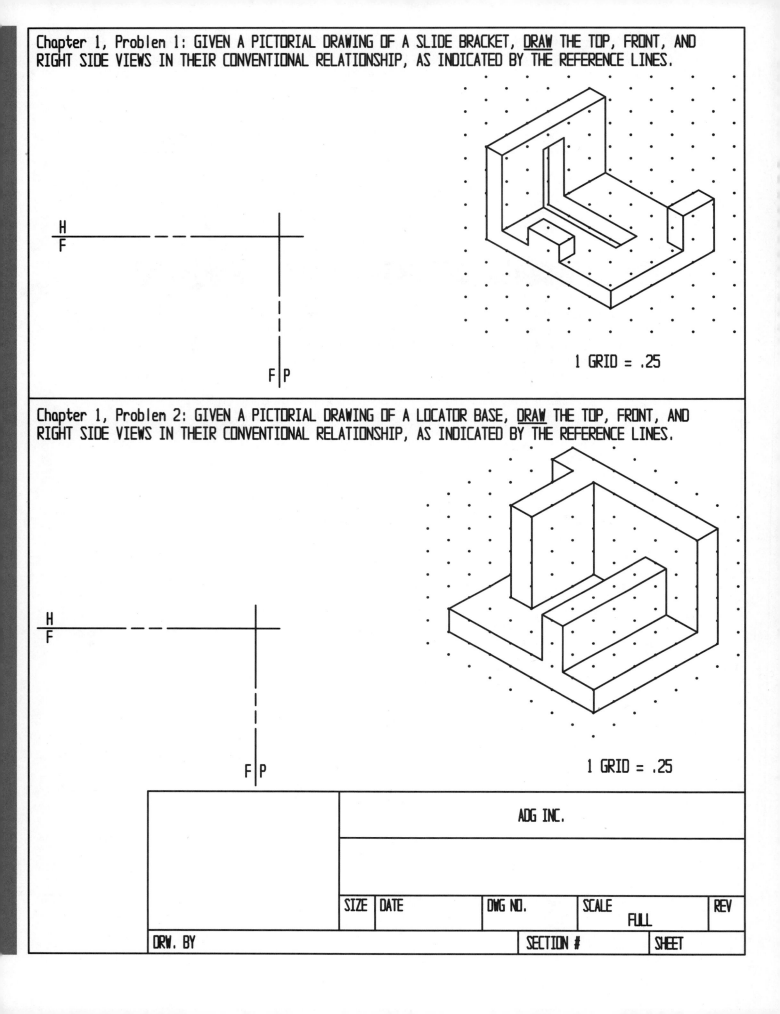

1 GRID = .25

Chapter 1, Problem 2: GIVEN A PICTORIAL DRAWING OF A LOCATOR BASE, DRAW THE TOP, FRONT, AND RIGHT SIDE VIEWS IN THEIR CONVENTIONAL RELATIONSHIP, AS INDICATED BY THE REFERENCE LINES.

1 GRID = .25

H
F

F P

ADG INC.

SIZE	DATE		DWG NO.		SCALE		REV
					FULL		

| DRW. BY | | | SECTION # | | SHEET | |

Chapter 1, Problem 3: GIVEN THE PICTORIAL DRAWING OF A STOP BLOCK, DRAW THE TOP, FRONT, LEFT SIDE, AND RIGHT SIDE VIEWS IN THEIR CONVENTIONAL RELATIONSHIP. USE REFERENCE LINES AND LABEL THEM APPROPRIATELY.

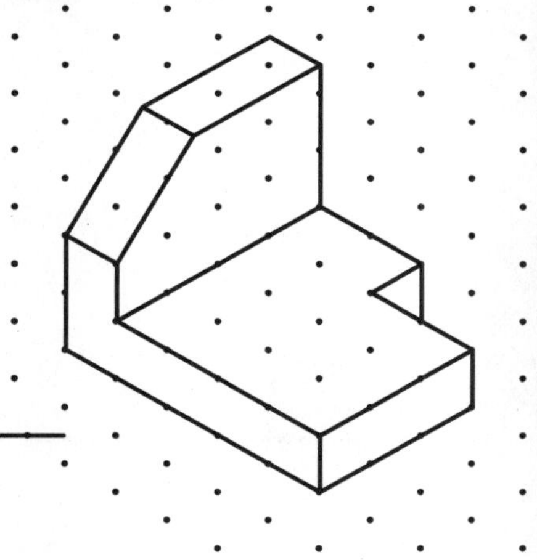

1 GRID = .25"

Chapter 1, Problem 4: GIVEN THE PICTORIAL DRAWING OF A WEDGE BLOCK, DRAW THE TOP, FRONT, AND RIGHT SIDE VIEWS. USE REFERENCE LINES AND LABEL THEM APPROPRIATELY.

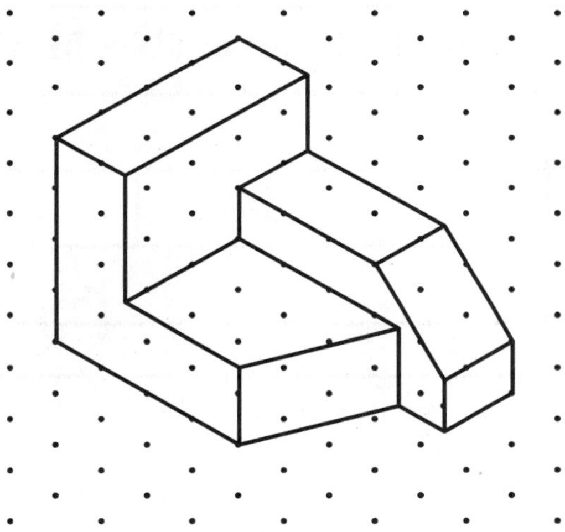

1 GRID = .25"

	ADG INC.			
SIZE	DATE	DWG NO.	SCALE FULL	REV
DRW. BY		SECTION #	SHEET	

Chapter 1, Problem 5: GIVEN A PICTORIAL DRAWING OF A CLIP, DRAW THE TOP, FRONT, AND RIGHT SIDE VIEWS. LABEL REFERENCE LINES CORRECTLY.

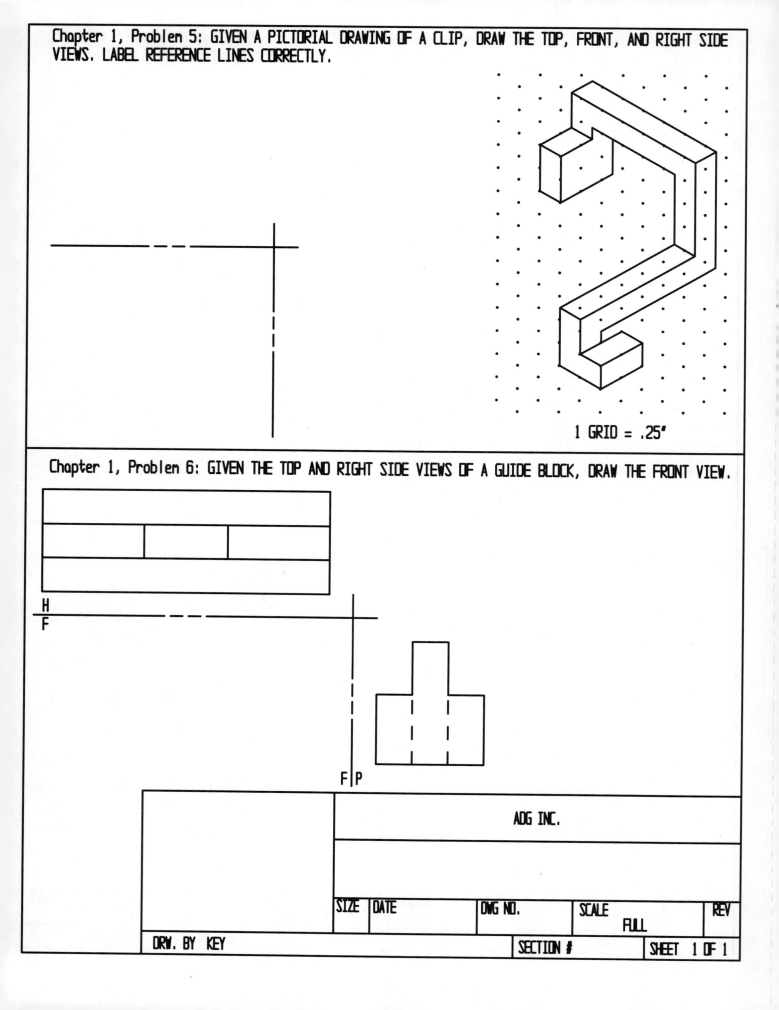

1 GRID = .25"

Chapter 1, Problem 6: GIVEN THE TOP AND RIGHT SIDE VIEWS OF A GUIDE BLOCK, DRAW THE FRONT VIEW.

H
F

F P

	ADG INC.			
SIZE	DATE	DWG NO.	SCALE FULL	REV
DRW. BY KEY		SECTION #		SHEET 1 OF 1

Chapter 1, Problem 7: GIVEN THE TOP AND THE FRONT VIEWS OF THE INDEX FINGER, DRAW THE ISOMETRIC DRAWING.

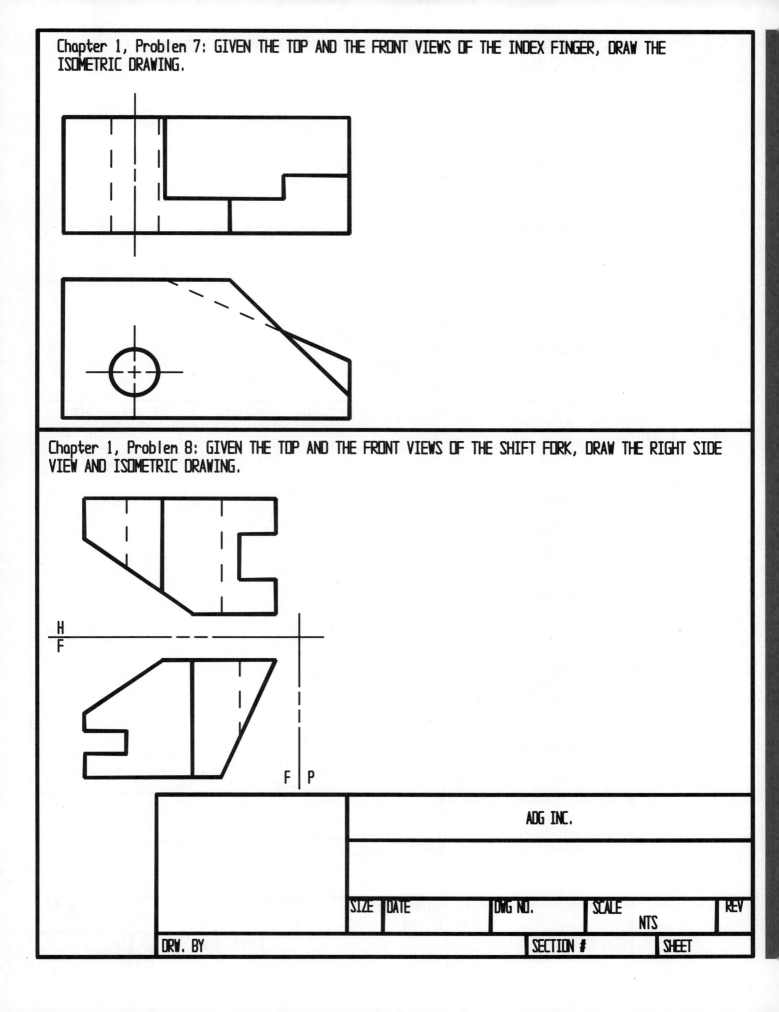

Chapter 1, Problem 8: GIVEN THE TOP AND THE FRONT VIEWS OF THE SHIFT FORK, DRAW THE RIGHT SIDE VIEW AND ISOMETRIC DRAWING.

H
F

F | P

ADG INC.

SIZE	DATE		DWG NO.	SCALE		REV
				NTS		
DRW. BY				SECTION #	SHEET	

Chapter 1, Problem 9: GIVEN THE RIGHT SIDE AND THE FRONT VIEWS OF THE LOCATOR BRACKET, DRAW THE TOP VIEW AND ISOMETRIC DRAWING.

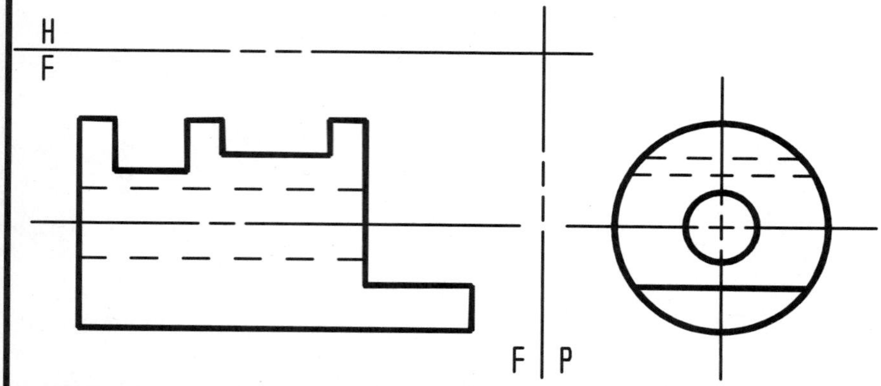

H
F

F | P

Chapter 1, Problem 10: GIVEN THE TOP AND THE FRONT VIEWS OF THE STOP BLOCK, ADD THE RIGHT SIDE VIEW AND ISOMETRIC.

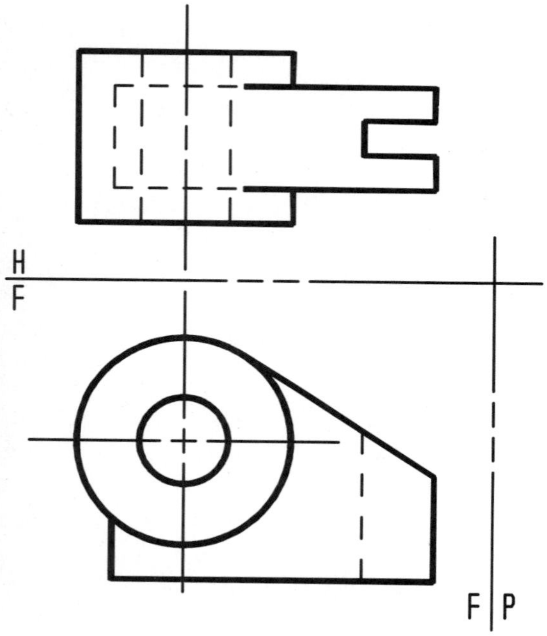

H
F

F | P

	ADG INC.				
SIZE	DATE	DWG NO.	SCALE NTS		REV
DRW. BY			SECTION #	SHEET	

Chapter 1, Problem 11: MEASURE THE FOLLOWING LINES USING THE APPROPRIATE SCALE. MEASURE FROM THE OUTSIDE OF THE LINE TO THE OUTSIDE OF THE LINE. LETTER THE ANSWER ON TOP OF THE LINE.

SCALE
3/16" = 1'0"   EXAMPLE: 7'9" |——————————————|

1/2" = 1'0" |————————————————————————|

3/4" = 1'0" |——————————————————————|

1/8" = 1'0" |————————————————————|

1-1/2" = 1'0" |——————————|

1" = 1'0" |——|

1" = 30' |——————————|

1" = 1' |——|

1" = 200' |————————————————————————————————|

1" = 40' |——————————————————————|

1" = 5' |————————————————————|

LETTER WHICH SCALE WAS USED TO MEASURE THE LINE.

EXAMPLE: 1" = 5' | 4.4' |——————|

                 | 3'9" |——————|

                 | 460.0' |——————————|

                 | 10'0" |————————————————————————|

                 | 4'0-1/4" |——————————————————————————|

                 | 2.8' |——————————————————|

                 | 1.5' |——————|

	ADG INC.			
	SCALES			
	SIZE	DATE	DWG NO.	SCALE FULL	REV
DRW. BY			SECTION #		SHEET

CHAPTER 2

Points and Lines in Space

Whether the object being studied is the Statue of Liberty, or the scaffolding surrounding it, it is still made up of points, lines, and planes.

After completing this chapter, you will be able to:

▸ *Locate a point or a line in a view, given locational information in two other views.*

▸ *Explain foreshortening.*

▸ *Describe how you will know if a line is in true length.*

▸ *Name three principal line types, and explain the differences between them.*

▸ *Describe the precise location of a line in space, in relationship to the principal projection planes.*

You are now well versed in the methods of representing three-dimensional solids on a single sheet of paper. All objects are made up of points, lines, and planes. Two points can define a line. Lines can define surfaces. Finally, surfaces combine to form an object. In orthographic drawings the entire object must be represented in all views. However, in descriptive geometry when solving for a particular measurement, the entire three-dimensional object does not have to be represented. Only the points, lines, or planes needed to solve the problem may be shown.

▼ LOCATING A POINT IN SPACE

Every point is located with three dimensions—height, width, and depth (Chapter 1). With the increased usage and capabilities of computer-aided-design and computer-aided-manufacturing (CAD/CAM), it is necessary to understand the three-axis coordinate system used in manufacturing.

When a three-dimensional (3D) numerical database is entered into a 3D CAD package, the computer becomes a vital link between design and production by transferring the 3D database to a receiving unit software, such as a computer numerical controlled (CNC) machine.

The three-axis coordinate system X, Y, Z—where **X equals width, Y equals depth,** and **Z equals height**—will be used in this text. (See Figure 2–27 on page 35.) Y equals depth and Z equals height coordinates may be the reverse of what you learned in another course, so be careful.

In Chapter 1, you learned that an origin or starting point, may be located anywhere on the object or near it. The figures in this section will place the origin where the three planes meet; therefore, X (width) will be left of the profile plane, Y (depth) will be behind the frontal plane, and Z (height) will be below the horizontal plane (Figures 2–1, 2–2 and 2–3).

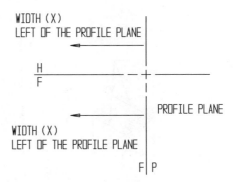

Figure 2–1 Width (X)—left of the profile plane.

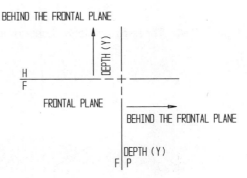

Figure 2–2 Depth (Y)—behind the frontal plane.

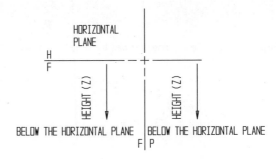

Figure 2–3 Height (Z)—below the horizontal plane.

To locate a point three-dimensionally, each measurement must be made parallel to an isometric or orthographic axis line (reference line, fold line, etc.). See Figure 2–4.

Figures 2–5, 2–6, and 2–7 illustrate the two-dimensional location of a point in each principal plane, and the line of sight utilized for each view. Solving descriptive geometry problems requires a methodical approach and good visualization skills. This text is based on a reliable set of instructions that will help you systemically solve descriptive geometry problems. For this reason, the line of sight used to observe each principal plane is very important and must be completely under-

stood. If you're still unsure how the line of sight works, sketch the three views as shown in the following figures on a piece of scrap paper, fold the sketch along the reference lines 90 degrees to form a box, and study the placement of the lines of sight.

Example

Refer to Figure 2–5. The two-dimensional location of point 1 in the front view shows only height and width; therefore, only the measurements X and Y will be used. Remember to study the line of sight for the front view.

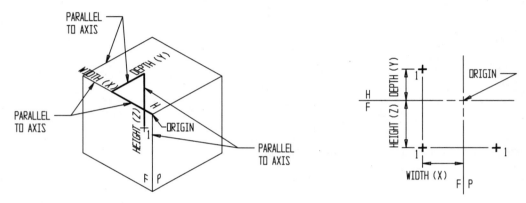

Figure 2–4 Three-dimensional location of point 1 at X, Y, Z.

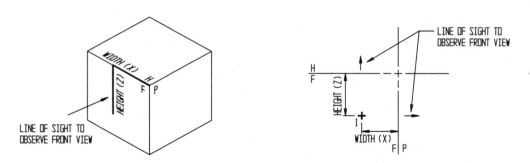

Figure 2–5 Two-dimensional location of a point in the front view.

Example

Refer to Figure 2–6. The two-dimensional location of point 1 in the top view shows only width and depth; therefore, only the measurements X and Z will be used. Remember to study the line of sight for the top view.

Example

Refer to Figure 2–7. The two-dimensional location of point 1 in the right side view shows only height and depth; therefore, only the measurements Y and Z will be used. Remember to study the line of sight for the right-side view.

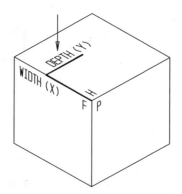

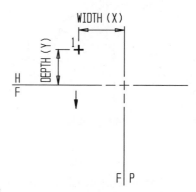

Figure 2–6 Two-dimensional location of a point in the top view.

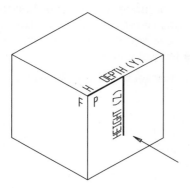

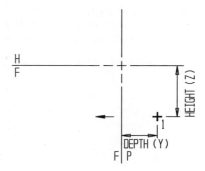

Figure 2–7 Two-dimensional location of a point in the profile view.

EXERCISE

Using drawing instruments, try the following problems:

Problem: In Figure 2–8, locate point A only in the front view. Sketch in the lines of sight for viewing the front view. In addition, answer the following questions to the nearest $\frac{1}{16}$". Draw you answer on Figure 2–8.)

1. Point A is how far to the left of the profile projection plane?

    X=

2. Point A is how far behind the frontal projection plane?

    Y=

3. Point A is how far below the horizontal projection plane?

    Z=

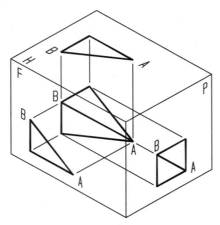

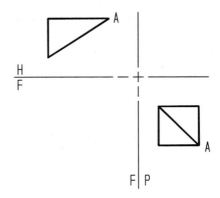

Figure 2–8 Practice problem.

Answer

1. You reviewed in Chapter 1 that properly aligned adjacent views share a dimension. The horizontal plane and frontal plane share width (distance left of the profile plane). In the top view, draw in the line of sight necessary to view the front view. Draw a projection line from point A, parallel to the line of sight and perpendicular to the H/F reference line, and extend it into the frontal plane.

2. The profile plane and frontal plane share height (distance below the horizontal plane). In the right-side view draw in the line of sight necessary to view the front view. Draw a projection line from point A, parallel to the line of sight and perpendicular to the F/P reference line and extend it into the frontal plane.

3. The location of point A is where the projection lines intersect in the front view.

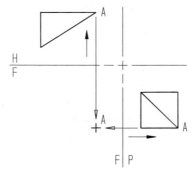

Figure 2–9 Locating point A the front view.

The distances must be measured perpendicular to the reference lines.

Point A is ¼" left of the profile projection plane.

Point A is 9⁄16" behind the frontal projection plane.

Point A is 1¹⁄16" below the horizontal projection plane.

Written as: X,Y,Z –¼,9⁄16,–1¹⁄16 (Page 35, Locating a Line in Space, will explain why the X and Z coordinates are negative.)

Problem: In Figure 2–10, locate only point B in the right-side view. Sketch in the line of sight for viewing the right-side view. In addition, answer the following questions to the nearest ¹⁄₁₆".

1. Point B is how far to the left of the profile projection plane?

    X=

2. Point B is how far behind the frontal projection plane?

    Y=

3. Point B is how far below the horizontal projection plane?

    Z=

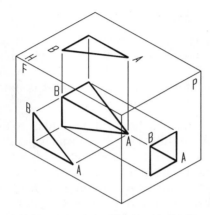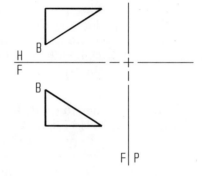

Figure 2–10 Practice problem.

Answer

1. Remember from Chapter 1 that adjacent views share a dimension. The profile plane and frontal plane share height (distance below the horizontal plane). In the front view, draw in the line of sight necessary to view the right-side view. Draw a projection line from point B, parallel to the line of sight and perpendicular to the F/P reference line, and extend it into the profile plane.

2. In Chapter 1 you reviewed that related views share a dimension, but that dimension must be transferred. The profile plane and horizontal plane share the depth dimension (behind the frontal plane). Transfer the depth of point B, using dividers, from the horizontal plane to the profile plane.

3. The location of point B is where the projection lines intersect in the front view.

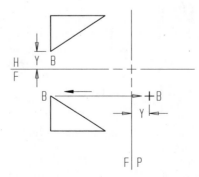

Figure 2–11 Locating point B i
the right-side view.

The distances must be measured perpendicular to the reference lines.

Point B is 7/8" to the left of the profile projection plane.

Point B is 3/16" behind the frontal projection plane.

Point B is 5/16" below the horizontal projection plane.

Written as: X,Y,Z –3/4, 3/16, –5/16 (Page 35, Locating a Line in Space will explain why the X and Z coordinates are negative.)

▼ LINE

A line is a straight path between two points. Lines are used to construct an object or part. Each line will appear **true length** (actual length of the line), as a **point view** (end view of the line), or **foreshortened** (line is viewed between true length and point view positions, making it appear shorter than it really is) in each view.

For ease of learning, only one line will be examined first. To visualize the examples in Figure 2–12, imagine the lines to be the center line of a pencil. Hold your pencil in the following positions and verify in which view the lines appear true length (TL). For example, for line 7-8 hold one end of the pencil in one hand and the other end in the other hand. Start by holding the pencil in front of you, in a horizontal position. The front view tells you point 8 is to the right of point 7. Therefore, your left hand represents point 7 and your right hand represents point 8. The front view also tells you height; therefore, move your right hand higher than your left hand. The top view shows the third dimension—depth.

Point 8 is farther behind the frontal plane than point 7; therefore, move your right hand away from you.

Determining if a line is true length in a principal plane is the first step in solving the majority of the descriptive geometry problems. This is true regardless of what you are solving for; for example, actual distance, shortest distance, or true shape.

There are several schools of thought as to how to locate the true-length line. One is, if a line is parallel to the reference line, it will be true length in the adjacent view. This method does not always prove true, as shown in Figure 2–13. However, the method used in this text, basing all steps around the line of sight, will always use the same steps to solve the majority of descriptive geometry problems. For example, once the correct line of sight is drawn, the reference line will always be drawn perpendicular to the line of sight. Therefore, the understanding and placement of the line of sight is critical.

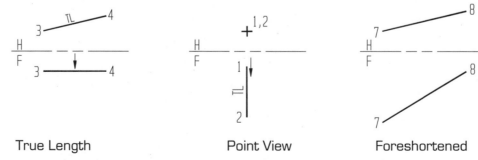

True Length   Point View   Foreshortened

Figure 2–12 Lines. Hold your pencil to represent the given views and verify in which view the lines appear true length.

The line of sight for locating a true length line in a principal view is as follows:

A. Locate the **true-length line**.

   1. Is the line true length in a principal view?

   a. If the line of sight is perpendicular to the object line in one view, it will appear true length in the adjacent view and can be measured (Figures 2–13 and 2–14).

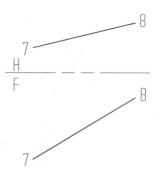

Figure 2–15 Oblique line.

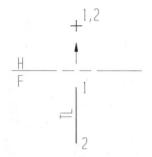

Figure 2–13 Point view of a true length line.

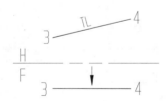

Figure 2–14 True-length line.

   2. If the above rule does not apply, it is an oblique line and cannot be measured in a principal view (Figure 2–15).

Each line has its own **line type** or name. **Whatever principal plane the line is true length in is the line type.** For example, if the line is true length in the front view (frontal plane), then the line type is **frontal**. Sometimes a line appears true length in more than one principal view and will have another name such as vertical or horizontal (level). If the line appears foreshortened—not true length or point view—in any of the principal views, the line type is **oblique**.

Examine the wedge shown in Figure 2–16. Pay special attention to lines A-B, A-C, C-D, and A-D. Line A-C (Figure 2–17) is **vertical**, that is, straight up and down in relation to the horizontal projection plane. Because it is vertical and parallel to the line of sight for the horizontal projection plane, it appears as a point in the top view and true length in the front and right-side view.

In contrast, line A-B (Figure 2–18) is **horizontal (level)**, that is, parallel to the horizontal projection plane. Because it is parallel to the horizontal plane and perpendicular to the line of sight in the adjacent view (front view), line A-B appears true length in the top view and as a point in the front view.

When viewing the front view of lines A-D and C-D (Figure 2–19), the lines are perpendicular to the line of sight in the top view, resulting in the true length of both lines in the front view. The line type for lines A-D and C-D is **frontal**, because the lines are true length in the front view (frontal plane). Line A-D is also called an **inclined line** because it is parallel to frontal plane, but not the horizontal plane.

Line C-D (Figure 2–20) is perpendicular to the line of sight for the top view; therefore it also appears true length in the top view. Because line A-D is sloping in the front view and not perpendicular to the line of sight for the top view, it is foreshortened in the top view.

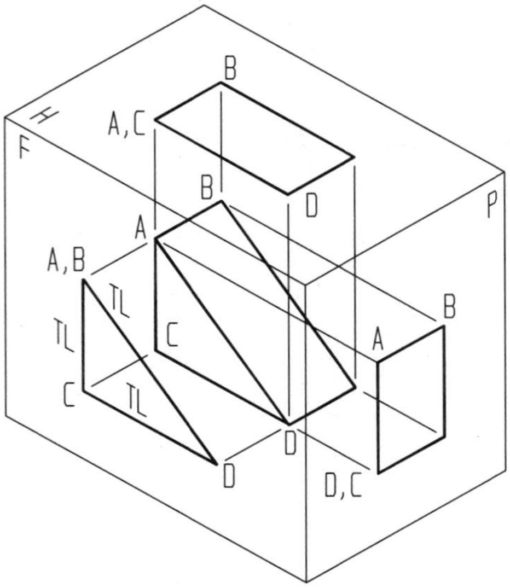

Figure 2-16 Wedge.

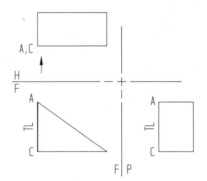

Figure 2-17 Line A-C is a vertical line.

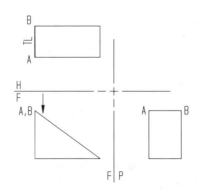

Figure 2-18 Line A-B is a horizontal (level) line.

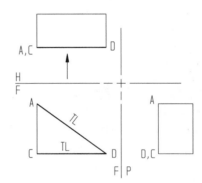

Figure 2-19 Lines A-D and C-D are true length in front view. Line A-D is inclined.

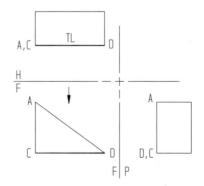

Figure 2-20 Line C-D is true length in top view.

EXERCISE

Problem: How do lines A-C, A-B, A-D, and C-D appear in the right-side view?

Answer

1. Line A-C appears true length in the right-side view, because the line of sight to view the right-side view is perpendicular to A-C in the adjacent view (front view) (Figure 2–21).

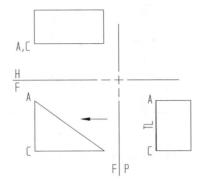

Figure 2–21　True length in right-side view.

2. Line A-B appears true length in the right-side view, because line A-B appears as a point view in the adjacent view (Figure 2–22).

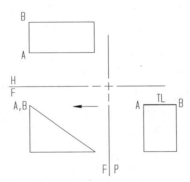

Figure 2–22　True length in right-side view.

3. Line A-D appears foreshortened in the right-side view, because the line of sight to view the right-side view is not perpendicular to A-D in the adjacent view (front view) (Figure 2–23).

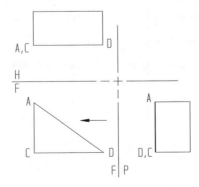

Figure 2–23　Line A-D foreshortened in right-side view.

4. Line C-D appears as a point in the profile, because the line of sight to view the right-side view is parallel to C-D in the adjacent view (front view) (Figure 2-24).

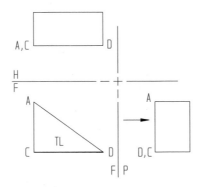

Figure 2-24 True length line in front view.

5. The line type of lines A-B and A-C is **profile**, because the lines are true length in the right-side view (profile plane).

Problem: Measure, to the nearest ⅟32", lines A-B, C-D, E-F, and G-H in Figure 2-25. List the line type of each line.

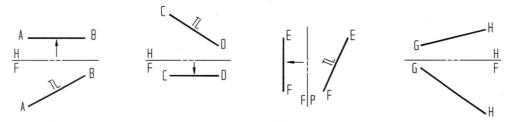

Figure 2-25 Measure the true length and state the line type for each line.

Answer

1. Line A-B = 1⅟16". Line type is frontal.
2. Line C-D = ⅝". Line type is horizontal or level.
3. Line E-F= ⅝". Line type is profile.
4. Line G-H = Cannot be measured as is, because it is oblique.

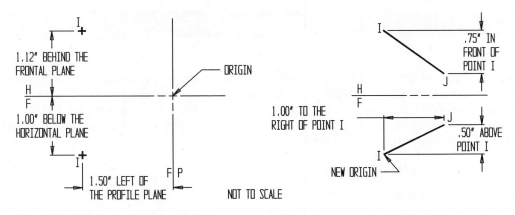

Figure 2–26 Locating a line in space.

▼ LOCATING A LINE IN SPACE

Each endpoint of the line can be located with three dimensions and then connected with a straight line.

Example
Line I-J (Figure 2–26)

Point I is: 1.5" left of the profile plane, 1.12" behind the frontal plane, and 1" below the horizontal plane. (Written as 1.5,1.12,1, from Exercise, page 36.)

Point J is: 1, −.75, .5 from point I. Notice the origin has been changed to point I.

The problems in Chapters 4 through 11 may be completed three-dimensionally on a 3D CAD system. Some CAD systems have more than one coordinate system, so it is also important that you enter the appropriate numbers for X, Y, and Z. When entering these coordinates into a CAD system some numbers will be entered as a negative number. As mentioned in the beginning of this chapter, this text will use the CNC coordinate axis orientation shown in Figure 2–27.

The CNC axis system is positive toward the right profile plane (to your right), behind the frontal plane (away from you), and toward the top horizontal plane (up). For example, point I is behind the frontal plane so it will be a positive coordinate. Point J uses point I as an origin and is in front of point I (traveling in front of the frontal plane), so it will be a negative coordinate.

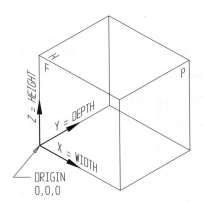

Figure 2–27 CNC axis orientation.

EXERCISE

Problem: Figure 2–28 includes a pictorial and an orthographic drawing of a locator block illustrating the concepts discussed in this chapter. Lines A-B, B-C, and A-C are singled out. Answer the following questions. *Hint:* highlighting the lines you are working with a different colored marker for each line helps visualization. Remember, measure to the nearest 1/16" and when a scale is not given, use full scale.

1. In which view is line A-B true length?

2. In which view is line B-C true length?

3. In which view is line A-C true length?

4. How far behind the frontal projection plane are points A, B, and C?

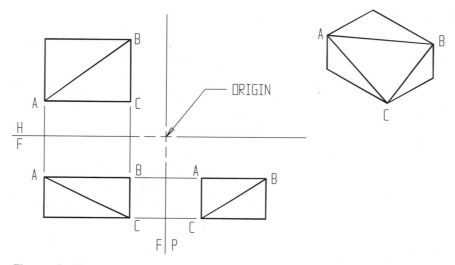

Figure 2–28 Review problem.

5. How far below the horizontal projection plane are points A, B, and C?

6. How far to the left of the profile plane are points A, B, and C?

7. What are the 3D coordinates for point B, if point A is the origin?

8. What are the 3D coordinates for point C, if point A is the origin?

9. Which point is the farthest behind the frontal plane?

10. What is the true length of line A-B?

Answer

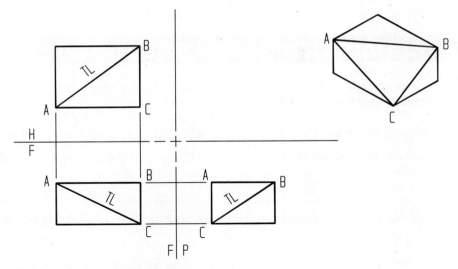

Figure 2–29 Answer to review problem.

1. Top.
2. Right side.
3. Front.
4. A ⅜"
   B 1¹⁄₁₆"
   C ⅜"

5. A ⁷⁄₁₆"
   B ⁷⁄₁₆"
   C ⅞"
6. A 1¼"
   B ⅜"
   C ⅜"

7. X = ⅞"
   Y = 1¹⁄₁₆"
   Z = 0
8. X = ⅞"
   Y = 0
   Z = –⁷⁄₁₆"

9. B
10. 1⅛

CHAPTER 2 STUDY QUESTIONS

The Study Questions are intended to assess your comprehension of chapter material. Please write your answers to the question in the space provided.

1. If you are given a three-view (front, top, right side) orthographic drawing of an object:

   a. In which view(s) are you able to see the distance that the object lies below the horizontal plane?

   _____

   b. In which view(s) do you see the distance that the object lies behind the frontal plane?

   _____

   c. In which view(s) do you see the distance that the object lies to the left of the right profile plane?

   _____

2. Explain the term foreshortened.

   _____

   _____

   _____

3. Measure the true length lines shown below. Also, sketch in the appropriate line of sight to view the true length line and label the true length line "TL".

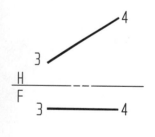 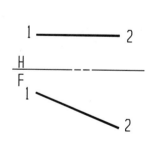 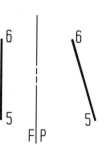 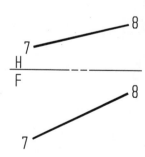

4. In the figure above, if the coordinates for point 7, are 0,0,0, what are the coordinates for point 8?

   _____

The problems in this chapter are more theoretical than most of the others in this text. Your ability to visualize points and lines as the building blocks of objects and other engineering situations is very important.

The following problems may be drawn using drawing instruments on the page provided, created using a 2D or 3D CAD system, or a combination of drawing board and CAD.

Chapter 2, Problem 1: POINTS A, B, C, D, E, AND F ARE LOCATED IN SPACE AS
SHOWN BY THE GIVEN HORIZONTAL AND FRONTAL PROJECTIONS. EXAMINE
THE POINTS CAREFULLY, AND ANSWER THE FOLLOWING QUESTIONS.

A. IS POINT A ABOVE POINT C? _____

B. IS POINT A BEHIND POINT B? _____

C. IS POINT C ABOVE POINT D? _____

D. WHICH POINT IS CLOSEST TO THE FRONTAL PLANE? _____

E. WHICH TWO POINTS ARE AT THE SAME ELEVATION (HEIGHT)? _____

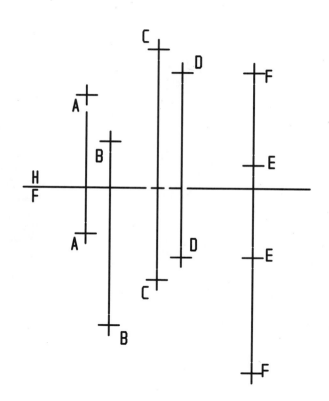

	ADG INC.			
	POINT LOCATION			
SIZE DATE	DWG NO.	SCALE FULL	REV	
DRW. BY		SECTION #	SHEET	

Chapter 2, Problem 2: DRAW THE TOP, FRONT, AND RIGHT SIDE VIEWS OF LINE AB
DESCRIBED BELOW, AND ANSWER QUESTION 1.

A IS TO THE LEFT OF B.
A AND B ARE EACH 1 INCH BELOW THE HORIZONTAL PLANE.
A IS 1/2 INCH BEHIND THE FRONTAL PLANE AND 2 INCHES LEFT OF THE RIGHT PROFILE PLANE.
B IS 1 INCH BEHIND THE FRONTAL PLANE AND 1/2 INCH LEFT OF THE RIGHT PROFILE PLANE.

1. WHAT IS TRUE LENGTH OF LINE AB? _____

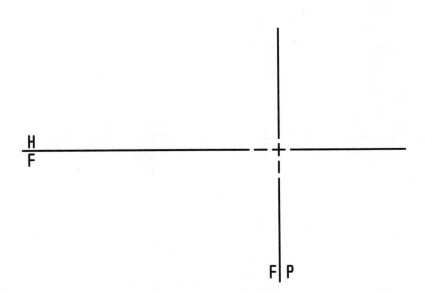

		ADG INC.		
		TRUE LENGTH		
	SIZE	DATE	DWG NO.	SCALE FULL	REV
DRW. BY			SECTION #		SHEET

Chapter 2, Problem 3:   DRAW THE TOP, FRONT, AND RIGHT SIDE PROJECTIONS OF LINE XY
DESCRIBED BELOW, AND ANSWER QUESTION 1.

X IS TO THE LEFT OF Y.
X AND Y ARE EACH 1/4 INCH BEHIND THE FRONTAL PLANE.
X IS 3/8 INCH BELOW THE HORIZONTAL PLANE AND 2-1/4 INCHES LEFT OF THE RIGHT PROFILE PLANE.
Y IS 15/16 INCH LEFT OF THE RIGHT PROFILE PLANE.
LINE XY MAKES AN ANGLE OF 45° WITH THE HORIZONTAL PLANE.

1. WHAT IS TRUE LENGTH OF LINE XY?   _____

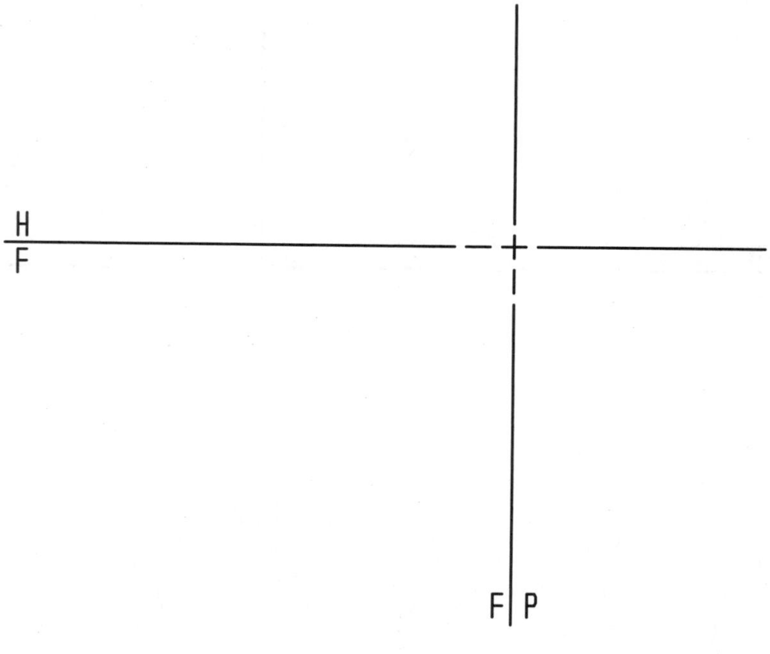

$\dfrac{H}{F}$

F | P

		ADG INC.		
		TRUE LENGTH		
SIZE	DATE	DWG NO.	SCALE FULL	REV
DRW. BY		SECTION #	SHEET	

Chapter 2, Problem 4: MOLDED PARTS FROM AN INJECTION PRESS FALL ONTO A CONVEYOR AT POINT D. THE CONVEYOR HAS TO DROP THE PARTS INTO A BIN AT POINT E WHICH IS 3 FEET TO THE RIGHT, 7'6" BEHIND, AND 5 FEET ABOVE POINT D.

LOCATE POINT E IN THE NECESSARY VIEWS.

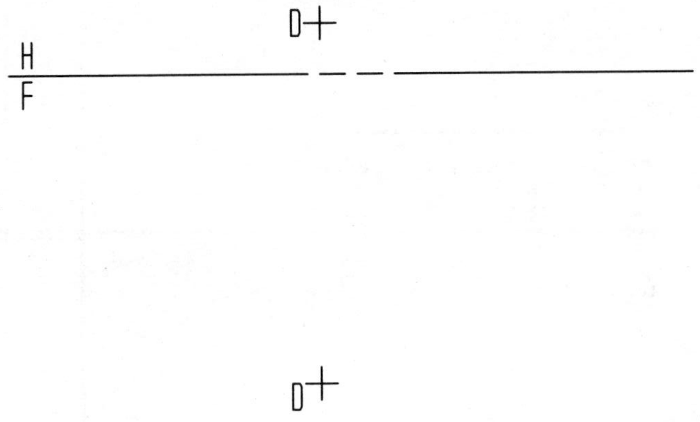

D+

H
—
F

D+

Chapter 2, Problem 5: A VERTICAL MAST, XY, IS ANCHORED BY THREE GUY WIRES, XA, XB, AND XC.
ANCHOR A IS 13 FEET WEST AND 4 FEET NORTH OF MAST XY AND IS AT AN ELEVATION OF 140 FEET.
ANCHOR B IS 12 FEET EAST AND 7 FEET NORTH OF MAST XY, AND IS AT AN ELEVATION OF 146 FEET.
ANCHOR C IS 5 FEET EAST AND 10 FEET SOUTH OF THE MAST AND IS AT AN ELEVATION OF 138 FEET.

LOCATE LINES XA, XB, AND XC IN THE NECESSARY VIEWS.

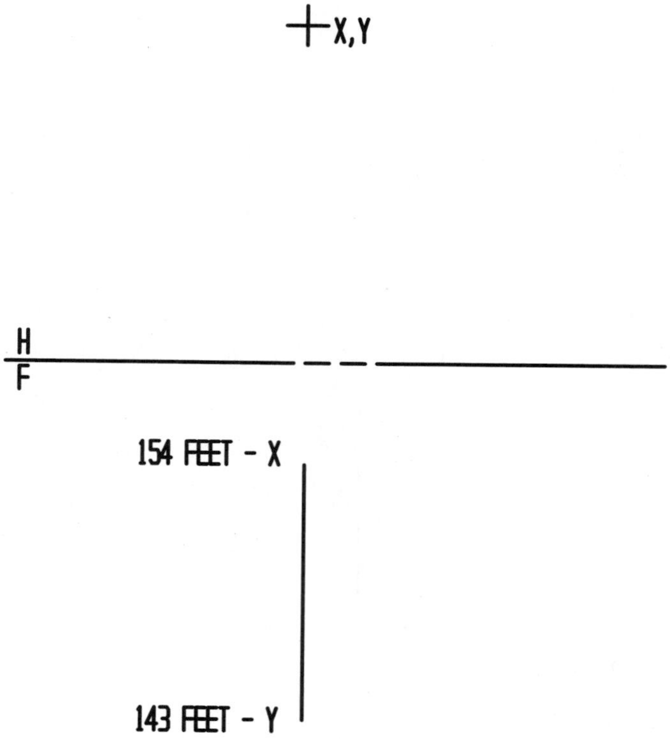

$+$ X,Y

$\dfrac{H}{F}$ ————— — — ——————

154 FEET – X

143 FEET – Y

	ADG INC.		
	MAST GUY WIRES		
SIZE DATE	DWG NO.	SCALE 1/8"=1'0"	REV
DRW. BY		SECTION #	SHEET

CHAPTER 3

Auxiliary Views

Cutting the ends of the I-beams on the ground before lifting them into place saves on the cost of renting a crane. To accomplish this, auxiliary views were necessary to determine at what angles to cut the I-beams.

This chapter will focus only on the procedure for drawing primary and secondary auxiliary views, rather than when, where, or why they are needed. The Helpful Hint section reinforces material covered, gives a different approach to the information, and includes practice problems. After completing this chapter, you will be able to:

▸ *Describe an auxiliary view.*

▸ *Explain the difference between primary and secondary auxiliary views.*

▸ *List the four fundamental views.*

▸ *Locate top-adjacent, front-adjacent, and side-adjacent auxiliary views using primary views.*

▸ *Locate principal views of an object using auxiliary views.*

You have learned that it is possible for an observer to move to different positions to view an object. In addition, you have seen how to draw the different views in their proper relationship on a sheet of paper. Sometimes it is necessary to view objects from various angles. Views projected on any projection plane other than the principal planes are **auxiliary views.** A **primary auxiliary view** is found by projecting onto a plane that is adjacent and thus perpendicular to one of the six principal planes of the orthographic box. A **secondary auxiliary view** is found by projecting onto a plane that is adjacent and thus perpendicular to a primary auxiliary view.

Most industrial drawings require dimensions. Almost all objects drawn are bound by lines and planes. However, lines or planes cannot be dimensioned in a view unless they appear true size in that particular view (see Chapter 2). A piece of steel rod cannot be dimensioned in any view unless it is drawn true length. A piece of steel plate cannot be dimensioned unless it is shown true or actual shape and size; therefore, you must know how to find the necessary view of the object that will allow you to dimension it where it shows true size and shape.

Finding the necessary view often requires the use of **auxiliary views** (primary, secondary, etc.). Examples of such a situation could be a guy wire, bridge truss, or a framework for an engine mount.

▼ PRIMARY AUXILIARY VIEWS

The steps for drawing an auxiliary view are the same as drawing a principal view. The only difference between drawing a principal view and an auxiliary view is the placement of the line of sight. Figures 3–1 through 3–9 illustrate this by comparing the steps to completing a principal view (right-side view) and a primary auxiliary view (front-adjacent) for a truncated pyramid. Look at the truncated pyramid in Figure 3–1.

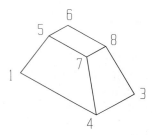

Figure 3–1   Isometric of a truncated pyramid.

For this demonstration, a right-side view and a front-adjacent auxiliary view will be drawn. Because both views are adjacent to the front view, locate the line of sight for viewing the front view (Figure 3–2).

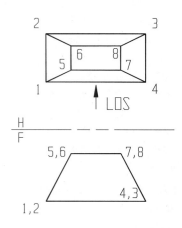

Figure 3–2   Line of sight for viewing front view.

I. Determine the line of sight.

This step requires a decision—where to place the **line of sight** (LOS). This step is important, because the rest of the steps are based on this step. The line of sight for auxiliary views may be specifically or randomly placed, depending on the desired results (Figure 3–3).

Chapter 2 reviews the specifically placed lines of sight used in locating principal views. In this chapter, the sections on Primary Auxiliary Views, and Secondary Auxiliary Views use randomly placed lines of sight to show you that no matter where the line of sight is placed, the rest of the steps of procedure are always the same. Chapter 4 introduces the specifically placed lines of sight used for finding particular information, such as true length lines.

Remember this chapter is focusing only on *how* to draw primary and secondary auxiliary views, not when, where, or why they are needed.

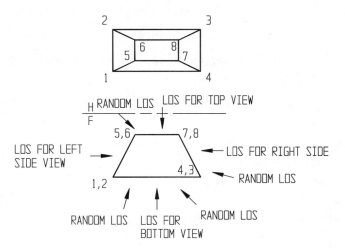

Figure 3–3   Specifically and randomly placed lines of sight (LOS).

In Figure 3–4, one line of sight is placed to view the right side of the object (Figure 3–4(A)) and one line of sight will be randomly placed to display, at the

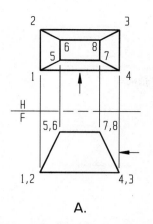

A.

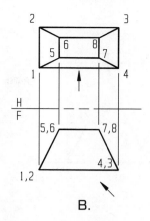

B.

Figure 3–4 Step I. Determine line of sight. (A) Line of sight for right-side view. (B) Randomly placed line of sight. This line of sight will show the bottom and right side of the object.

same time, the bottom and right side of the object (Figure 3–4(B)).

II. Draw the reference line perpendicular to the line of sight and label to show the two adjacent planes (Figure 3–5).

III. Draw the projection lines parallel to the line of sight (Figure 3–6).

IV. Transfer the points of measurement to the new view from the related view and label lightly (Figure 3–7).

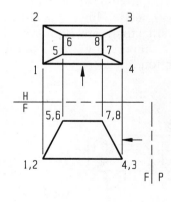

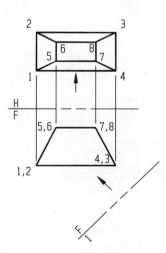

Figure 3–5 Step II. Reference line drawn perpendicular to the line of sight.

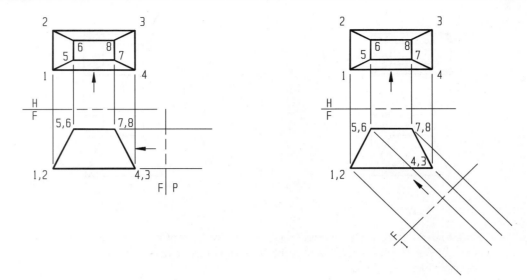

Figure 3–6 Step III. Projection lines drawn parallel to line of sight.

In other words:

(1) Go back to the reference line perpendicular to the last line of sight (related view—two views back). In this case, reference line H/F, which is perpendicular to the line of sight to see the front view.

(2) Measure the distance to be transferred. The distance is parallel to the last line of sight and is measured from the reference line to an endpoint. In Figure 3–7, the distance being transferred is true depth.

(3) Transfer distance to auxiliary view. The distance will be transferred to the reference line per-

pendicular to the current line of sight (F/?) along the appropriate projection line parallel to the current line of sight. In other words, any views adjacent to the front view will show true depth.

(Here's another way to think of it. Every object has three dimensions. The view being observed (front view) shows two dimensions (height and width). These dimensions are projected to the adjacent view (right-side view and front-adjacent view) using projection lines. The third dimension (depth) is missing; therefore, it must be transferred from the related view (top view) to the new view along the projection lines.)

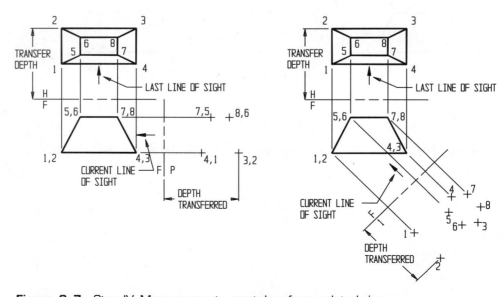

Figure 3–7 Step IV. Measurements are taken from related view.

V. Connect the points and label true-length lines, edge views, and/or true shapes.

Remember visibility. No matter what, the perimeter (outside lines) of the view is always visible (Figure 3–8). Visibility of interior lines has to be established (Figure 3–9). The line of sight for the auxil-

iary view indicates that you are looking at the bottom and right side; therefore, those lines will be visible (object) lines in view 1. The left side and top of the truncated pyramid are the farthest from the line of sight; therefore, their lines will be drawn as hidden lines in view 1 (Figure 3–9(B)).

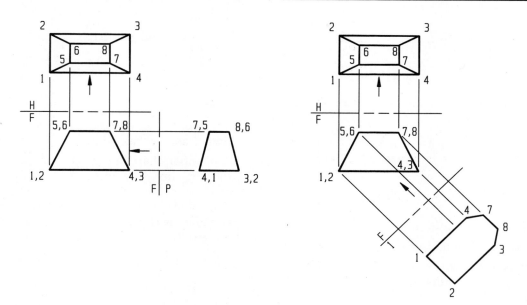

Figure 3–8 Step V. Perimeter of view will always be visible.

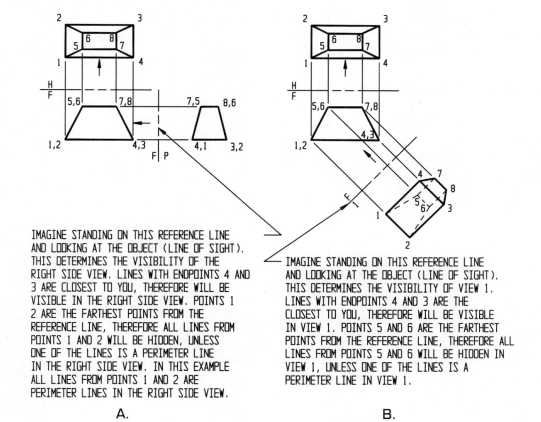

IMAGINE STANDING ON THIS REFERENCE LINE AND LOOKING AT THE OBJECT (LINE OF SIGHT). THIS DETERMINES THE VISIBILITY OF THE RIGHT SIDE VIEW. LINES WITH ENDPOINTS 4 AND 3 ARE CLOSEST TO YOU, THEREFORE WILL BE VISIBLE IN THE RIGHT SIDE VIEW. POINTS 1 2 ARE THE FARTHEST POINTS FROM THE REFERENCE LINE, THEREFORE ALL LINES FROM POINTS 1 AND 2 WILL BE HIDDEN, UNLESS ONE OF THE LINES IS A PERIMETER LINE IN THE RIGHT SIDE VIEW. IN THIS EXAMPLE ALL LINES FROM POINTS 1 AND 2 ARE PERIMETER LINES IN THE RIGHT SIDE VIEW.

A.

IMAGINE STANDING ON THIS REFERENCE LINE AND LOOKING AT THE OBJECT (LINE OF SIGHT). THIS DETERMINES THE VISIBILITY OF VIEW 1. LINES WITH ENDPOINTS 4 AND 3 ARE THE CLOSEST TO YOU, THEREFORE WILL BE VISIBLE IN VIEW 1. POINTS 5 AND 6 ARE THE FARTHEST POINTS FROM THE REFERENCE LINE, THEREFORE ALL LINES FROM POINTS 5 AND 6 WILL BE HIDDEN IN VIEW 1, UNLESS ONE OF THE LINES IS A PERIMETER LINE IN VIEW 1.

B.

Figure 3–9 Determine visibility for inside points of view.

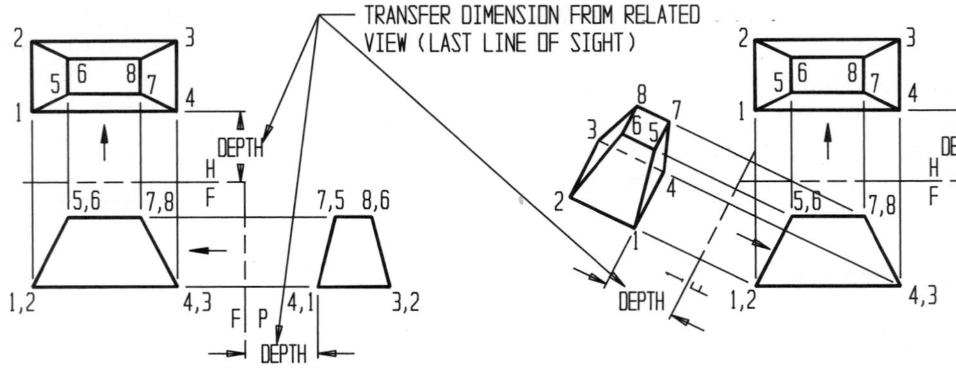

Figure 3-10 Front-adjacent auxiliary view.

This text encompasses the front-adjacent auxiliary, side-adjacent auxiliary, and top-adjacent auxiliary views.

A front-adjacent auxiliary view (Figure 3–10):

■ Always projects from the front view

■ Always utilizes the frontal reference plane

(Measurements are taken from a F/? reference line and transferred to a F/? reference line—for example, from F/H to F/1.)

■ Always shows true depth

(Remember that the projection lines for the front-adjacent auxiliary view transfer two dimensions, height and width, from the front view. The third dimension, depth, must be transferred from the related view (top view) to the new view (front-adjacent auxiliary view).)

A side-adjacent auxiliary view (Figure 3–11):

■ Always projects from the side view

■ Always utilizes the profile reference plane

(Measurements are taken from a P/? reference line and transferred to a P/? reference line—for example, from P/F to P/1.)

■ Always shows true width

(Remember that the projection lines for the side-adjacent auxiliary view transfer two dimensions (depth and height) from the side view. The third dimension (width) must be transferred from the related view (front view) to the new view (side-adjacent auxiliary view).)

A top-adjacent auxiliary view (Figure 3–12):

■ Always projects from the top view

■ Always utilizes the horizontal reference plane

(Measurements are taken from a H/? reference line and transferred to a H/? reference line—for example, from H/F to H/1.)

■ Always shows true height.

(Remember that the projection lines for the top-adjacent auxiliary view transfer two dimensions (width and depth) from the top view. The third dimension (height) must be transferred from the related view (front view) to the new view (top-adjacent auxiliary view).)

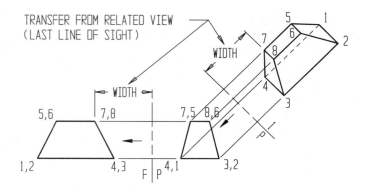

Figure 3-11 Side-adjacent auxiliary view.

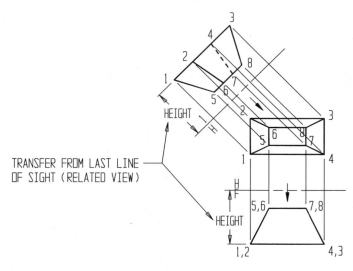

Figure 3-12 Top-adjacent auxiliary view.

EXERCISE

Problem: Let's try another top-adjacent auxiliary view of the truncated pyramid together.

First the line of sight necessary to see the top view is added (Figure 3–13).

I. Determine the line of sight.

The line of sight has been randomly placed to display the back and right side of the object at the same time (Figure 3–14).

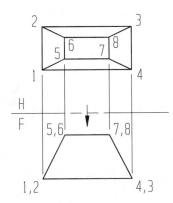

Figure 3–13   Practice   problem for top-adjacent auxiliary.

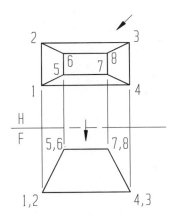

Figure 3–14   Determine the line of sight.

For the rest of the steps, sketch the solution on the figures. The correct answer is shown in the following figure.

II. Draw the reference line perpendicular to the line of sight and label (Figure 3–15).

III. Draw the projection lines parallel to the line of sight (Figure 3–16).

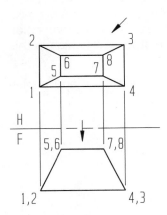

Figure 3–15   Draw the reference line perpendicular to the line of sight.

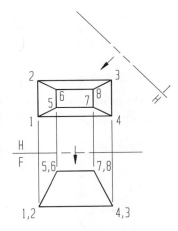

Figure 3–16   Draw the   projection lines parallel to the line of sight.

IV. Transfer the points of measurement to the new view from the related view and label lightly (Figure 3–17).

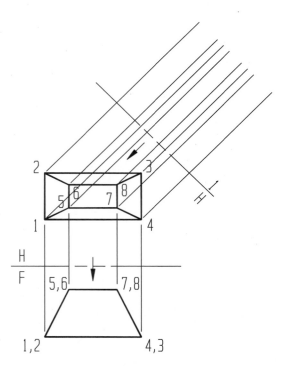

Figure 3–17 Transfer the measurements from the related view.

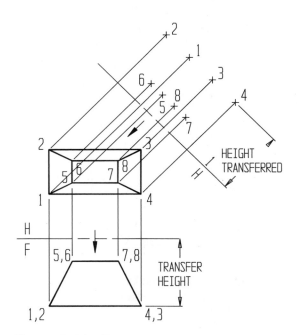

Figure 3–18 Connect the points and **label** true-lenght lines, edge views, and/or true-shape planes.

In other words, go back to the reference line perpendicular to the last line of sight (back two views—in this case, the line of sight to see the top view in Figure 3–13). Measure the distance parallel to the last line of sight from the reference line to the endpoint.

Here's another way to think of it. Every object has three dimensions. The view being observed (top view) shows two dimensions (width and depth). These dimensions are projected to the adjacent view (top-adjacent view) using projection lines. The third dimension (height) is missing; therefore, it must be transferred from the related view (which in this case is the front view) to the new view.

V. Connect the points and **label** true-length lines, edge views, and/or true-shape planes (Figure 3–18). Remember visibility.

The perimeter (outside lines) of the view is shown in (Figure 3–19).

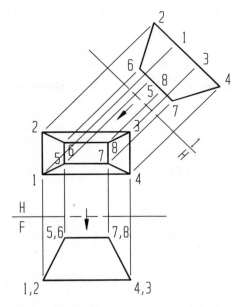

Figure 3-19   The perimeter (outside lines) of the view are visible.

Answer

Figure 3-20 shows finished results. The line of sight indicates that you are looking at the back and right side of the object, therefore those lines will be visible (object) lines in view 1. The left side and front of the truncated pyramid are the farthest from the line of sight; therefore, their lines will be drawn as hidden lines, unless they are perimeter lines, in view 1.

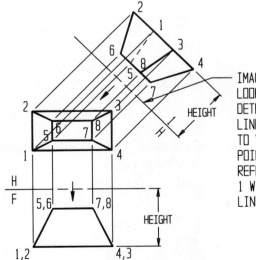

IMAGINE STANDING ON THIS REFERENCE LINE AND LOOKING AT THE OBJECT (LINE OF SIGHT). THIS DETERMINES THE VISIBILITY OF VIEW 1. LINES WITH ENDPOINTS 3 OR 8 ARE CLOSEST TO YOU, THEREFORE WILL BE VISIBLE IN VIEW 1. POINT 1 IS THE FARTHEST POINT FROM THE REFERENCE LINE, THEREFORE ALL LINES FROM POINT 1 WILL BE HIDDEN, UNLESS IT IS A PERIMETER LINE IN VIEW 1 - SUCH AS LINE 1-2.

Figure 3-20   Top-adjacent auxiliary view.

▼ SECONDARY AUXILIARY VIEWS

A secondary auxiliary view is found by projection onto a plane that is adjacent and therefore perpendicular to a primary auxiliary view.

A secondary auxiliary view:

■ Always projects from a primary auxiliary view

■ Always utilizes (measures from) the primary reference plane

■ Always shows distance from related view

The steps are the *same* as those used in drawing a principal view or a primary auxiliary view. To save time and space, the primary auxiliary view is already established in Figure 3–21.

I. Determine **line of sight** (Figure 3–21). In this case, the line of sight is randomly placed to see the bottom and right side.

II. Draw the **reference line** perpendicular to the line of sight and label (Figure 3–22).

III. Draw **projection lines** parallel to the line of sight (Figure 3–23).

IV. Transfer **points of measurement** to the new view from the related view and lightly label (Figure 3–24).

V. Connect the points and **label** true-length lines, edge views, and/or true-shape planes (Figure 3–25).

The perimeter of the secondary auxiliary view is shown in Figure 3–25.

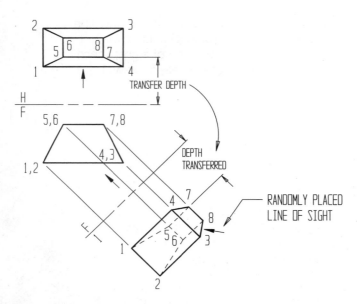

Figure 3–21 Line of sight for secondary auxiliary view.

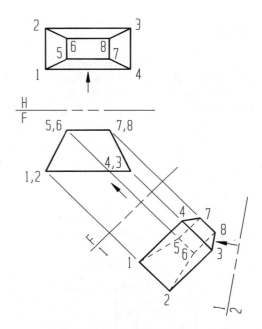

Figure 3–22 Reference line perpendicular to the line of sight.

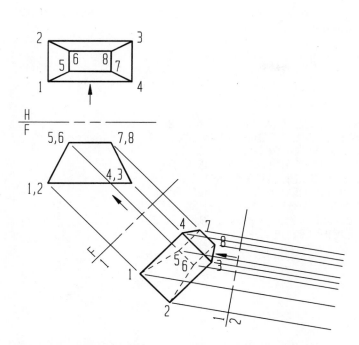

Figure 3-23 Draw the projection lines parallel to the line of sight.

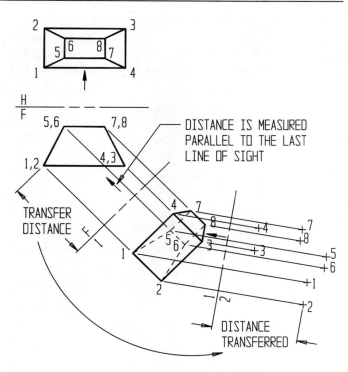

Figure 3-24 Transfer distance from related view.

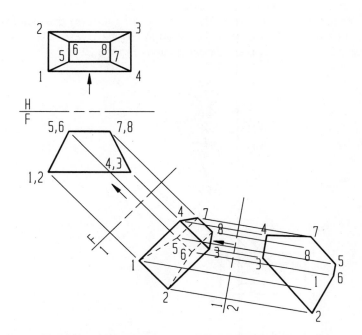

Figure 3-25 Perimeter of the secondary auxiliary view.

Take the time to review Figures 3–8 through 3–12 and study the visibility of the new views. Notice, the perimeter of the object is *always* visible. Visibility of interior lines has to be established. The endpoint closest to the line of sight will be visible in the new view. The endpoint farthest from the line of sight will be hidden in the new view.

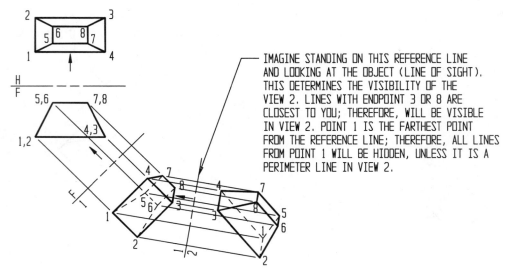

IMAGINE STANDING ON THIS REFERENCE LINE AND LOOKING AT THE OBJECT (LINE OF SIGHT). THIS DETERMINES THE VISIBILITY OF THE VIEW 2. LINES WITH ENDPOINT 3 OR 8 ARE CLOSEST TO YOU; THEREFORE, WILL BE VISIBLE IN VIEW 2. POINT 1 IS THE FARTHEST POINT FROM THE REFERENCE LINE; THEREFORE, ALL LINES FROM POINT 1 WILL BE HIDDEN, UNLESS IT IS A PERIMETER LINE IN VIEW 2.

Figure 3–26   Visibility.

EXERCISE

Problem: Try sketching the secondary auxiliary view on Figures 3–28 through 3–33. Remember the steps are the same as those used for locating principal views and primary auxiliary views. The random line of sight is given in Figure 3–27. Start sketching on Figure 3–28. The answer is shown on the following view, Figure 3–29.

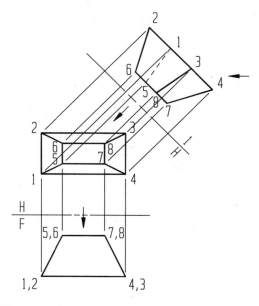

Figure 3–27   Secondary auxiliary view practice problem.

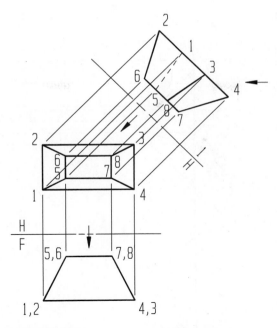

Figure 3-28 Draw the reference line perpendicular to the line of sight.

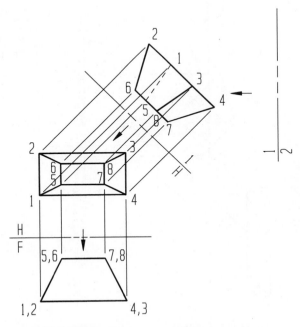

Figure 3-29 Draw the projection lines parallel to the line of sight.

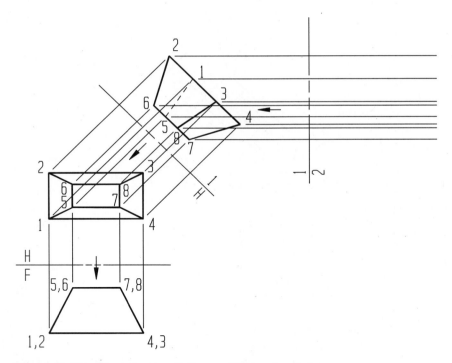

Figure 3–30 Transfer the distances from related view.

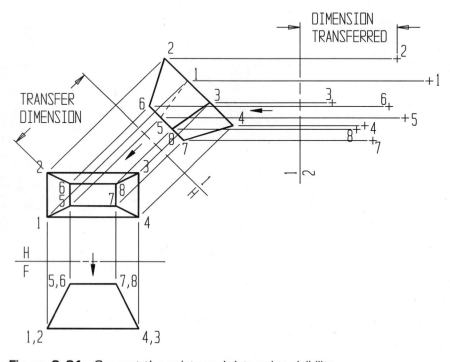

Figure 3–31 Connect the points and determine visibility.

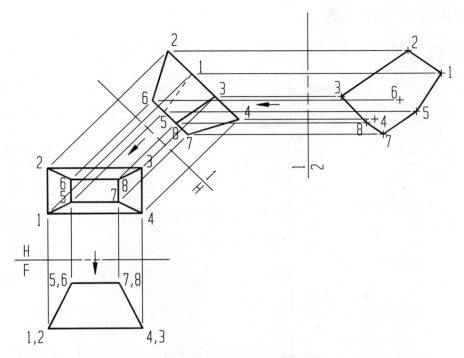

Figure 3-32 Perimeter of the secondary auxiliary view.

Answer

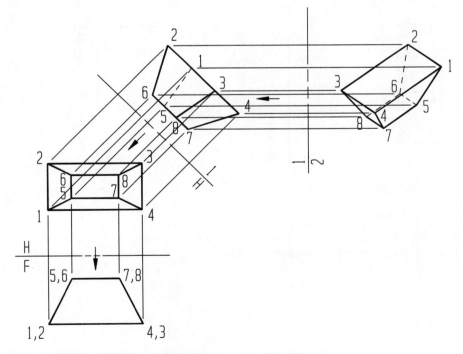

Figure 3-33 Visibility of the secondary auxiliary view.

▼ FUNDAMENTAL VIEWS

Almost all objects or situations a drafter may have to draw, such as the length of a guy wire, the shortest distance between two struts, or the true size of a gazebo truss, may be drawn through the use of one or more of the **four fundamental views**. A thorough working knowledge of the four fundamental views will be the basis for solving all problems using descriptive geometry, and is described in detail in the following chapters.

The four fundamental views are:

■ The true length (TL) of a straight line
■ The point view of a line
■ The edge view (EV) of a plane
■ A plane in its true size (TS) and shape

The way to look at the object (setting up a line of sight) is determined by knowing which one of the fundamental views is needed. Each of the four basic types of views has its own line of sight and must be used in sequence. (These are the specific lines of sight referred to in Chapter 3, page 46.) This section will only show you examples of each type. Deciding which line(s) of sight to use will be discussed in later chapters.

I. Determine **line of sight**.

  A. Locate **true-length line**. This is used to find the length of a line. The line may represent the center line of a pipe, cable, etc. (Figure 3–34).

  1. Is the true-length line in one of the **principal planes**?

    a. If the line of sight is perpendicular to the object line in one view, it will appear true length in the adjacent view and can be measured.

  2. If the above rule does not apply, it is an oblique line and cannot be measured as in a principal view.

    a. Draw a **primary auxiliary** view to find the true length.

      (1) Draw the line of sight perpendicular to the oblique line in either principal view.

  B. Draw the **point view of a true-length line**. This is used to find shortest distances between a known point and a line, parallel lines, skewed lines, etc. (Figure 3–35). An example could be the clearance between two conveyor ducts. Use your pen-

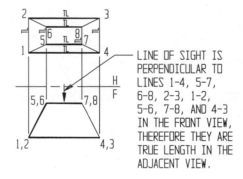

LINE OF SIGHT IS PERPENDICULAR TO LINES 1-4, 5-7, 6-8, 2-3, 1-2, 5-6, 7-8, AND 4-3 IN THE FRONT VIEW, THEREFORE THEY ARE TRUE LENGTH IN THE ADJACENT VIEW.

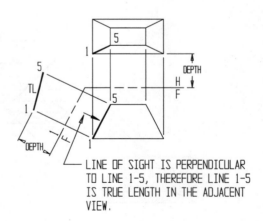

LINE OF SIGHT IS PERPENDICULAR TO LINE 1-5, THEREFORE LINE 1-5 IS TRUE LENGTH IN THE ADJACENT VIEW.

Figure 3–34 True length line in principal view and front-adjacent auxiliary view. Line of sight is perpendicular to the object line.

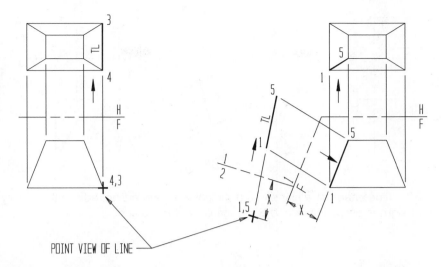

Figure 3–35   Point view of a true length line in a principal view and a front-adjacent auxiliary view. Line of sight is parallel to the true length line.

cil to represent a line. Hold it up and look perpendicular to it—it is now true length. Turn your pencil so you are looking parallel to it, now you see the point view of the line.

1. The line of sight must be drawn parallel to the true-length line.

C. Draw an **edge view of a plane** (all points fall on the same line). This is used to find the shortest distance between a known point and a plane, two planes, dihedral angle, etc. (Figure 3–36). An example could be the angle between two windshields in an airplane.

1. An edge view is found in any view where a line in the plane appears as a point.

D. Draw the **true shape of a plane**. This is used to find a specific location, dimensions, area, etc., of a plane (surface) (Figure 3–37). For example, the true size of a hip roof is needed to calculate the number of shingles needed.

1. The line of sight must be drawn perpendicular to the edge view of a plane.

If the above lines of sight are followed, the steps of procedure listed on page 46–49 for drawing an auxiliary view will always be the same.

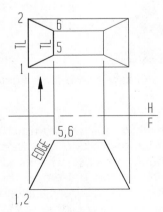

Figure 3–36   Edge view of a plane in a principal view. Line of sight is parallel to a true length line in the plane.

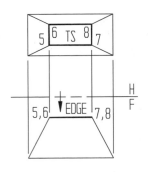
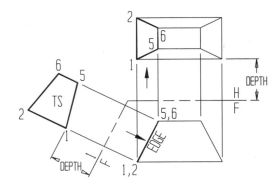

Figure 3–37 True shape of a plane in a principal view and a front-adjacent auxiliary view. Line of sight is perpendicular to the edge view of the plane.

I. Determine **line of sight**.

II. Draw the **reference line** perpendicular to the line of sight and label.

III. Draw **projection** lines parallel to the line of sight.

IV. Transfer **points of measurement** to the new view from the related view and lightly label.

V. Connect the points and **label** true-length lines, edge views, and/or true-shape planes.

Because steps II– IV are based on the location of the line of sight, step I is the most critical step. Until the descriptive geometry steps of procedure have been memorized, refer to the review page (Appendix B) when solving descriptive geometry problems.

Figures 3–34 through 3–37 show examples of the four fundamental views. Note the placement of the line of sight for each problem. No matter which line of sight is chosen, steps II–V are always the same. How to determine which line of sight to use will be discussed in more detail in the following chapters. Remember, the answers may be found in a principal view, primary auxiliary view, or secondary auxiliary view.

HELPFUL HINTS

To help visualize which dimensions to transfer, photo copy or trace, and cut out Figures 3–39, 3–41, 3–43, and 3–45. Fold 90 degrees on reference lines, and study the dimensions. To simplify what is happening only one line will be used. Imagine it to represent the center line of your pencil. Try holding your pencil in the same position as the line and viewing it. Do not throw these models out, we will be referring back to them at a later date.

A front-adjacent auxiliary view (Figure 3–38):

■ Always projects from the front view
■ Always utilizes the frontal reference plane for measurements
■ Always shows true depth

To set up for a front-adjacent auxiliary view, you must first look at the front view. Figure 3–38(A) shows the line of sight in the top view necessary to see the front view. Notice it is parallel to depth measurements, which is why you cannot see depth in the front view. The front view shows width and height; therefore, the dimension missing from the top view—depth—must be transferred to the front-adjacent auxiliary view (Figure 3–38 (B)). That is the last line of sight where you would be standing to view the front view (related view, two views back). Cut and fold Figure 3–39. Pick up the model and move it around until you have looked parallel to each line of sight.

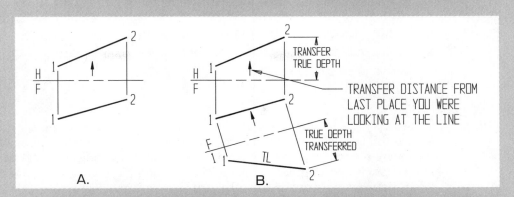

Figure 3-38 (A) Line of sight to view the front view. (B) Front-adjacent auxiliary view.

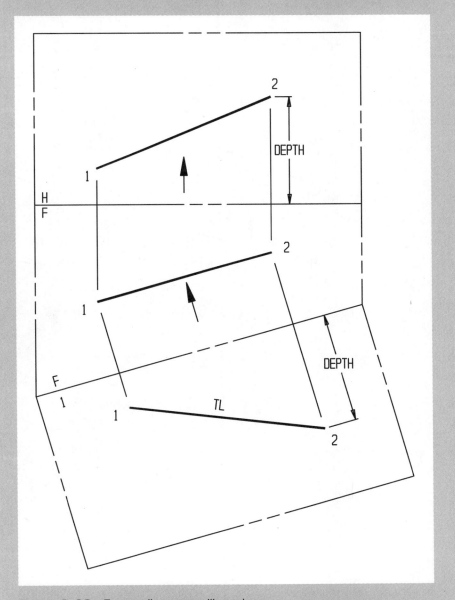

Figure 3-39 Front-adjacent auxiliary view.

A top-adjacent auxiliary view (Figure 3–40):

■ Always projects from the top view
■ Always utilizes the horizontal reference plane for measurements
■ Always shows true height

To set up for a top-adjacent auxiliary one must first look at the top view. Figure 3–40 (A) shows the line of sight in front view necessary to see the top view. Notice it is parallel to height measurements, which is why you cannot see height in the top view. The top view shows width and depth; therefore, the missing dimension—height—must be transferred from the front view to the top-adjacent auxiliary view (Figure 3–40(B)). That is the last line of sight where you would be standing to view the top view (related view, two views back). Cut and fold Figure 3–41. Pick up the model and move it around until you have looked parallel to each line of sight.

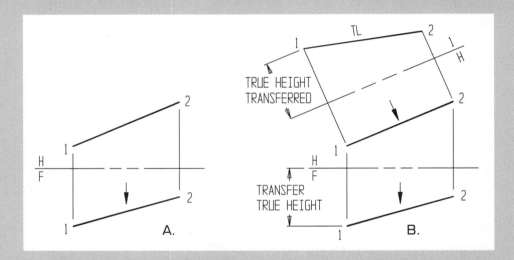

Figure 3–40   (A) Line of sight to view top view. (B) Top-adjacent auxiliary view.

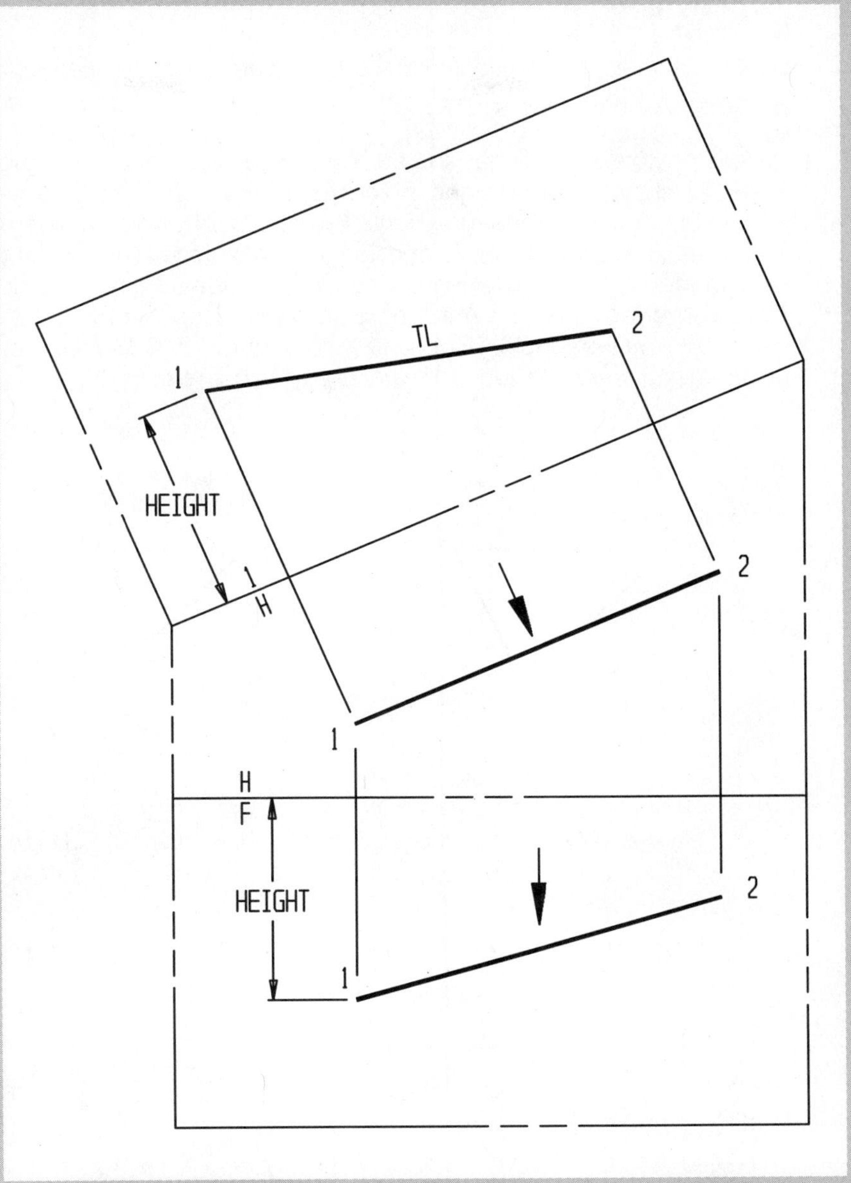

Figure 3–41 Top-adjacent auxiliary view.

A side-adjacent auxiliary view (Figure 3–42):

■ Always projects from the side view
■ Always utilizes the profile reference plane for measurements
■ Always shows true width

To set up for a side-adjacent auxiliary view, you must first look at the side view. Figure 3–42(A) shows line of sight in the front view necessary to see the side view. Notice it is parallel to width measurements, which is why you cannot see width in the side view. The side view shows depth and height; therefore, the missing dimension—width—must be transferred from the front view to the side-adjacent auxiliary view (Figure 3–42(B)). That is the last line of sight where you would be standing to view the side view (related view, two views back). Cut and fold Figure 3–43. Pick up the model and move it around until you have looked parallel to each line of sight.

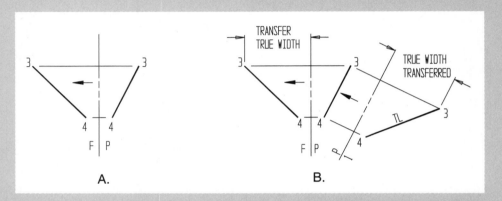

Figure 3–42 (A) Line of sight to view side view. (B) Side-adjacent auxiliary view.

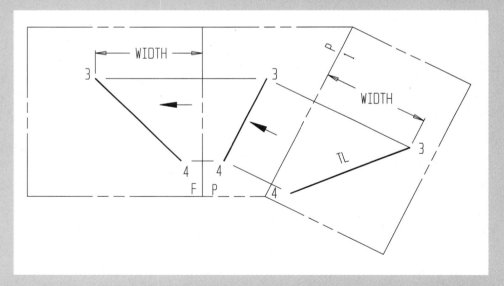

Figure 3–43 Side-adjacent auxiliary.

A secondary auxiliary (Figure 3–44):

■ Always projects from a primary auxiliary view
■ Always utilizes the primary reference plane for measurements
■ Always shows true distance from related view

To set up for a secondary auxiliary view one must first look at the auxiliary view. Figure 3–44(A) shows line 1-2 true length in a top-adjacent auxiliary view. Figure 3–44(B) shows the line of sight parallel to the true length line in the top-adjacent auxiliary view, which will result in seeing the point view of line 1-2 in the secondary auxiliary view. The rule is the same for transferring the missing dimension for the secondary auxiliary: go back to the last line of sight (related view, two views back) and transfer the missing distance. Cut and fold Figure 3–45. Pick up the model and move it around until you have looked parallel to each line of sight.

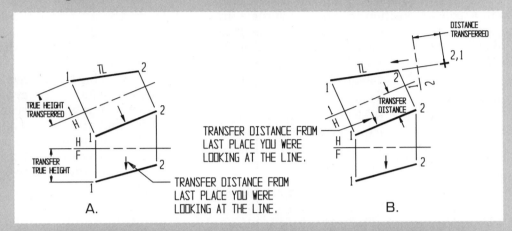

Figure 3–44   (A) Primary auxiliary view. (B) Secondary auxiliary view.

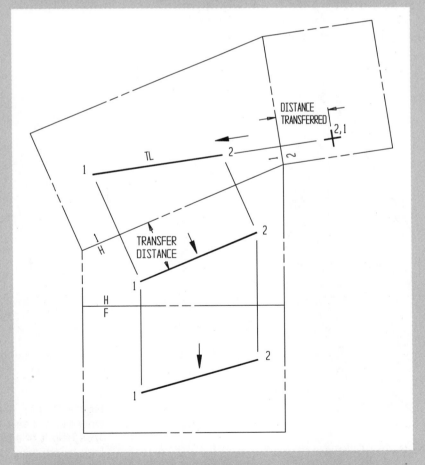

Figure 3–45   Secondary auxiliary view.

▼ LOCATING PRINCIPAL VIEWS OF AN OBJECT USING AUXILIARY VIEWS

Sometimes information must be transferred back to a principal view. For example, if the shortest distance for a new pipe was found in an auxiliary view, the new pipe's location must be shown in the principal views. (This is explained in Chapters 5 and 7.) A part also might be in a skewed position in a principal view. In order to draw the part in the principal views, the true shape and size of the object must be drawn in an auxiliary view (Figure 3–46) and transferred back to the principal views (Figures 3–47 through 3–50).

This section will focus on only how to transfer information back to a related view. How to set up the inclined axis for the part in the principal views (Figure 3–46) will be covered in Chapter 4.

Example

Customers were complaining that after 2000 hours of run time on Model 37625, an arm starts to slip to the right. The arm eventually binds, resulting in down time for the machine while maintenance is being completed. The design team recommends a ½" stop on the shaft be added to keep the arm from slipping. As a member of the design team, one of your responsibilities is to update the assembly drawing.

The drawing shows the shaft is at an oblique angle (Figure 3–46) so, the stop will also be shown askewed. To locate the exact position in the principal views draw the true shape of the part first (Figure 3–47). The face of the stop will be drawn true shape in the view that shows the end view of the shaft. (See Chapters 4 and 6.) The set screw hole has been omitted to simplify the drawing. The arrows on the projection lines show the direction of travel for transferring dimensions.

If the principal views are given, and you are completing auxiliary view 2, transfer the information from the last related view. Figure 3–48(A) shows the point A distance being transferred to auxiliary view 2 from the related view—top view. If you can transfer the information one way, you can transfer it back. Figure 3–48(B) shows the point 2 distance being transferred to the top view from the related view—auxiliary view 2. The arrows on the projection lines are to aid you in the direction of travel.

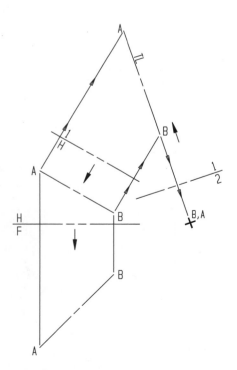

Figure 3–46 Oblique axis given.

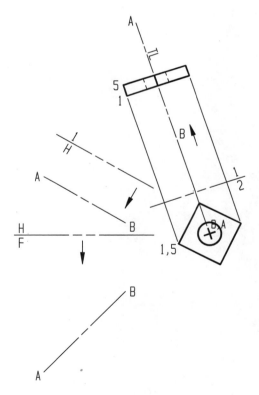

Figure 3–47 True shape of stop drawn where axis is shown as a point view (view 2). True height of stop shown where axis is true length (view 1).

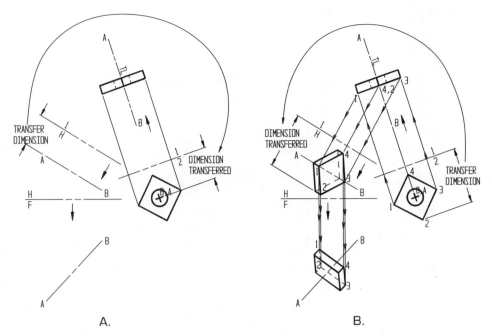

Figure 3–48 Transfer dimensions to related view. (A) Point A transferred to view 2. (B) Point 2 transferred to top view.

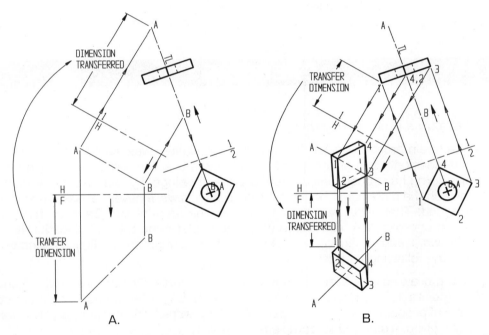

Figure 3–49 Transfer distance from related view. (A) Point A transferred to view 1. (B) Point 1 transferred to front view.

You are always working with three views: new view, adjacent view, and related view. Figure 3–49(A) shows the point A distance being transferred from front view to the related view—auxiliary view 1. If you can transfer the information one way, you can transfer it back. Figure 3–49(B) shows point 1 being transferred to the front view from the related view—

auxiliary view 1. The arrows on the projection lines show the direction of travel.

The same is done for the circle, except it must be segmented first. View 2, in Figure 3–50, shows the circle segmented into 12 equal parts. To simplify drawing, only points 6 and 7 are transferred back to the front view.

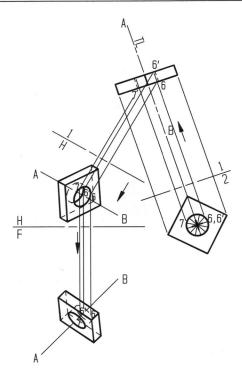

Figure 3–50 Divide hole and transfer dimensions to related view.

EXERCISE

Problem: Let's try a problem together.

As a quick fix to a motorized wheel chair lift for a van, a plate was added for support. Unfortunately, the solution was not thought out and now a brace must be welded to the steel support plate. You have been asked to update the drawings, by showing the weld location for the brace on the plate. Oh, Oh! The original drawings were never revised when the steel plate was added. Therefore, you must first draw the steel plate. (Because true shapes have not been covered, the true shape of the steel plate and the weld location are given in auxiliary views 1 and 2, Figure 3–51.) Sketch the solution on the figures. Answers will be shown in the following figures.

To simplify the drawing, the steel plate is broken into parts. First, transfer the front face of the plate (points 1,2,3,4) to the top view (Figure 3–51). (Remember, you must still follow the steps of procedure covered on page 46–49.) Steps I, line of sight, and II, reference line H/1, are given. You need to complete steps III–IV.

III. Draw **projection lines** parallel to the line of sight.

IV. Transfer **points of measurement** to the new view from the related view and lightly label.

V. Connect the points and **label** true-length lines, edge views, and/or true-shape planes.

Transfer the front face of the plate (1-2-3-4) to the front view (Figure 3–52). Again, step I (line of sight to view top view) and step II (reference line H/F) are given. Complete steps III–V.

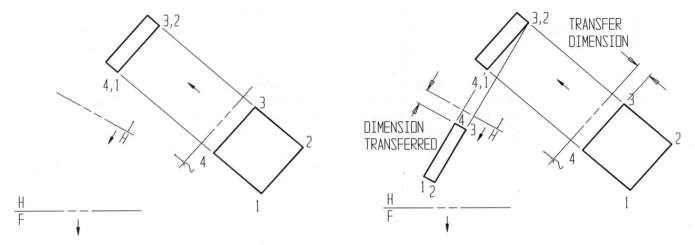

Figure 3–51 True shape of steel plate.

Figure 3–52 Front face of steel plate transferred to top view.

III. Draw **projection lines** parallel to the line of sight.

IV. Transfer **points of measurement** to the new view from the related view and lightly label.

V. Connect the points and **label** true-length lines, edge views, and/or true-shape planes.

Number the back face of the steel plate in views 1 and 2 (Figure 3–53). In view 2, start with number 5 located directly under point 1. Locate points 6-8 in a counterclockwise direction.

Transfer the back face of the steel plate (5-6-7-8) to the top view and complete the view (Figure 3–54).

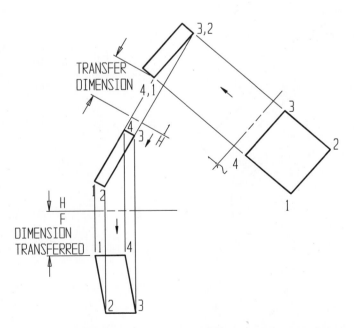

Figure 3–53 Front face of steel plate transferred to front view.

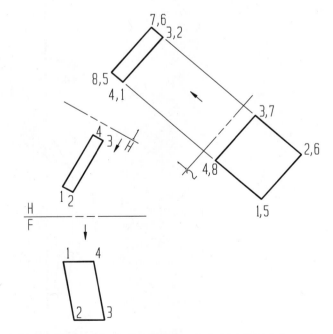

Figure 3–54 Back face in views 1 and 2 numbered.

Remember visibility. All perimeter lines of the object will always be visible. The interior lines must be analyzed. Imagine standing on reference line H/1, looking at the 1st auxiliary view. The points closest to you (4 and 1) will be visible in the adjacent view. The points farthest from you (7 and 6) will be hidden.

Transfer the back face of the steel plate (5-6-7-8) to the front view and complete the view (Figure 3–55). To view the visibility for the front view, standing on reference line H/F look at the top view. Points 1 and 2 are closest to the H/F reference line; therefore, they will be visible. Points 7 and 8 are the farthest from the H/F reference line; therefore, they will be hidden (Figure 3–56).

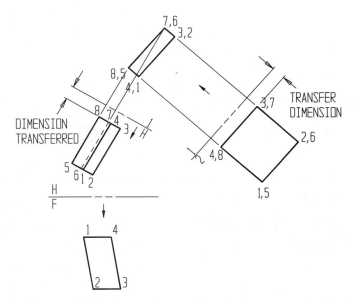

Figure 3–55 Back face of steel plate transferred to top view.

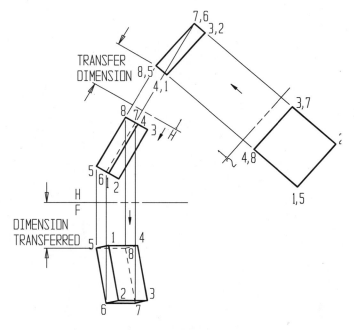

Figure 3–56 Back face transferred to front view.

The weld location has been located in views 1 and 2. Transfer the weld location to the front view (Figure 3-57).

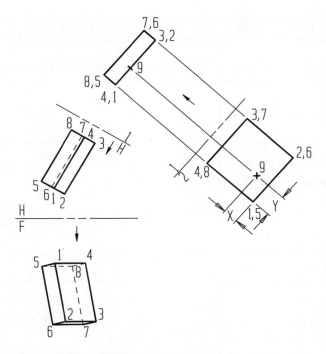

Figure 3-57 Weld location in views 1 and 2.

Answer

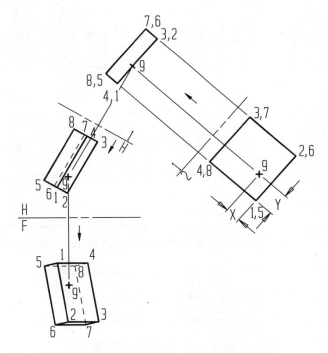

Figure 3-58 Weld location in principal views.

CHAPTER 3 STUDY QUESTIONS

The Study Questions are intended to asses your comprehension of chapter material. Please write your answers to the questions in the space provided.

1. What is a primary auxiliary view?

    _____

    _____

2. What is the purpose of an auxiliary view?

    _____

    _____

3. Describe the difference between a primary and secondary auxiliary view.

    _____

    _____

4. The following three auxiliary views are samples of top-adjacent, front-adjacent, and side-adjacent. Label each auxiliary view with the correct name.

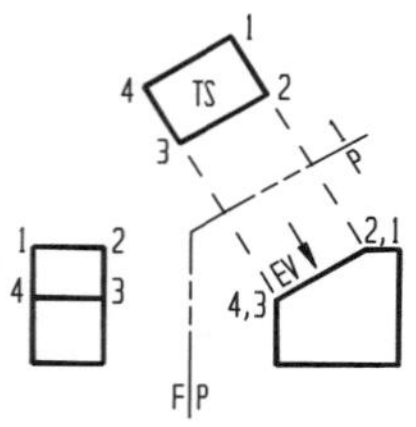

    A 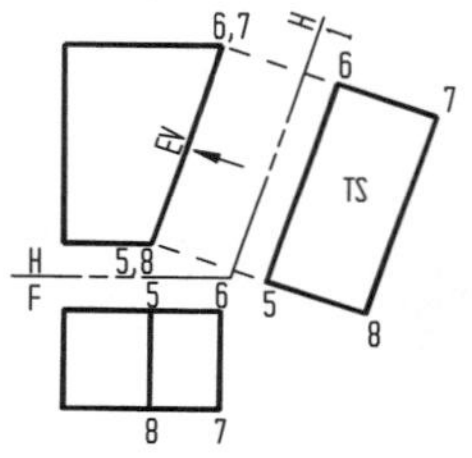 _____   B 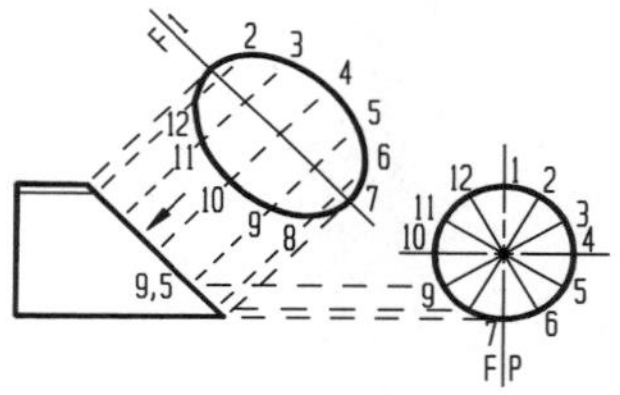 _____   C _____

5. What true dimension of the object is shown in the front-adjacent views?

    _____

    _____

6. List the five basic steps of procedure for drawing primary auxiliary views.

    _____

    _____

    _____

    _____

    _____

7. Draw the primary and secondary auxiliary views to the object below. Use the random lines of sight given. Draw the primary auxiliary reference line F/1 through the given "+" mark labeled F/1. Draw the secondary auxiliary reference line 1/2 through the given "+" mark labeled one-half.

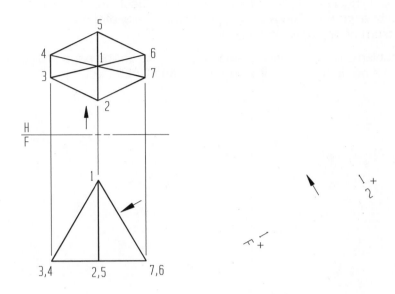

The problems contained in this chapter are meant to provide practice in the use of descriptive geometry principles. They are designed to test your understanding of the basic fundamentals and of the theoretical methods used, and to teach you to apply the principles while finding solutions to a variety of engineering problems. Many of the problems are derived from the broad spectrum of engineering settings.

The following problems may be drawn using instruments on the page provided, created using a 2D system, or a combination of drawing board and CAD.

Chapter 3, Problem 1: USING THE GIVEN LINES OF SIGHT, DRAW THE APPROPRIATE PRIMARY AUXILIARY VIEWS.

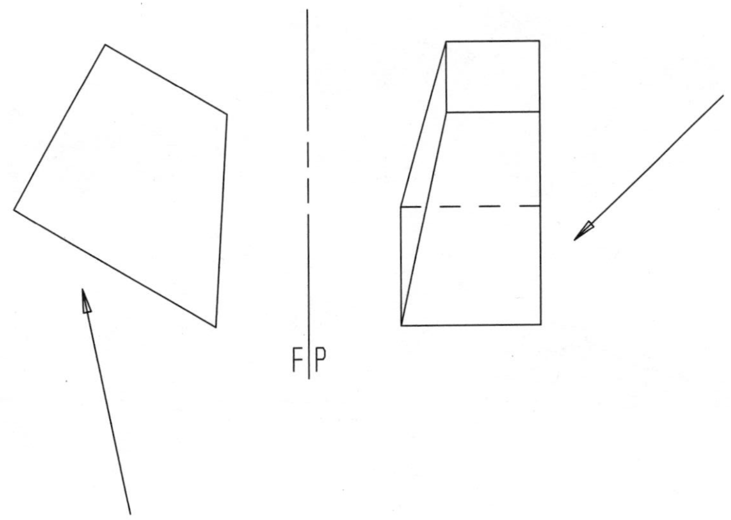

F/P

ADG INC.

WEDGE

SIZE	DATE		DWG NO.		SCALE	REV
					1" = 2"	
			SECTION #		SHEET	

DRW. BY

Chapter 3, Problem 2: USING THE GIVEN LINES OF SIGHT, DRAW THE APPROPRIATE PRIMARY AUXILIARY VIEWS.

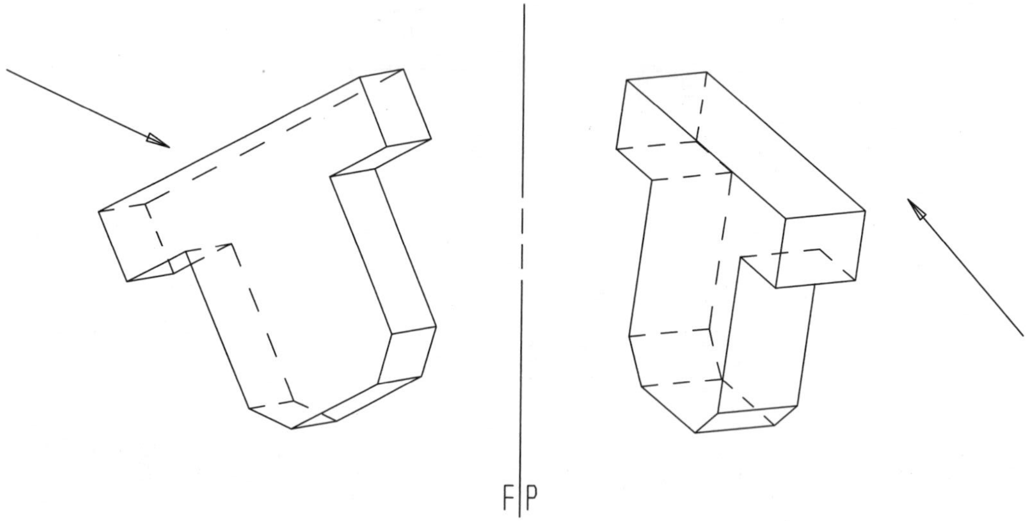

F/P

ADG INC.

CAVITY RETAINER

SIZE	DATE	DWG NO.	SCALE	REV
			FULL	
		SECTION #	SHEET	

DRW. BY

Chapter 3, Problem 3: USING THE GIVEN LINES OF SIGHT,
DRAW THE APPROPRIATE PRIMARY AND SECONDARY AUXILIARY VIEWS.

H
F

ADG INC.

DOOR STOP

SIZE	DATE	DWG NO.	SCALE	REV
			1" = 2"	
		SECTION #	SHEET	

DRW. BY

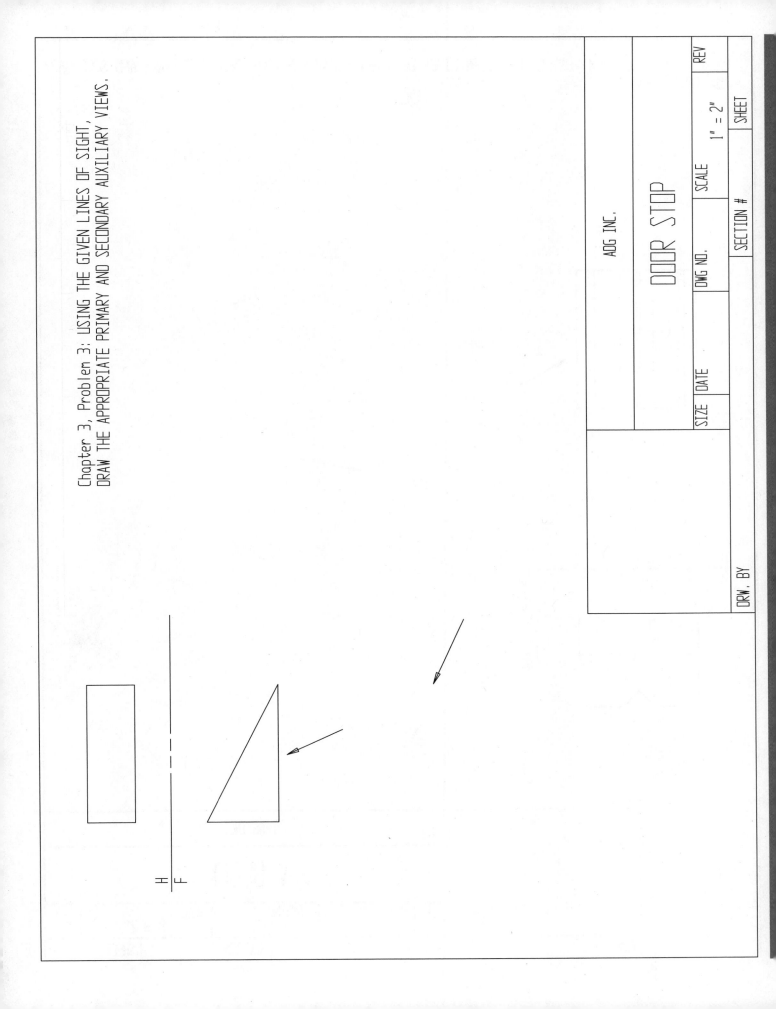

H
F

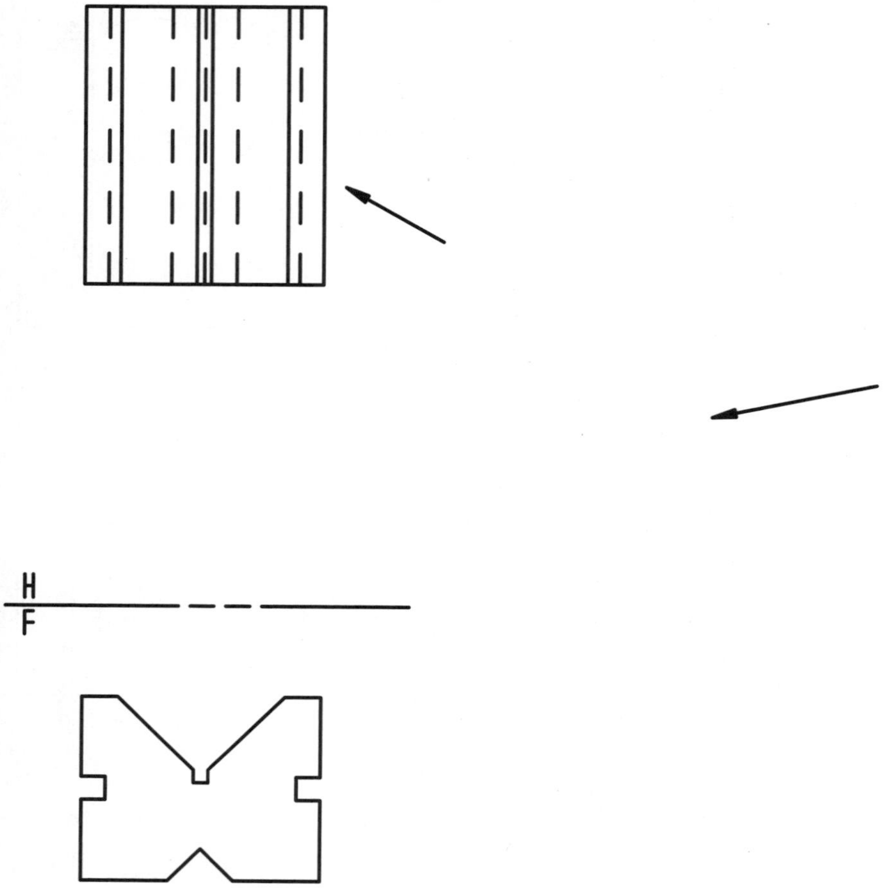

	ADG INC.			
	V BLOCK			
SIZE	DATE	DWG NO.	SCALE 1" = 2"	REV
DRW. BY			SECTION #	SHEET

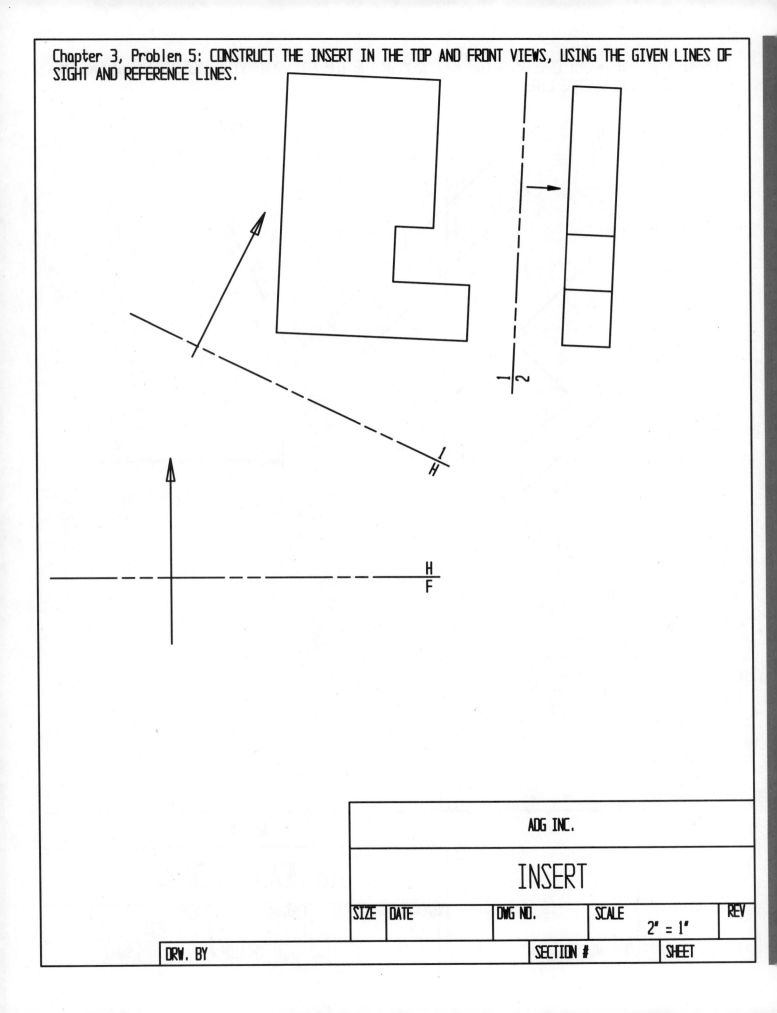

		ADG INC.		
		INSERT		

SIZE	DATE	DWG NO.	SCALE	REV
			2" = 1"	

| DRW. BY | | | SECTION # | SHEET |

Chapter 3, Problem 6: CONSTRUCT GIB HEAD KEY IN THE TOP AND FRONT VIEWS, USING THE GIVEN LINES OF SIGHT AND REFERENCE LINES.

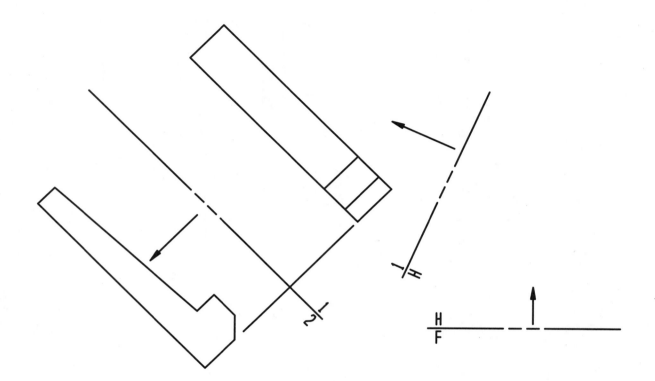

	ADG INC.			
	GIB HEAD TAPER KEY			
SIZE	DATE	DWG NO.	SCALE FULL	REV
DRW. BY		SECTION #		SHEET

CHAPTER 4

Lines

How much line is necessary to rig the Brig Niagara?

After completing this chapter, you will be able to:

▸ *Identify the type of a line.*

▸ *Locate the true length of a line.*

▸ *Locate the point view of a line.*

▸ *Define and locate the bearing and slope of a line.*

As mentioned in Chapter 3, at times specific lines of sight are needed in order to view:

■ The true length (TL) of a straight line
■ Point view of a true length line
■ Edge view (EV) of a plane
■ A plane in its true size (TS) and shape

This chapter will encompass when, where, why, and how to determine the true length of a line and the point view of a line not displayed in a principal view, as well as slope and bearing of a line.

▼ TYPES OF LINES

In descriptive geometry, there are several **line types**, or classifications. Before the majority of descriptive geometry problems can be solved, the line type must be resolved. **Line type is determined by the principal plane in which the line appears true length.** A **vertical line** (Figure 2–17, page 32) has the unique properties of being true length in all elevation views, such as front and right side views. If the line is not true length in a principal plane, it is called an **oblique line**. (See Figure 4–1. To make it easier to view the lines being referred to, the rest of the object has been omitted.)

Example

The line type for line A-B, in Figure 4–2, is **frontal**, because it is viewed true length in the frontal plane.

The line of sight is perpendicular to the object line in one view (top view); therefore, the line appears true length in the adjacent view (front view).

Example

The line type for line C-D, in Figure 4–3, is **horizontal**, because it is viewed true length in the horizontal plane. The line of sight is perpendicular to the object line in one view (front view); therefore, the line appears true length in the adjacent view (top view).

Example

The line type for line E-F, in Figure 4–4, is **profile**, because it is viewed true length in the profile plane. The line of sight is perpendicular to the object line in one view (front view); therefore, the line appears true length in the adjacent view (side view).

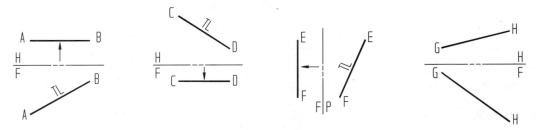

Figure 4–1   Line types: frontal, horizontal, profile, oblique.

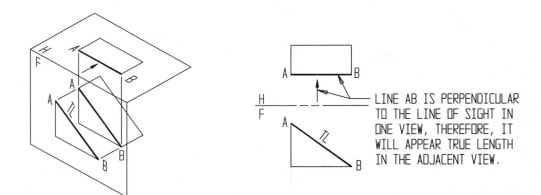

LINE AB IS PERPENDICULAR TO THE LINE OF SIGHT IN ONE VIEW, THEREFORE, IT WILL APPEAR TRUE LENGTH IN THE ADJACENT VIEW.

Figure 4–2   Line type—frontal.

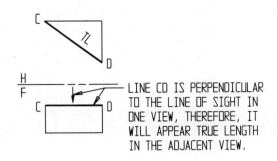

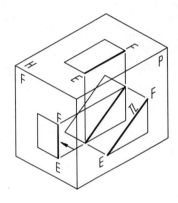

Figure 4-3 Line type—horizontal.

LINE CD IS PERPENDICULAR
TO THE LINE OF SIGHT IN
ONE VIEW, THEREFORE, IT
WILL APPEAR TRUE LENGTH
IN THE ADJACENT VIEW.

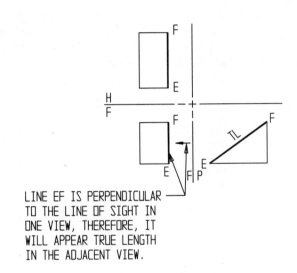

LINE EF IS PERPENDICULAR
TO THE LINE OF SIGHT IN
ONE VIEW, THEREFORE, IT
WILL APPEAR TRUE LENGTH
IN THE ADJACENT VIEW.

Figure 4-4 Line type—profile.

▼ TRUE LENGTH OF A LINE IN AN AUXILIARY VIEW

In building or engineering a length of steel tubing, member of a truss, guy wires, highway, mining shaft, or conveyor line may be needed. Sometimes the length can easily be found by actually measuring the distance or reading the dimension given in a principal view. But, sometimes this cannot be done for one reason or another. This could be because it may not be convenient or possible to measure an actual part, or the dimension may be missing from the views. In these cases, auxiliary views are completed to find the missing information.

To find the **true length** of a cable, pipe, conveyor line, etc., you must be looking perpendicular to it. Always check for the true length in the principal views first. (See Chapter 2.)

Review

I. Determine **line of sight**.

  A. Locate **true-length line.**

    1. Is the true-length line **in** one of the **principal planes**?

      a. If the line of sight is perpendicular to the object line in one view, it will appear true length in the adjacent view and can be measured.

    2. If the above rule does not apply, it is an oblique line and cannot be measured as is.

An **oblique line** is a straight line that is not parallel to any of the six planes of the orthographic box. When a line is oblique, it will not show true length in any of the principal views. In each of these views, the line will be foreshortened. Finding the true length of the line requires the use of an auxiliary view. A line will appear true length where the line of sight is perpendicular to the line in the adjacent view.

EXERCISE

Problem: Figure 4–5 illustrates how to find the true length of a piece of steel rod that is part of an engine hoist and is drawn foreshortened in the top and front views. If the line of sight is drawn perpendicular to the rod, it will make auxiliary plane 1 parallel to the rod; therefore, the auxiliary view of the rod will appear true length. (The reference lines between views other than the primary views are labeled with consecutive numbers.) Notice where distances H_1 and H_2 (height 1 and height 2) were found, and how they were transferred to the auxiliary view. Any point on a line or object will appear the same distance below the reference lines in all elevation views that are related to the top view. Also note that this auxiliary view is an auxiliary elevation view, since it is perpendicular to the horizontal projection plane (Figure 4–6).

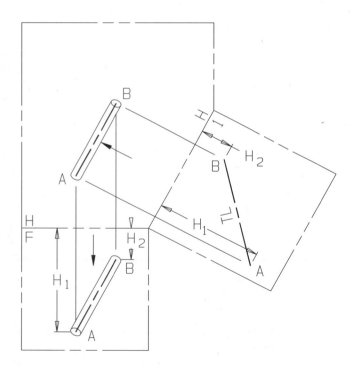

Figure 4–5 True length of line A-B found in auxiliary view.

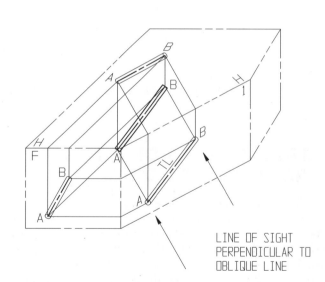

Figure 4–6 Auxiliary projection plane parallel to oblique line.

Remember, this is where the systematic steps of procedure studied in Chapter 3 will help you. Figures 4–7 through 4–14 will walk you through the steps. Sketch the solution on the figure. The correct answer will be shown on the next figure.

I. Determine line of sight for steel rod (A-B).

    A. Locate **true-length line**.

        1. Is the true-length line **in** one of the **principal planes**?

            a. If the line of sight is perpendicular to the object line in one view, it will appear true length in the adjacent view and can be measured.

The steel rod is not true length in a principal view (step I.A.1.a.); therefore, step I.A.2., listed below, must be followed.

        2. If the above rule does not apply, it is an oblique line and cannot be measured as is.

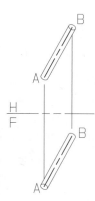

Figure 4–7 Oblique line. Sketch in the line of sight to view the top view.

            a. Draw a **primary auxiliary** view **to find the true length** (Figure 4–7). Drawing a top-adjacent or a front-adjacent auxiliary view will result in the same answer. For this problem, complete a top-adjacent auxiliary view. At this time, lightly draw in the line of sight needed to view the top view. Go ahead and draw on this page. The correct answer is shown on the next figure.

                (1) Draw the line of sight perpendicular to the oblique line in either principal view (Figure 4–8). This step allows the observer to move until his line of sight is perpendicular to the object line. The line of sight may be placed in either principal plane or either side of the line, but try to avoid overlapping views. Now rotate yourself so you are looking perpendicular to the oblique line in the top view and lightly draw in the line of sight. Remember, the rest of the steps are based around this step, so do not skip it.

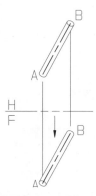

Figure 4–8 Draw the line of sight perpendicular to oblique line to complete a top-adjacent auxiliary view.

II. Draw the **reference line** perpendicular to the line of sight and label (Figure 4–9). (The reference lines between views other than the primary views are labeled with consecutive numbers.)

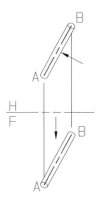

Figure 4–9 Draw the reference line perpendicular to the line of sight.

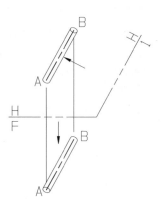

Figure 4–10 Draw projection lines parallel to the line of sight.

III. Draw **projection lines** parallel to the line of sight (Figure 4–10).

IV. Transfer **points of measurement** to the new view from the related view and label (Figure 4–11).

    A. Top-adjacent auxiliary:

        1. Always projects from the top view

        2. Always utilizes the horizontal reference plane for measurements

        3. Always shows true height

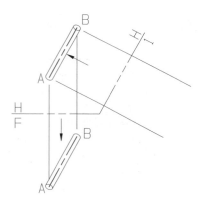

Figure 4–11 Transfer points of measurements from related view.

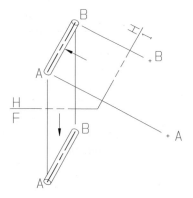

Figure 4–12 Connect the points and label TL.

V. Connect the points and **label** true-length lines, edge views, and/or true-shape planes (Figure 4–12).

Answer

Figure 4–13 shows the true length of the rod. To simplify and speed up the drawing process, the pipe, cable, shaft, etc., is usually represented as an object line or a center line. The same answer would have been found if the front-adjacent auxiliary view had been completed.

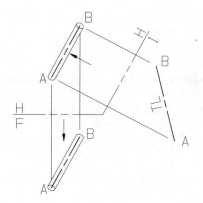

Figure 4–13 True length of rod.

Problem: Figure 4–14 shows a rafter of a hip roof in its true length on an inclined auxiliary projection plane. View line A-B from the front view. The same answer would be found if the line was viewed from the top view.

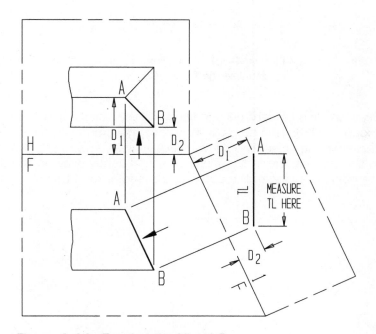

Figure 4–14 True length of line A-B.

Remember, this is where the systematic steps of procedure will help you. Figures 4–15 through 4–21 will walk you through the steps. Sketch the solution for each step on the figure. The correct answer will be shown in the next figure.

I. **Determine line of sight** of line A-B in the hip roof.

   A. Locate **true-length line**.

      1. Is the true-length line **in** one of the **principal planes**?

         a. If the line of sight is perpendicular to the object line in one view, it will appear true length in the adjacent view and can be measured.

Line A-B in the hip roof is not true length in a principal view (step I.A.1.a.); therefore, step I.A.2. must be followed.

   2. If the above rule does not apply, it is an oblique line and cannot be measured as is.

      a. Draw a **primary auxiliary** view **to find the true length**. (Figure 4–15). The line of sight may be placed in either principal plane or either side of the line, but try to avoid overlapping views. Use the front view. At this time, lightly draw in the line of sight needed to view the front view.

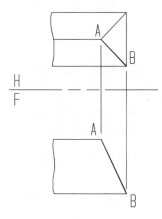

Figure 4–15 Draw line of sight to visualize the front view.

   (1) Draw the line of sight perpendicular to the oblique line in either principal view (Figure 4–16). This step allows you to move until your line of sight is perpendicular to the object line. Now rotate yourself so you are looking perpendicular to the oblique line in the front view and lightly draw in the line of sight. Remember, the rest of the steps are based around this step, so do not skip it.

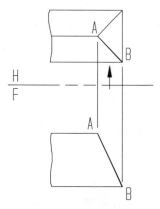

Figure 4–16 Draw line of sight perpendicular to oblique line.

  II. Draw the **reference line** perpendicular to the line of sight and label (Figure 4–17). (The reference lines between views other than the primary views are labeled with consecutive numbers.)

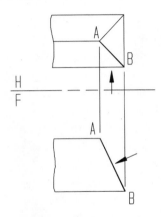

Figure 4-17 Draw reference line
perpendicular to line of sight.

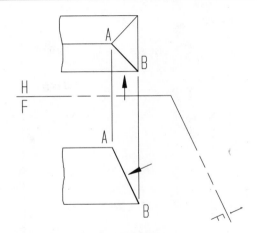

Figure 4-18 Draw projection lines parallel
to the line of sight.

III. Draw **projection lines** parallel to the line of sight (Figure 4-18).

IV. Transfer **points of measurement** to the new view from the related view and label (Figure 4-19).

    A. Front-adjacent auxiliary:

        1. Always projects from the front view

        2. Always utilizes the frontal reference plane for measurements

        3. Always shows true depth

V. Connect the points and **label** true-length lines, edge views, and/or true-shape planes (Figure 4-20).

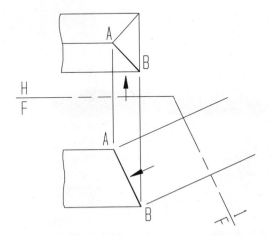

Figure 4-19 Transfer points of measurements
from related view.

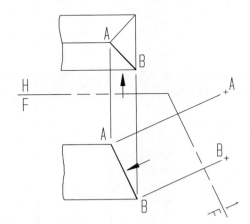

Figure 4-20 Connect the points and label TL.

Answer

Figure 4–21 shows the true length of line A-B in the hip roof.

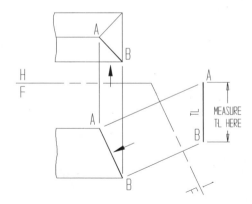

Figure 4–21 True length of line A-B.

HELPFUL HINTS

When solving a problem organize yourself first by asking the following questions before you start to draw.

What am I **solving for**?

What is **given**?

What views are given?

What type of line do I have?

What other information do I know?

What do I **need to solve** this problem?

EXERCISE

Problem: Complete Figure 4–22 to find the true length of line C-D. The problem may be solved by completing a top-adjacent view or a front-adjacent view; only one or the other view is necessary. Try completing both adjacent views for the practice. In the previous figures, the reference lines have intersected so you could cut and fold the figure to create a 3D model. In reality, the reference line may be placed anywhere, as long as it does not interfere with any other entity. For this exercise, place the reference line approximately ½" away from line C-D. If you are not sure where to start, answer the questions following the figure first.

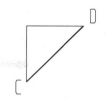

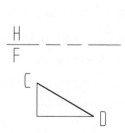

Figure 4–22 Practice problem.

Answer the following questions:

What am I **solving for**?

What is **given**?

What views are given?

What type of line do I have?

What other information do I know?

What do I **need to solve** this problem?

Answer

What am I **solving for**?

True length of line C-D

What is **given**?

What views are given?

■ Top and front

What type of line do I have?

■ The answer is oblique, because:

(Use steps of procedure I.A.1.a and b, and I.A.2. to answer this question.)

I. Determine **line of sight**.

  A. Locate **true-length line**.

    1. Is the true-length line **in** one of the **principal planes**?

      a. If the line of sight is perpendicular to the object line in one view, it will appear true length in the adjacent view and can be measured.

    2. If the above rule does not apply, it is an oblique line and cannot be measured as is.

What other information do I know?

■ None

What do I **need to solve** this problem?

■ Auxiliary view is needed because:

(Use steps of procedure I.2.a.1 to answer this question.)

---

2. If the above rule does not apply, it is an oblique line and cannot be measured as is.

a. Draw a **primary auxiliary** view **to find the true length**.

(1) Draw the line of sight perpendicular to the oblique line in either principal view.

Decide what type of auxiliary drawing will be done—top-adjacent, front-adjacent, or side-adjacent. Any primary auxiliary view will result in the correct answer. For this example a top-adjacent auxiliary drawing was decided on. Go back and try the problem, then come back and check the answer (Figure 4–23).

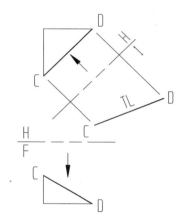

Figure 4–23   True length of line C-D.

▼ BEARING OF A LINE

The **bearing** (direction) of a line (true length or oblique) is the compass reading or angle of a line. Because bearing is used solely for map direction, it is read using a compass. If a hiker is lost (I've lost my bearing, what direction should I go?), he will use a compass to decide on the correct direction to travel. Because a compass is held horizontally—parallel to the earth—and we are on top of the earth, the bearing is found (read) only in the top orthographic view. The angle the line makes in the top view is read using a protractor from a north-south line (Figure 4–24). North is assumed to be directed toward the top of the page, unless otherwise specified.

The direction of the line (bearing) is read from the high end of the line toward the low end, unless an origin is given, such as in Figure 4–24. By stating "what is the bearing from A to B?", A became the origin. The bearing for line A-B, in Figure 4–24, is N63°E. The origin overrides any information given in an elevation view. If an origin is not specified, for example—"What is the bearing of line 1-2?", the bearing is read from the high end of the line toward the low end (Figure 4–25). Remember, height can only be determined in an elevation view (front, side, or top-adjacent view). Imagine placing a ball on the line and watching the direction it rolls—this shows the low end of the line. Try visualizing this by holding your pencil up and moving it into the position shown in Figure 4–25. The bearing for line 1-2 is N63°W.

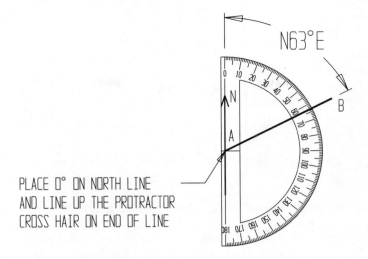

Figure 4-24 Bearing of the line A-B, from A to B is N63°E.

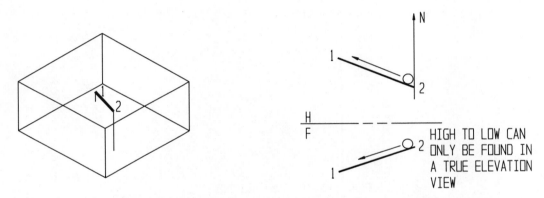

Figure 4-25 Bearing direction high to low. Notice point 2 is higher than point 1.

EXERCISE

Problem: Identify the highest endpoint of each line in the following practice problem (Figure 4-26).

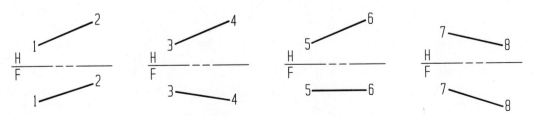

Figure 4-26 Identify the highest endpoint of each line.

Answer

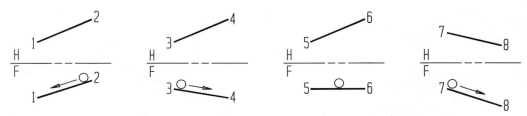

Figure 4–27 Endpoints 2, 3, and 7 are the highest endpoints. Line 5-6 is level.

There are two ways bearing may be read: azimuth or compass bearing.

1. **Azimuth bearings** are measured with respect to north in a clockwise direction up to 359° (360° should be written as "due north"). (See Figure 4–28.) Azimuth is used in aerial and marine navigation and in angular computations to avoid confusion that might be caused by reference to the four points of a compass.

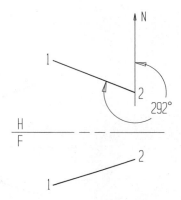

Figure 4–28 Azimuth: N292°.

2. **Compass bearing** is measured with respect to north and south (Figure 4–29). Compass bearings always begin with a north or south reading. The angle with north and south are measured toward east or west, with a maximum of 89°. Write 0° as due north, 90° as due east, 180° as due south, and 270° as due west.

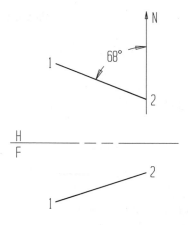

Figure 4–29 Compass bearing: N68°W.

Problem: A north arrow was not given in the previous problems, so it was assumed to be toward the top of the paper or away from the front view. Figure 4–30 is similar to that problem above, but with a specific north arrow given. First a line is drawn parallel to the given north arrow line, through either the highest end of the line or the origin. Find the azimuth and compass bearing of line 1-2 on Figure 4–30.

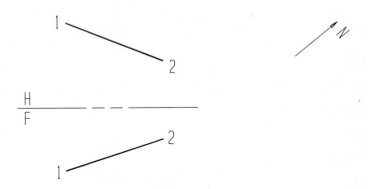

Figure 4–30 North arrow given.

Answer

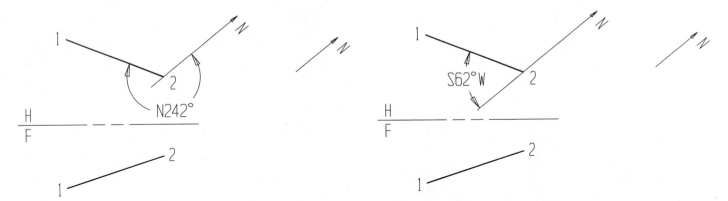

Figure 4–31 Azimuth with north arrow given: N242°.

Figure 4–32 Compass bearing with north arrow given: N62°W.

Problem: Figure 4–33 is a representation of a topographical map. The curved lines represent elevation—feet above sea level. Figures 4–33 through 4–35 show four cross country skiers leaving from point A. They are all leaving from point A, so A will be the origin, whether they are skiing up or downhill. What is the compass bearing for each skier? What is the azimuth for each skier?

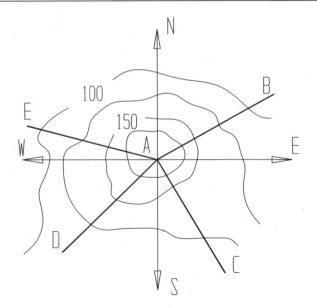

Figure 4–33 Topographic map.

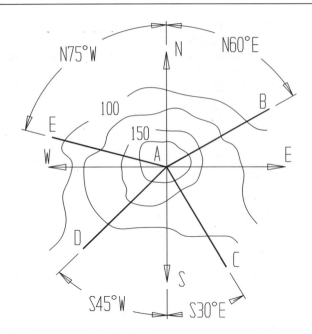

Figure 4–34 Compass bearing of four lines.
A-B: N60°E, A-C: S30°E, A-D: S45°W, A-E: N75°W.

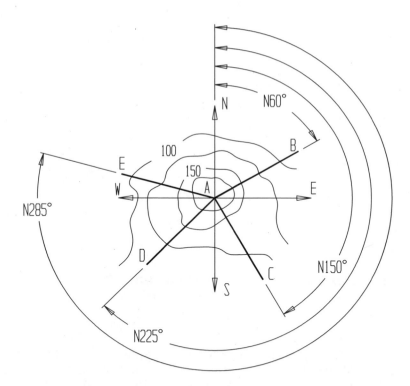

Figure 4–35 Azimuth reading of four lines. A-B: N60°,
A-C: N150°, A-D: N225°, A-E: N285°.

Answer

Students often confuse the steps to finding the bearing of a line with those to finding the slope of a line. It is important to remember that the bearing of a line is measured in the top view and is independent of its slope. This means the line's bearing is fixed regardless of whether the line is level or has a steep slope. For example, if you were cross-country skiing in a N30°E direction, it would not matter if you were skiing up or down hill—you would still be skiing N30°E. This N30°E can only be read in the top view, not in an elevation view, such as a top-adjacent view (Figure 4–36).

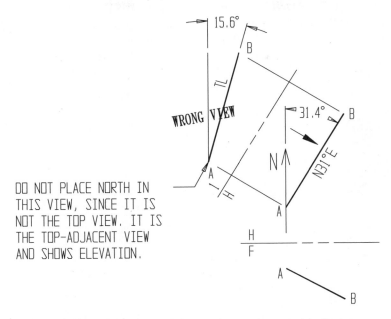

DO NOT PLACE NORTH IN
THIS VIEW, SINCE IT IS
NOT THE TOP VIEW. IT IS
THE TOP-ADJACENT VIEW
AND SHOWS ELEVATION.

Figure 4–36   Measure bearing in top view—not top-adjacent.

HELPFUL HINTS

Make up a reference note card, to set beside you when working on homework stating the following:

> **Bearing of a Line**
>
> 1. Used solely for a map direction.
> 2. Any type of line (true lenth or oblique),
> 3. Found (read) in top view only.
> 4. Measure toward low end, unless an origin is given.(Low end is found only by looking in an elevation view—front, side, or top-adjacent.)
> 5. North is assumed to be directed toward the top of the page, unless otherwise specified.
> 6. Independent of its slope. (The bearing (direction) of the line is fixed regardless of whether the line is level, or it has a steep slope. Do not measure the angle of the line in the auxiliary view.)

Referring to this note card often is the easiest way to learn the steps.

▼ SLOPE OF A LINE

While the bearing of a line is the line's direction in the top view, the **slope** of a line is the angle between a true length line and a horizontal plane. Figure 4–37 shows a building with a wheelchair ramp. The angle between the ground (horizontal plane) and the ramp is the slope. If the wheelchair ramp slope is too steep, it is difficult to wheel or push a wheelchair up the ramp or to control it coming down the ramp. Since we have not studied planes, we will study only line 1-2 of the ramp and one line of the roof.

In orthographic drawings, the slope can only be measured when the line appears true length in a view created adjacent to the horizontal plane (reference line H/?, such as H/F or H/1). In the special case of a frontal line, the true slope is seen in the front view, such as a wheelchair ramp. But in the case of oblique lines (line 3-4 in Figures 4–38 and 4–39) the true slope of the line can be seen only in an elevation view that shows its true length. To prove this to yourself, be sure to read the helpful hints section.

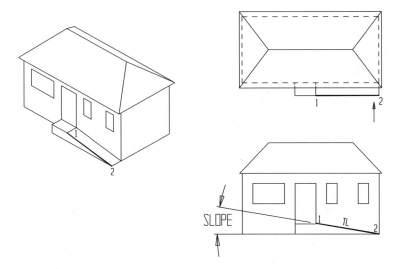

Figure 4–37 Slope of a line.

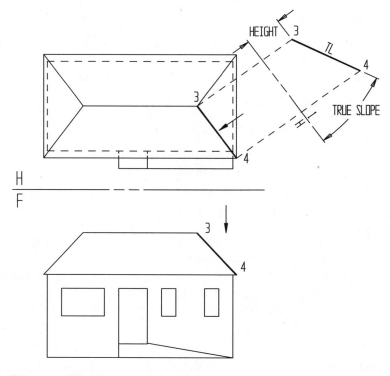

Figure 4–38 Slope of line 3-4.

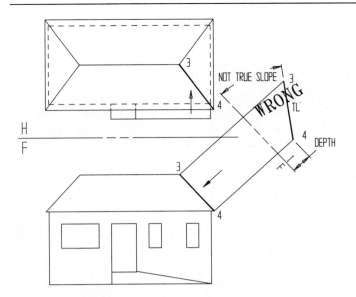

Figure 4–39 Slope is only found when the line is true length in an elevation view.

Slope may be expressed as slope angle or percent grade. Architectural drafting and structural engineering also require a means of specifying inclination.

Slope Angle

The angle itself measured in degrees. This is the most commonly used expression for slope. In this text all slope angles will be expressed as a positive number. However, you may work for a company that also uses negative readings. Figure 4–40 shows slope angle in a principal elevation view. Figure 4–41 shows slope in an elevation auxiliary view.

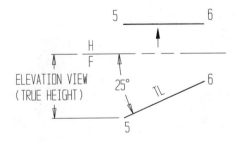

Figure 4–40 Slope angle of line 5-6 is 25°.

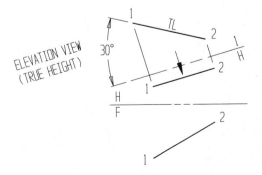

Figure 4–41 Slope angle of line 1-2 is 30°.

Percent Grade

Civil engineering, railroad and highway work, express slope in percent grade. Road signs stating a percent, such as 5%, are warning trucks that they are approaching a steep hill (grade) and to slow down and down shift. Percent grade is found by evaluating the following expression: percent grade = rise/run × 100. Rise being a vertical distance, and run being a horizontal distance.

Observe also that the grade is the tangent of the slope angle multiplied by 100. Percent grade will be expressed in both positive and negative numbers. If the line is uphill, in direction of travel, it is said to be positive. If the line is downhill, it will be negative. If an origin is not given, read the line in the front view from left to right. For example, lines 1-2 and 5-6, in Figures 4–42 and 4–43 respectively, do not have an origin. Study the front views. Both lines are traveling uphill (read line left to right).

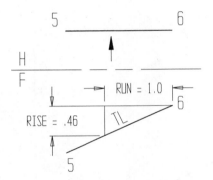

Figure 4–42 Percent grade of line 5-6 is 46%.

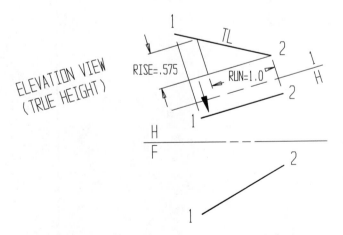

Figure 4–43 Percent grade of line 1-2 is 57.5%.

The following page demonstrates how to read the percent grade. It is easiest to read the run (parallel to the H/? reference line) as 100%, then measure the rise (perpendicular to the run).

Example

The percent problem below is shown twice. Figures 4–44 and 4–45 use 20 scale, and Figures 4–46 and 4–47 use 40 scale. The 40 scale is preferred, because the smaller divisions eliminate some of the guess work.

Part I

1. Locate a true-length line in elevation view.
2. Construct line parallel to H/1 reference line through one endpoint of the true-length line.
3. Using any scale, mark off the run—10 equal divisions on line drawn in step 2. (The scale may be any size, but spaces must be equal.) Each mark will represent 10 units totaling 100 units.

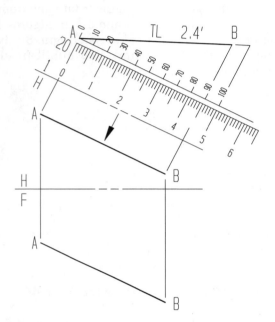

Figure 4–44

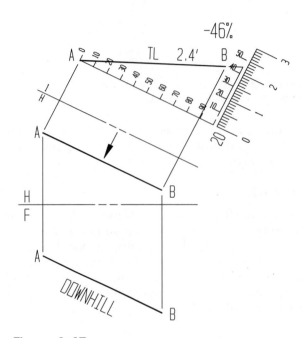

Figure 4–45

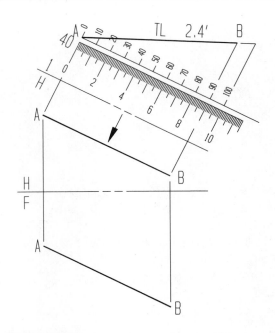

Figure 4–46

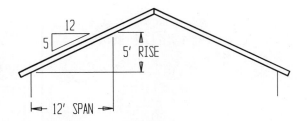

Figure 4–48 Pitch—5:12.

6. If the line is uphill, it is positive. If it is downhill, it will be a negative number. If an origin is not given, read the line in the front view from left to right. (–46%)

Other Means of Specifying Inclination

1:XX Where 1 is the rise and XX is the horizontal distance.

The unit 1 must be drawn first and then the horizontal distance.

In architectural drafting, slope can be seen as the pitch of a roof, expressed as the ratio of vertical rise to each 12 inch span. The graphical method of indicating **pitch** on a drawing is illustrated in Figure 4–48. In structural engineering, slope is the bevel of a beam and is indicated as shown in Figure 4–49.

Part II

4. Draw construction line perpendicular to H/1 reference line from the 10th division mark found in step 2 to the true length line found in step 1.

5. Using the scale used in step 3, measure the rise (distance of the perpendicular line). (46/100 = .46 or 46%)

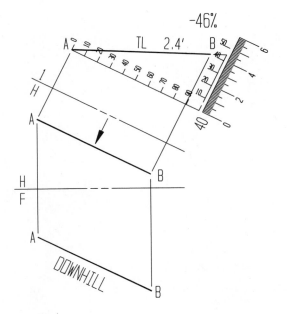

Figure 4–47

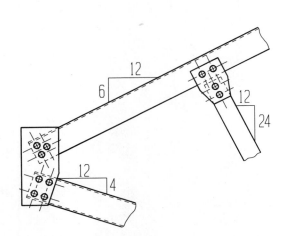

Figure 4–49 Pitch—4:12, 6:12, and 24:12.

HELPFUL HINTS

Make reference cards with the following information:

> **Slope of a Line**
>
> Must have:
>
> 1. A true-length line
> 2. In an elevation view (because the front view, side view or top-adjacent view shows one end of the line is higher than the other).
> 3. Create a construction line parallel to the edge view of the horizontal plane (H/? reference line, such as, H/F or H/1).
> 4. Measure the slope between the true-length line and the construction line created in step 3.

To help visualize what is happening, such as how the horizontal plane appears as an edge view, photocopy or trace Figure 4–50, cut it out, and fold each plane 90° on the reference lines. Set the box on a table and position yourself so the front view is eye level and your line of sight is level (parallel to the horizontal plane (floor) and perpendicular to the frontal plane). Walk around the box looking perpendicular to each plane. Notice when your line of sight is level (parallel to the floor) and perpendicular to true-length line, true slope can be seen. If your line of sight is not level, for example when looking at the top view, the line of sight is perpendicular to the floor. Can you see true slope?

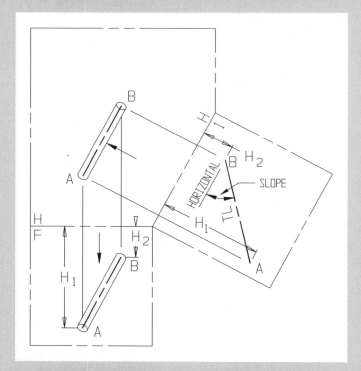

Figure 4–50 Cut out to demonstrate slope three-dimensionally.

Percent Grade of a Line

1. Using an engineer's scale, measure off 10 equal divisions parallel to the horizontal plane (run), with each unit equaling 10 units totaling 100 units. (Use any convenient scale.)

2. Measure the vertical distance (rise), using the same scale as in step 1.

3. An uphill line is positive and a downhill line is negative. If an origin is not given, read the line in the front view from left to right.

Bearing and slope are often used in the mechanical world without reference to their names, and only given in degrees or distances. For example, tunnel gates for an injection mold generally have slope, but not bearing (Figure 4–51). However, sometimes a tunnel gate will be at a compound angle (Figure 4–52).

▼ POINT VIEW OF A LINE

Thus far, we have used the auxiliary view to show the true length of a line (the first fundamental view). The second fundamental view is to show the point view of a true-length line. Point views of a line are very important in finding distances between parallel lines, perpendicular lines, skewed lines, etc., and will be covered in more detail in later chapters.

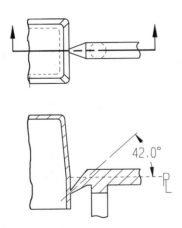

Figure 4–51 The slope of the tunnel gate is 42.0°.

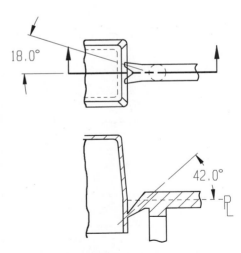

Figure 4–52 This tunnel gate has the same slope of 42.0°, but a bearing of 18.0°.

EXERCISE

Problem: To find a point view of a line, the line of sight must be parallel to the true length view of the line. This means you must look at the end of the line. Use your pencil to represent a line. Hold it up and look perpendicular to it. It is now true length. Turn your pencil so you are looking parallel to it, now you see the point view of the line. Looking at the hip roof displayed below (Figure 4–53), label which lines appear as point views in the front view? Right side view? Top view? Are those lines true length in the adjacent views?

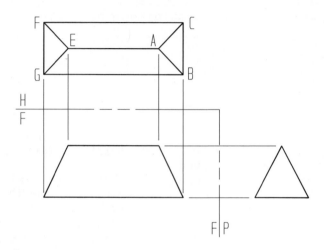

Figure 4–53   Hip roof.

Answer

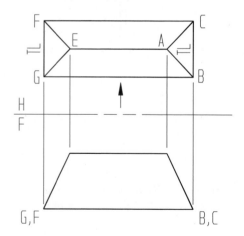

Figure 4–54   Front view.

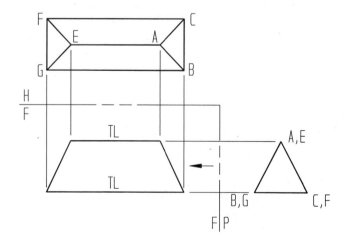

Figure 4–55   Right-side view.

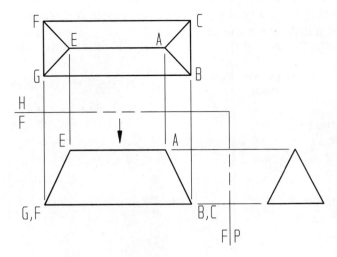

Figure 4–56   Top view.

Remember, a line must appear true length before a view may be drawn viewing the line as a point. Sometimes the answer is in a principal view, as shown in Figure 4–57. Sometimes an auxiliary view will be needed, as shown in Figure 4–58. Always analyze the problem first, to see if you may already have the answer.

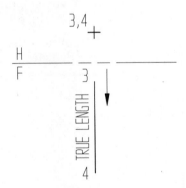

Figure 4–57   Point view of line 3-4.

Example
Solving for point view of line 3-4 in Figure 4–57:

What am I **solving for**?

What is **given**?

What **do I need** to solve this problem?

Answer

What am I **solving for**?

■ Point view of line 3-4

What is **given**?

■ True-length line in front view

■ Point view in top view

What is **needed**?

■ Nothing—problem is already solved

Example

Solving for the point view of a line, when the line is true length in a principal view (Figure 4–58).

What am I **solving for**?

What is **given**?

What do I **need to solve** the problem?

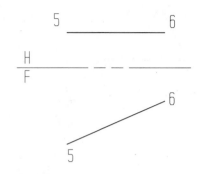

Figure 4–58 Find point view of line 5-6.

Answer

What am I **solving** for?

■ Point view of line 5-6

What is **given**?

■ Top and front views

■ True-length line in front view

What is **needed**?

■ Step I.A. is already given, so skip to step I.B. and finish steps II–V. Steps are listed below (Figure 4–59).

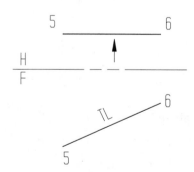

Figure 4–59 Line 5-6 is true length in the front view.

I. Determine **line of sight**.

   A. Locate **true-length line**.

      1. Is the true-length line **in** one of the **principal planes**?

a. If the line of sight is perpendicular to the object line in one view, it will appear true length in the adjacent view and can be measured.

B. Draw the **point view of a true-length line.**

   1. The line of sight must be drawn parallel to the true-length line (Figure 4–60).

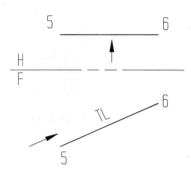

Figure 4–60 Line of sight is parallel to true length line.

II. Draw the **reference line** perpendicular to the line of sight and label.

III. Draw **projection lines** parallel to the line of sight.

IV. Transfer **points of measurement** to the new view from the related view and label.

V. Connect the points and **label** the true-length lines, edge views, and/or true-shape planes (Figure 4–61).

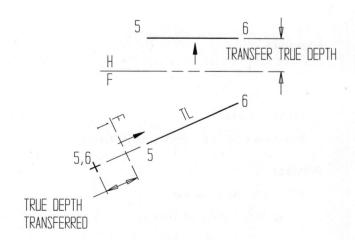

Figure 4–61 Transfer measurement from related view.

Notice the only new step was the placement of the line of sight!

Example

Solving for the point view of a line, when the line is oblique (Figure 4–62).

What am I **solving** for?

What is **given**?

What is **needed**?

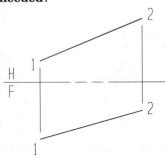

Figure 4–62 Find point view of line 1-2.

Answer

What am I **solving** for?

- Point view of a line

What is **given**?

- Top and front views

- Oblique line

What is **needed**?

- Solve for true-length line (skip to step I.A.2.)

- To draw the line of sight parallel to the true-length line to solve for point view (step I.B.)

I. Determine line of sight.

A. Locate **true-length line.**

1. Is the true-length line **in** one of the **principal planes**?

   a. If the line of sight is perpendicular to the object line in one view, it will appear true length in the adjacent view and can be measured.

2. If the rule above does not apply, it is an oblique line and cannot be measured as is.

   a. Draw a **primary auxiliary** view **to find the true length.**

(1) Draw the line of sight perpendicular to the oblique line in either principal view (Figure 4–63).

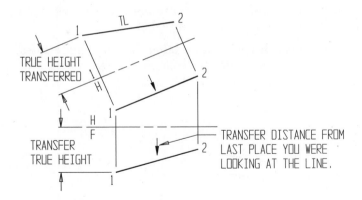

Figure 4–63 Must have true length of line 1-2 first.

II. Draw the **reference line** perpendicular to the line of sight and label.

III. Draw **projection lines** parallel to the line of sight.

IV. Transfer **points of measurement** to the new view from the related view and label.

V. Connect the points and **label** the true-length lines, edge views, and/or true-shape planes.

Step I.A. is solved, now solve for step I.B.

B. Draw the **point view of a true-length line.**

1. The line of sight must be drawn parallel to the true-length line (Figure 4–64).

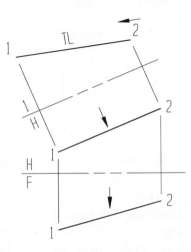

Figure 4–64 Line of sight parallel to true-length line.

II. Draw the **reference line** perpendicular to the line of sight and label.

III. Draw **projection lines** parallel to the line of sight.

IV. Transfer **points of measurement** to the new view from the related view and label.

V. Connect the points and **label** the true length lines, edge views, and/or true-shape planes.

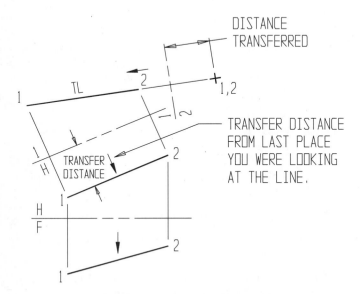

Figure 4–65 Transfer measurement from related view.

▼ LOCATING A LINE WHEN GIVEN BEARING, TRUE LENGTH, AND SLOPE

Sometimes a specific bearing, slope, and length of line is given. This known information is used to locate the line in the principal views. For example, wheelchair ramps are constructed with a maximum slope of 1:12, and if the rise of the ramp is higher than 30" a level platform must be installed. This information can be found in the principal views. However, if a specific bearing for the ramp is added the answer can no longer be found in a principal view, because the ramp is no longer true length in an elevation view.

1. Analyze the problem first. Ask yourself:
   ■ What am I **solving for**?
   ■ What is **given**?
   ■ What conditions are **needed** to locate each of the following: true length, slope, and bearing.
   ■ Ask which—true length, slope, or bearing—can be drawn with what is given first, second, and third.

2. Draw the true length, slope, and bearing in the proper order.

3. Transfer the second point of the line back to the principal views.

Example
Line 1-2 is 2" long, has a downward slope of 20° and a bearing of N60°E. Locate line 1-2 in the principal views. Point 1 is given (Figure 4–66).

Figure 4–66 Locate line 1-2 when given bearing, true length and slope.

Answer
What am I **solving for**?

   ■ Line 1-2 in the top and front views

What is **given**?

   ■ Endpoint 1 in the top and front views

   ■ True length distance, slope, and bearing information

Ask what conditions are **needed** to locate each of the following: true length, slope and bearing.

   ■ True length can be drawn in any view, as long as the line of sight is perpendicular to the object line in the adjacent view.

   ■ Slope can only be found where the object line is true length in an elevation view.

   ■ Bearing can be found of any type of line in the top view.

Ask whether true length, slope, or bearing can be drawn with what is given.

   ■ True length? No, you can't tell which view the line will be true length in. And you can't create a true-length line, because a line is needed in order to draw a line of sight perpendicular to it.

   ■ Slope? No, you need a true-length line.

   ■ Bearing? Yes, one end of the line is shown in the top view and the length and type of the line doesn't matter (Figure 4–67).

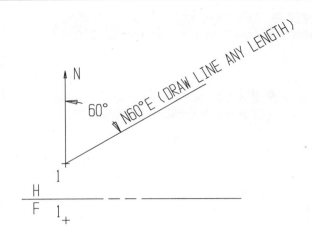

Figure 4–67 Locate bearing first.

Ask whether true length or slope can be drawn next.

■ True length? You can set up the line of sight and find the given point, but a distance to the second point is needed or slope of the line is needed.

■ Slope? Yes, you can set up the line of sight for true-length line in a true elevation view (top-adjacent view). While the true-length line is needed, the actual distance is not required. Be careful when locating upward and downward. Toward the top view is up!! That does not mean toward the top of the paper. (See up arrow in Figure 4–68. Copy or trace Figure 4–68 and fold 90° along the reference lines to prove to yourself which way is up.)

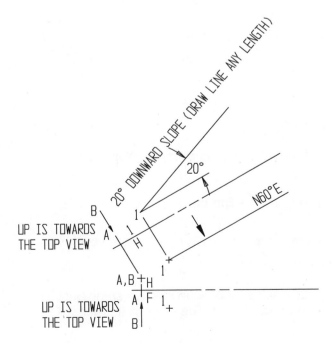

Figure 4–68 Locate slope next.

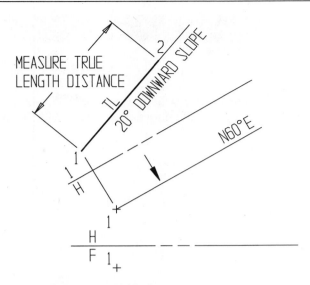

Figure 4–69 Locate true length of line 1-2.

Can true-length distance now be found?

■ Yes, true length can be measured on line showing slope (Figure 4–69).

Transfer the second point of the line back to the principal views. Projection lines used to transfer endpoints back are drawn with hidden lines and directional arrows to help you visualize the process. Check your downward slope. Is line 1-2 sloping uphill or downhill from point 1 in the front view? It is sloping downhill as per the instructions.

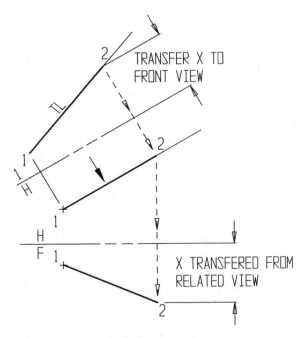

Figure 4–70 Transfer point 2 back to front view.

EXERCISE

Problem:

Let's try a practice problem. A new 8.5' conveyor line, line 1-2, is to be run at an uphill slope of 20 degrees in a N70°W direction. Show line 1-2 in the front view (Figure 4–71).

Figure 4–71 Endpoint 1 of line 1-2 given in top and front views.

First organize yourself by answering the following questions.

■ What am I **solving for?**

■ What is **given?**

■ Ask what conditions are **needed** to locate each of the following: true length, slope and bearing.

■ Ask which of true length, slope and bearing can be drawn with what is given first, second and third.

Answer

What am I **solving** for?

■ Line 1-2 in the top and front views

What is given?

■ Endpoint 1 in the top and front views

■ True-length distance, slope, and bearing information

Ask what conditions are needed to locate each of the following: true length, slope, and bearing.

■ True length can be drawn in any view, as long as the line of sight is perpendicular to the object line in the adjacent view.

■ Slope can only be found where the object line is true length in an elevation view.

■ Bearing can be found for any type of line in the top view.

Remember, the steps of procedures for true length, bearing, and slope are the same, just in a different order. Sketch the solution for each step on the figure. The correct answer will be shown in the following figure.

Ask whether true length, slope, or bearing can be drawn with what is given.

- True length? No, you can't tell which view the line will be true length in. And you can't create one, yet because there isn't a line to look at perpendicular.
- Slope? No, you need a true-length line.
- Bearing? Yes, one end of the line is shown in the top view and the length and type of the line doesn't matter (Figure 4–72).

If solving on a 3D CAD system, slope must be completed before bearing.

Figure 4–72   Draw in N70°W bearing.

Ask whether true length or slope can be drawn next.

- True length? You can set up the line of sight and find the given point, but a distance to the second point is needed or slope of the line is needed.
- Slope? Yes, you can set up the line of sight for true-length line in a true elevation view (top-adjacent view). While the true length line is needed, the actual distance is not required. Be careful when locating upward and downward. Toward the top view is up!! That does not mean toward the top of the paper.

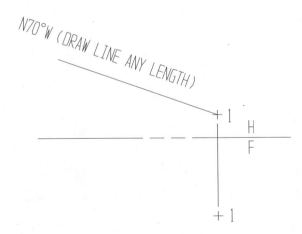

Figure 4–73   Locate 20° upward slope.

Can true-length distance now be found?
- Yes, true length can be measured on the line showing slope (Figure 4–74).

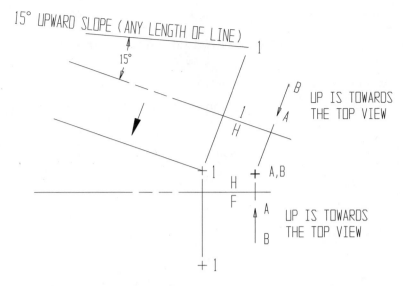

Figure 4–74 Locate true length of line 1-2.

Transfer the second point of the line back to the principal views.

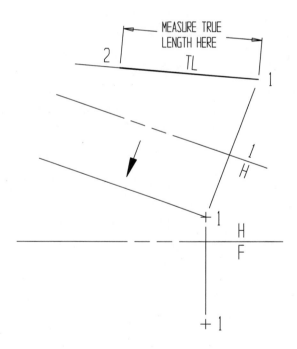

Figure 4–75 Transfer point 2 back to front view.

Answer

Finished problem. Check the upward slope. Is the line sloping uphill from point 1 in the front view?

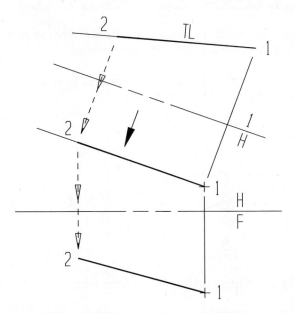

Figure 4–76 Line 1-2 in top and front view.

CHAPTER 4 STUDY QUESTIONS

The Study Questions are intended to assess your comprehension of chapter material. Please write your answers to the questions in the space provided.

1. Label the types of lines shown below. Also, sketch in the appropriate lines of sight to view the true-length lines and label the true-length lines TL.

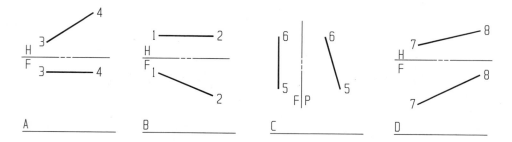

A _____   B _____   C _____   D _____

2. Where is the line of sight placed to see the true length of a line?

    _____

3. Locate and measure the true-length line for line 3-4. **Show construction work.**

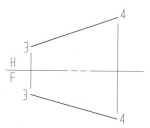

4. Define the bearing of a line.

    _____

    _____

5. List the conditions that must exist in order to measure the bearing of a line.

    _____

    _____

6. What is the compass bearing for the drawing below? **Show construction work.**

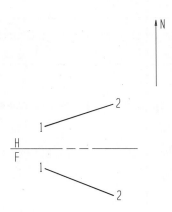

7. There are three figures shown below, each showing line 1-2 in the same position. One figure measures compass bearing, one figure measures azimuth, and one figure measures bearing incorrectly. What is the correct compass bearing for line 1-2?

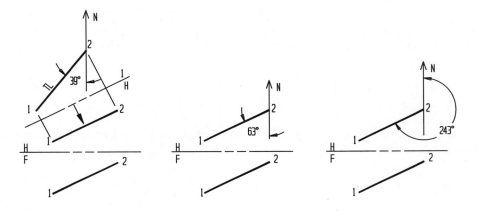

CHOOSE THE CORRECT COMPASS BEARING FOR LINE 1-2:

A. N 39° E   E. N 63° E   I. N 63°

B. N 39° W   F. N 63° W   J. S 63°

C. S 39° E   G. S 63° E   K. N 243°

D. S 39° W   H. S 63° W   L. S 243°

8. There are three figures shown in question number 7, each showing line 1-2 in the same position. One figure measures compass bearing, one figure measures azimuth, and one figure measures bearing incorrectly. What is the correct azimuth for line 1-2?

CHOOSE THE CORRECT COMPASS BEARING FOR LINE 1-2:

A. N 39° E   E. N 63° E   I. N 63°

B. N 39° W   F. N 63° W   J. S 63°

C. S 39° E   G. S 63° E   K. N 243°

D. S 39° W   H. S 63° W   L. S 243°

9. Define the true slope of a line.

_____

10. List the conditions that must exist in order to measure the slope of a line.

_____

_____

11. What is the slope angle for the drawing below. **Show construction work.**

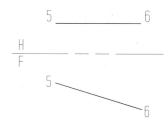

12. What is the percent grade for the drawing shown below. **Show construction work.**

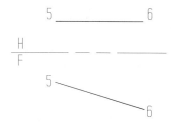

13. What is the true length, compass bearing, and slope angle for the following problem? **Show construction work.**

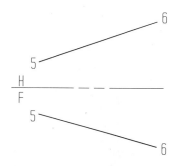

14. True   False   In order to obtain a point view of a line, your line of sight may be selected parallel to any view of the line.

15. What two lines of sight are necessary to locate a point view of the line.

_____

16. Locate the point view of line 1-2. **Show construction work.**

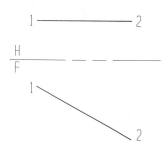

The problems in this chapter are based on a variety of industrial applications. In solving these problems you should try to utilize direct approaches that get to the heart of the problem. At this analysis stage of the design process, many problems use center lines and single lines to represent the problem situation. Yet, when necessary, relevant sizes are given.

The following problems may be drawn using instruments on the page provided, created using a 2D or 3D CAD system, or a combination of drawing board and CAD. Refer to appendix A for additional dimensions needed to solve problems three-dimensionally.

Chapter 4, Problem 1:   IN EACH OF THE FOUR PROBLEMS BELOW, THERE IS A VIEW WHICH IS MISSING.
COMPLETE THE MISSING VIEW. IF THE LINE IS TRUE LENGTH, LABEL IT TL IN THE APPROPRIATE
VIEW, DRAW IN THE LINE OF SIGHT TO VIEW IT TRUE LENGTH, AND MEASURE IT. LETTER THE LINE
TYPE AND ACTUAL LENGTH ON THE LINES PROVIDED.

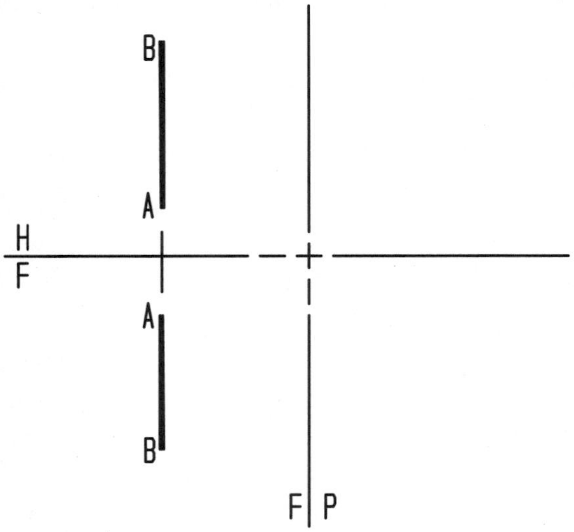

B

A

$\dfrac{H}{F}$

A

B

F | P

TYPE OF LINE _____

TRUE LENGTH _____

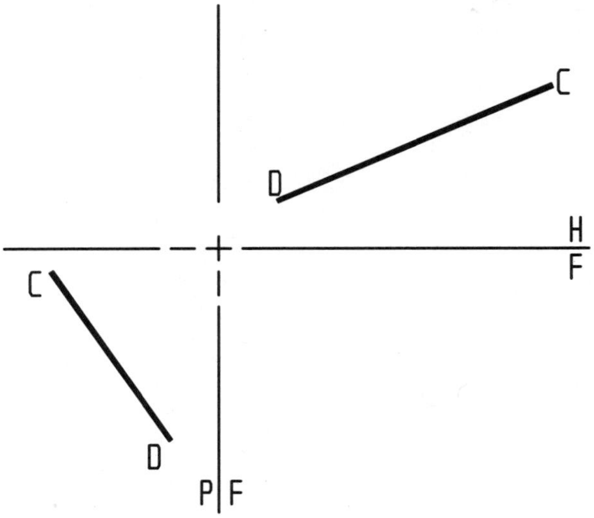

C

$\dfrac{H}{F}$

C

D

D

P | F

TYPE OF LINE _____

TRUE LENGTH _____

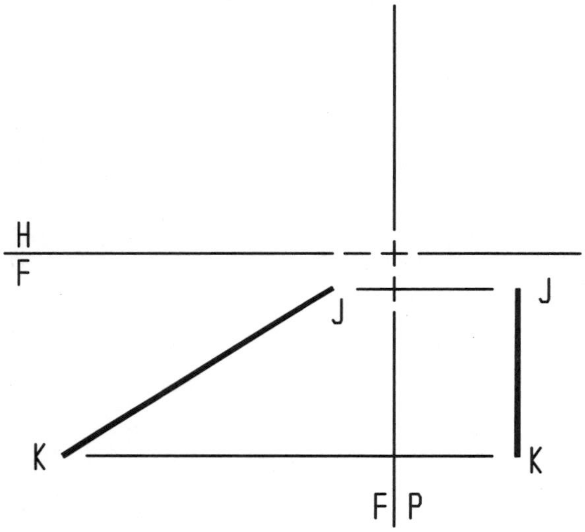

$\dfrac{H}{F}$

J   J

K   K

F | P

TYPE OF LINE _____

TRUE LENGTH _____

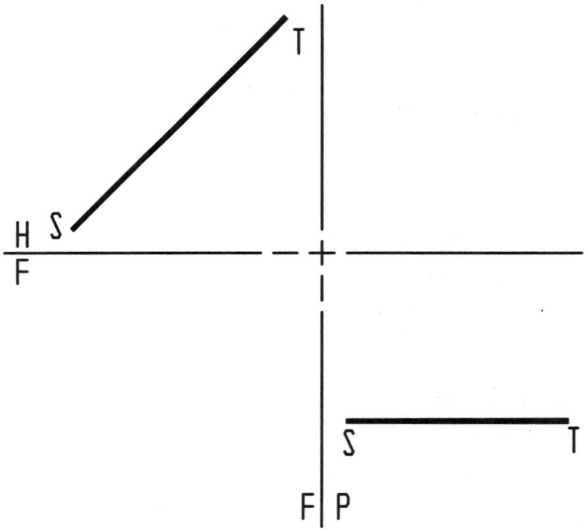

T

$\dfrac{H}{F}$ S

S   T

F | P

TYPE OF LINE _____

TRUE LENGTH _____

	ADG INC.			
	TRUE LENGTH				
	SIZE	DATE	DWG NO.	SCALE FULL	REV
	DRW. BY			SECTION #	SHEET

Chapter 4, Problem 2: TWO ORTHOGRAPHIC VIEWS OF FOUR LINES ARE GIVEN BELOW. COMPLETE THE MISSING VIEW. IF THE LINE IS TRUE LENGTH IN A PRINCIPAL VIEW: LABEL IT TL, DRAW THE LINE OF SIGHT NEEDED TO VIEW THE TRUE LENGTH LINE, AND MEASURE IT. LETTER THE LINE TYPE AND ACTUAL LENGTH ON THE LINES PROVIDED. DO NOT DRAW AN AUXILIARY VIEW TO LOCATE THE TRUE LENGTH LINE.

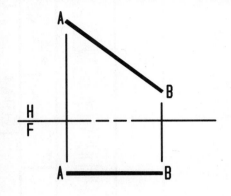

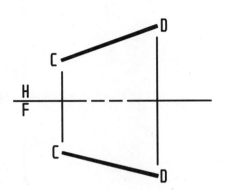

TYPE OF LINE _____

TYPE OF LINE _____

TRUE LENGTH _____

TRUE LENGTH _____

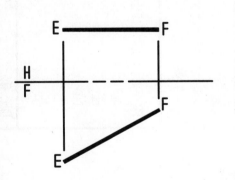

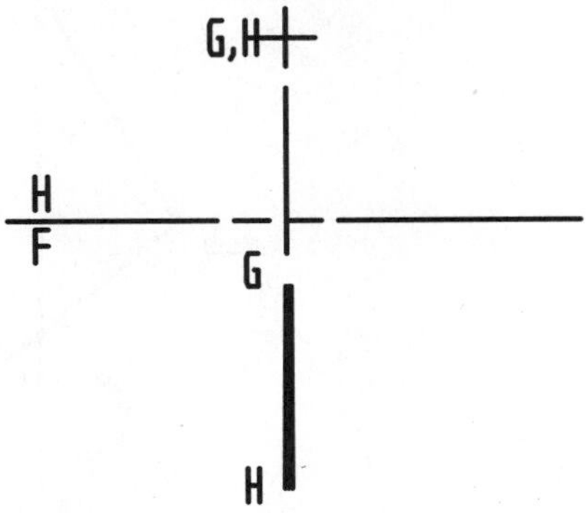

TYPE OF LINE _____

TYPE OF LINE _____

TRUE LENGTH _____

TRUE LENGTH _____

	ADG INC.
	LINE TYPE

SIZE	DATE	DWG NO.	SCALE FULL	REV
DRW. BY			SECTION #	SHEET

Chapter 4, Problem 3: THE PROPERTY LINE OF A BUILDING SITE IS SHOWN. FIND THE BEARING OF EACH SEGMENT OF THE PROPERTY LINE. WITHOUT AN ELEVATION VIEW, HIGH-LOW CANNOT BE DETERMINED. THEREFORE, USE THE FIRST LETTER OF EACH PROBLEM AS THE ORIGIN. FOR EXAMPLE, THE ORIGIN FOR LINE AB IS A.

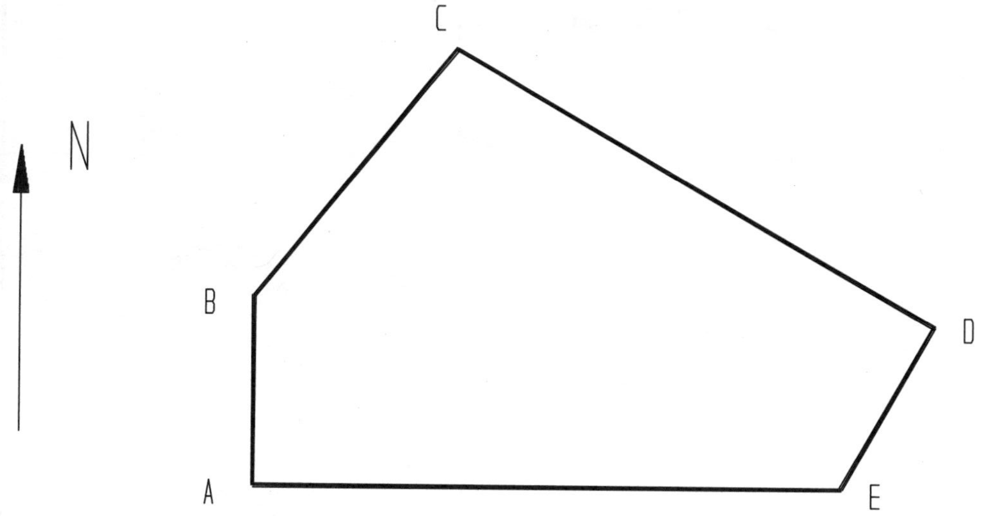

N

PROPERTY LINE BEARING

PROPERTY LINE	BEARING
AB	
BC	
CD	
DE	
AE	

DRW. BY

ADG INC.

PROPERTY LINES

SIZE	DATE		DWG NO.		SCALE	REV
					NTS	
			SECTION #			SHEET

Chapter 4, Problem 4: A SECTION OF CONCRETE WALL WITH A LEVEL TOP IS SHOWN. CORNERS AB AND CD ARE TO BE COVERED WITH STEEL ANGLE FOR PROTECTION AGAINST CHIPPING. FIND THE TRUE LENGTH AND SLOPE ANGLE OF CORNERS AB AND CD.

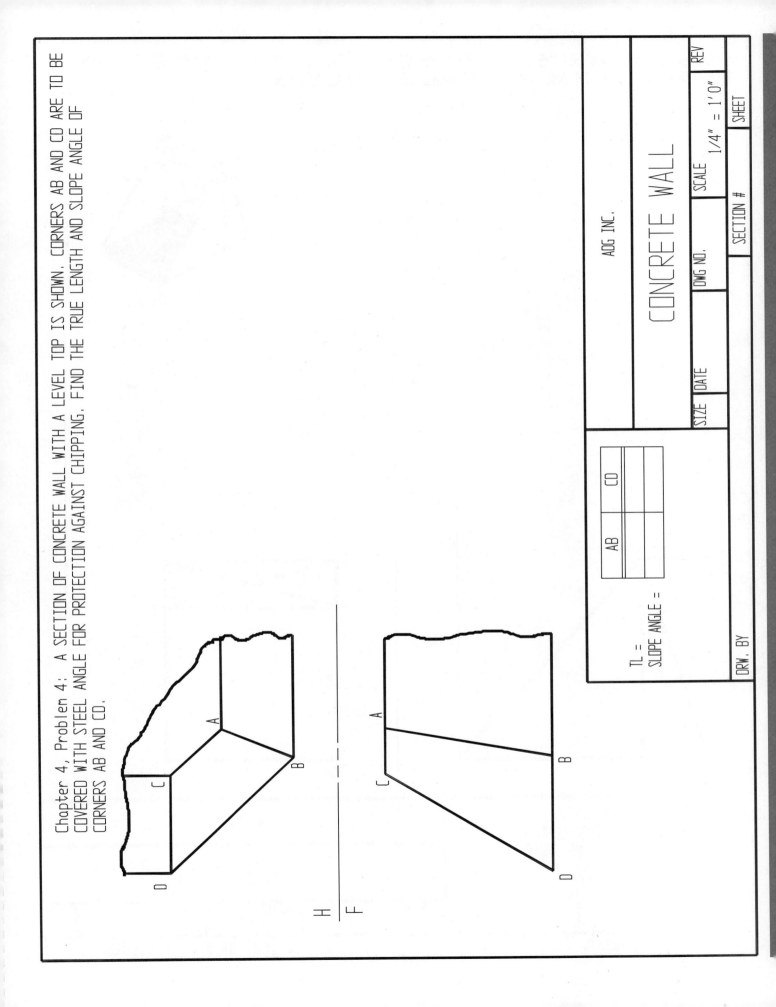

H
—————
F

ADG INC.

CONCRETE WALL

	AB	CD

TL =
SLOPE ANGLE =

SIZE	DATE		DWG NO.		SCALE	REV
					1/4" = 1'0"	
					SECTION #	SHEET

DRW. BY

Chapter 4, Problem 5: THE FRONT AND SIDE VIEWS OF THE END PANEL OF A BRIDGE ARE SHOWN.
FIND THE TRUE LENGTH AND THE SLOPE ANGLE OF DIAGONAL MEMBER XY.

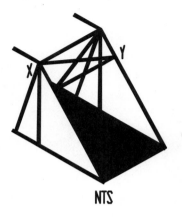

NTS

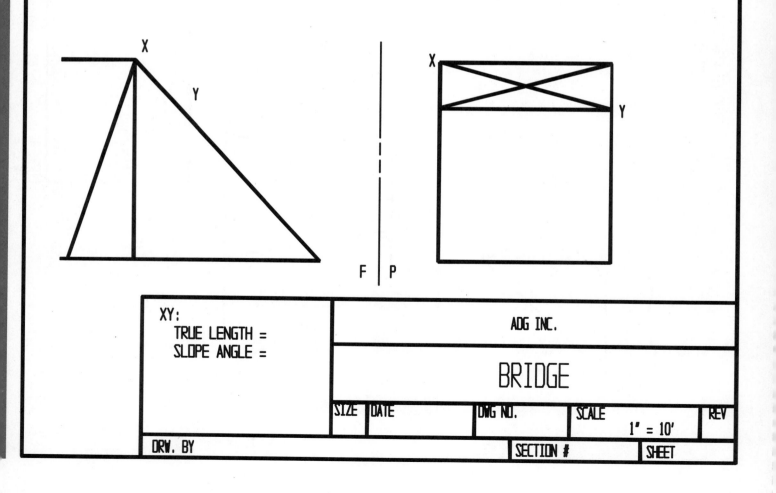

F P

XY:	ADG INC.
TRUE LENGTH =	
SLOPE ANGLE =	**BRIDGE**

SIZE	DATE	DWG NO.	SCALE	REV
			1" = 10'	
DRW. BY		SECTION #	SHEET	

Chapter 4, Problem 6: A DERRICK, SUPPORTING A HORIZONTAL DERRICK BOOM, YZ, IS MOUNTED ON A BARGE AS SHOWN. FIND THE TRUE LENGTH AND THE SLOPE ANGLE OF THE ANCHOR CABLE, WY, AND THE STRUCTURAL MEMBER, XY.

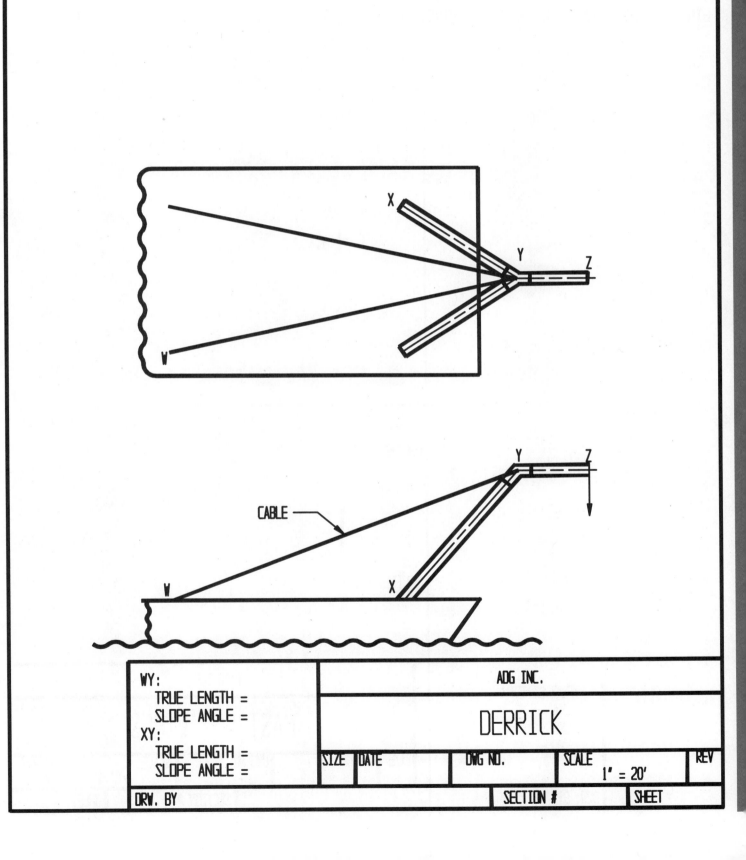

WY:		ADG INC.			
TRUE LENGTH =					
SLOPE ANGLE =		DERRICK			
XY:					
TRUE LENGTH =	SIZE DATE	DWG NO.	SCALE		REV
SLOPE ANGLE =			1" = 20'		
DRW. BY			SECTION #	SHEET	

Chapter 4, Problem 7: A VERTICAL MAST, XY, IS ANCHORED BY THREE GUY WIRES, XA, XB, AND XC.
ANCHOR A IS 13 FEET WEST AND 4 FEET NORTH OF MAST XY AND IS AT AN ELEVATION OF 140 FEET.
ANCHOR B IS 12 FEET EAST AND 7 FEET NORTH OF MAST XY, AND IS AT AN ELEVATION OF 146 FEET.
ANCHOR C IS 5 FEET EAST AND 10 FEET SOUTH OF THE MAST AND IS AT AN ELEVATION OF 138 FEET.
(PLATE 2-5)

FIND THE TRUE LENGTH OF EACH GUY WIRE.

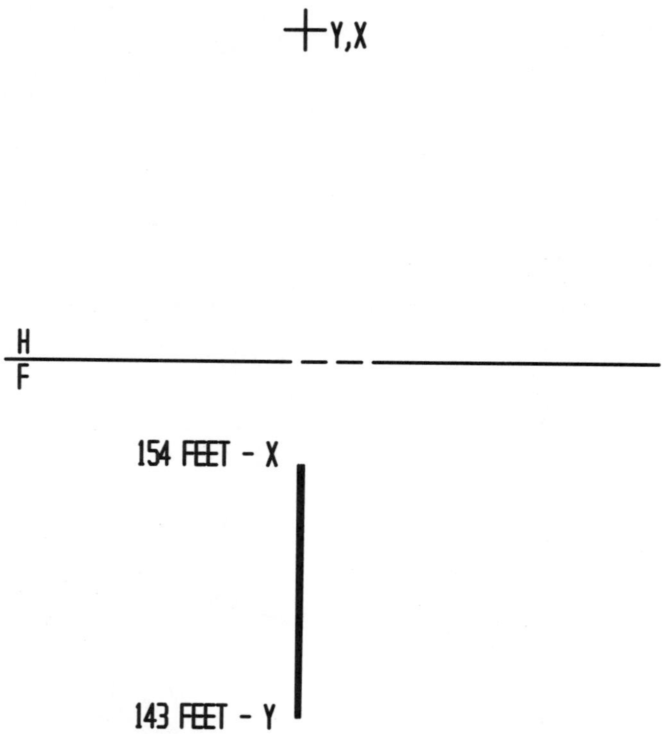

$+$ Y,X

$\dfrac{H}{F}$ ———————— — — ————————

154 FEET - X

143 FEET - Y

TRUE LENGTH:	ADG INC.			
XA = XB = XC =	MAST GUY WIRES			
	SIZE	DATE	DWG NO.	SCALE 1/8"=1'0"	REV
DRW. BY			SECTION #		SHEET

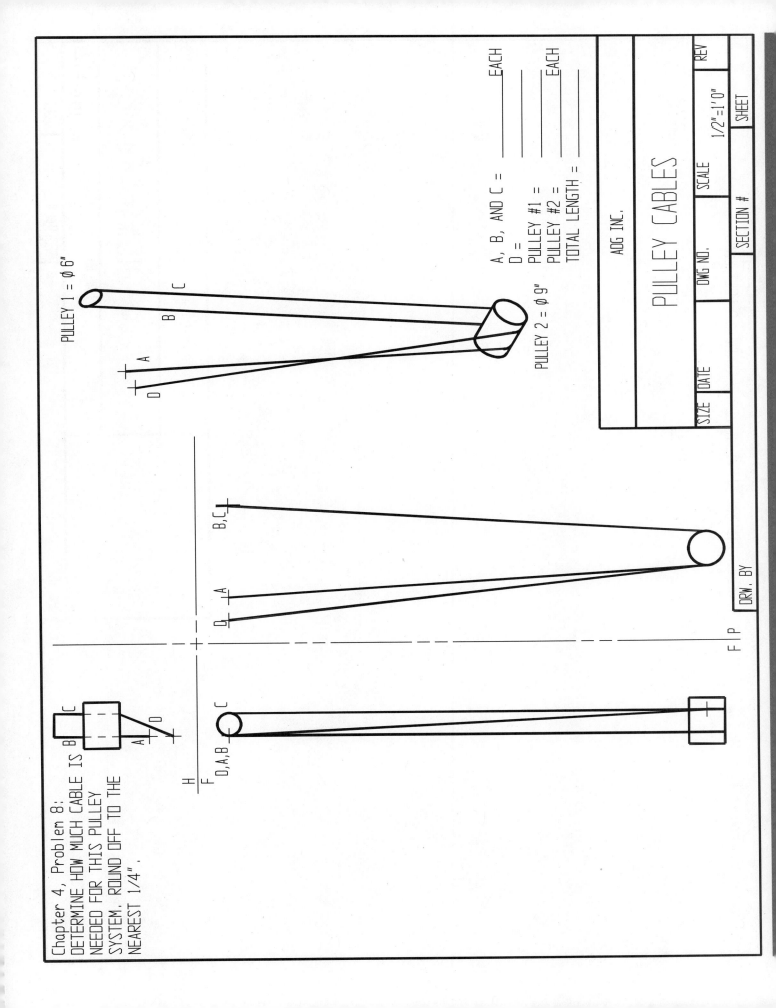

Chapter 4, Problem 8;
DETERMINE HOW MUCH CABLE IS
NEEDED FOR THIS PULLEY
SYSTEM. ROUND OFF TO THE
NEAREST 1/4".

PULLEY 1 = ⌀ 6"

PULLEY 2 = ⌀ 9"

A, B, AND C = _____ EACH
D = _____
PULLEY #1 = _____
PULLEY #2 = _____ EACH
TOTAL LENGTH = _____

ADG INC.

PULLEY CABLES

SIZE	DATE	DWG NO.	SCALE 1/2"=1'0"	REV
	DRW. BY	SECTION #	SHEET	

Chapter 4, Problem 9: DRAW TWO VIEWS OF A STRAIGHT HIGHWAY SECTION, AB, THAT IS 250 FEET LONG,
AND RUNS N47°W WITH A DOWNGRADE OF 6 PERCENT FROM POINT A.

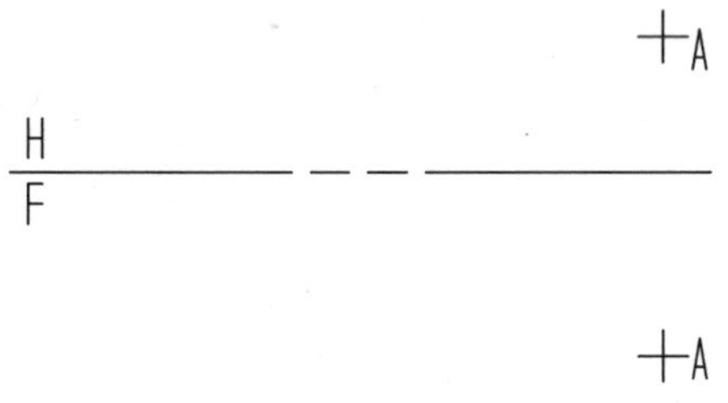

+A

H
—————————
F

+A

ADG INC.

HIGHWAY

SIZE	DATE	DWG NO.	SCALE	REV
			1" = 100'	
DRW. BY		SECTION #	SHEET	

Chapter 4, Problem 10: DRAW TWO VIEWS OF A MINE SHAFT, MN. BEGINNING AT THE BOTTOM OF THE MINE
SHAFT (POINT M), LINE MN IS A BEARING OF S35°W, AN UPHILL GRADE OF 15 PERCENT, AND 150 FEET IN LENGTH.

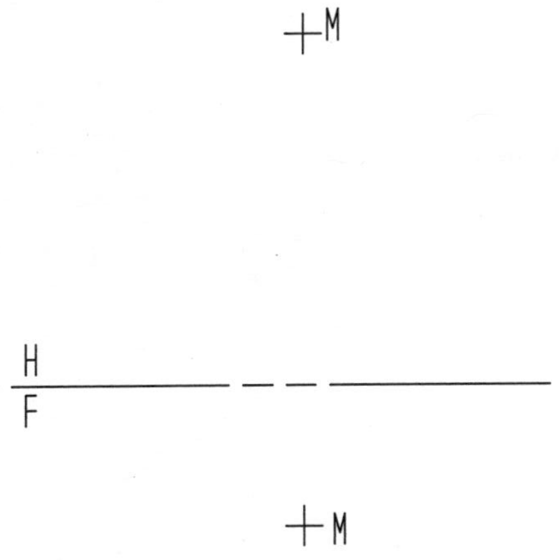

+ M

H
F

+ M

ADG INC.

MINE SHAFT

SIZE	DATE	DWG NO.	SCALE	REV
			1" = 100'	
	SECTION #		SHEET	

DRW. BY

Chapter 4, Problem 11: A CONVEYOR LINE SYSTEM IS CURRENTLY INSTALLED AS INDICATED BY THE
FOLLOWING INFORMATION.

SECTION	BEARING	SLOPE/GRADE	TRUE LENGTH
AB	N40°W	-20%	135'
BC	DUE WEST	0°	105'
CD	S30°W	15° DOWNWARD	150'

A REVISION IN PRODUCTION SEQUENCE CALLS FOR THE ELIMINATION OF THE PRESENT CONVEYOR SYSTEM AND
THE INSTALLATION OF A NEW CONVEYOR LINE DIRECTLY FROM A TO D. FIND THE TRUE LENGTH, BEARING, AND
SLOPE OF THE NEW LINE.

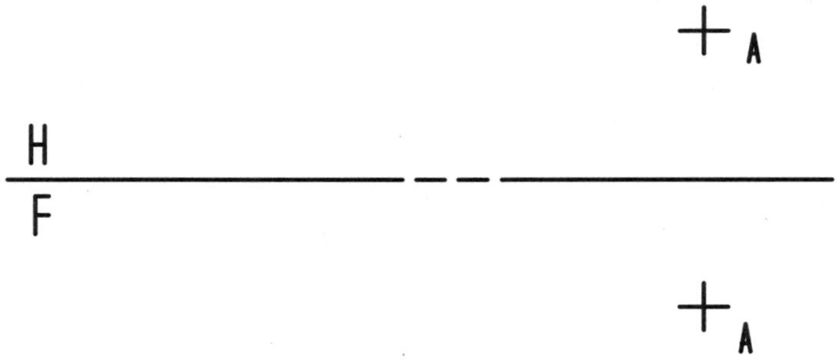

$+$ A

$\dfrac{H}{F}$ ———— – – ————

$+$ A

AD	ADG INC.
TRUE LENGTH =	
SLOPE ANGLE =	NEW CONVEYOR LINE
BEARING =	

SIZE	DATE	DWG NO.	SCALE 1" = 100'	REV

| DRW. BY | | SECTION # | SHEET 1 OF 1 |

Line Characteristics

When given a set of drawings for these pipes, how will you be able to identify which pipes intersect and which do not intersect? Which pipes are above or below other pipes?

After completing this chapter, you will be able to:

- *Determine if a point is on a line.*
- *Determine if lines are intersecting or nonintersecting.*
- *Determine if lines are parallel or not.*
- *Determine if lines are perpendicular or not.*
- *Construct a view that will show the clearance between two parallel lines.*
- *Find the shortest distance from a given point to a line.*
- *Find the shortest distance from two skewed lines.*

We know that everyday objects are made up of points, lines, and planes. This chapter will discuss the relationship between lines. Lines may be intersecting or nonintersecting. There are four typical line characteristics for two lines: parallel, perpendicular, intersecting, or skewed.

Having studied the concepts of orthographic projection and two of the four fundamental views, you are well versed in the basics of descriptive geometry. The purpose of this chapter is to build on those skills. You will be introduced to a variety of line problems that occur often in actual practice and that drafters and engineers should know how to solve. You will continue to use the same methods of logical thinking and three-dimensional visualization you applied to the problems in previous chapters.

▼ LOCATE A POINT ON A LINE

Sometimes a point is located on a line and sometimes it only appears to be located on the line. Figure 5–1 shows an example of each. Which example in Figure 5–1 shows the point located on the line and which one only appears to be on the line?

You are right, you do not have enough information to answer that question. Remember, each view shows only two dimensions. An adjacent view is needed to show the third dimension. By studying the front view in Figure 5–2, can you now tell which point is located on the line?

Figure 5–2 shows the same coordinates for lines A-C and D-E. The top views show points B and F are located at the same width and depth. The adjacent view (front view) shows point B's height the same as line A-C;

therefore, it is located on line A-C. Point F's height is higher than line D-E; therefore, it is not located on line D-E.

Will you need to know this information? Yes. For example, if you are drawing a new gas line connecting to an existing gas line, will it matter if the gas lines intersect? Yes; therefore, the connecting point between the two gas lines must be drawn on the line. Would you want the end of a hopper chute to run into or rest on a gas pipe? No; therefore, the end of the hopper must not be drawn on the line.

Let's compare a two-dimensional representation to a three-dimensional representation of the examples given above. Figure 5–3(A) shows the intersection of two gas lines. Figure 5–3(B) shows the gas line and an end of a hopper chute. Because an isometric shows all three dimensions at the same time, it is not difficult to see

Figure 5–1 Which point, B or F, is on the line?

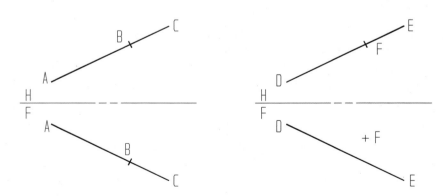

Figure 5–2 Point B is on line A-C. Point F is not on line D-E.

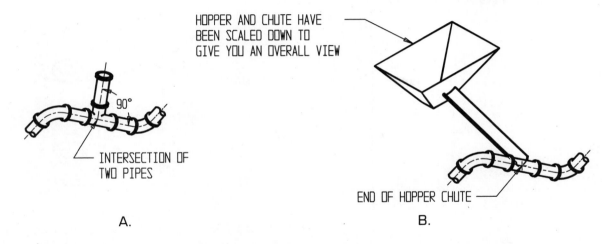

Figure 5–3   (A) Isometric of two pipes. (B) Isometric of a pipe and a hopper.

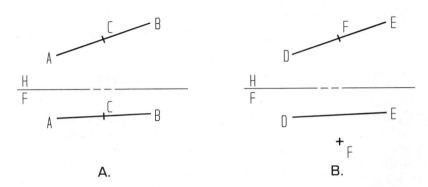

Figure 5–4   (A) The two pipes intersect. (B) The pipe and hopper do not intersect.

that the two gas lines intersect, but it is not as easy to determine if the hopper shoot and gas line touch.

Two-dimensional representation is another story. In Figure 5–4(A), line A-B represents the existing gas line and point C represents where the new gas line intersects the existing gas line. Point C shows the location of where to install the tee-joint. The gas line in Figure 5–3(B) is represented in Figure 5–4(B) as line D-E and the end of the hopper chute is represented as point F. Study Figure 5–4. Both top views look the same, because the top view shows two dimensions (width and depth) of the location. The answer is in the adjacent view (front view) because it shows the third dimension—height.

▼ INTERSECTING LINES

When lines intersect, the point of intersection is a point that lies on both lines. At least two views are necessary to test for intersection. Each view shows only two dimensions; the adjacent view shows the missing dimension. If the crossing point in adjacent views line up on the same projection line, the two lines intersect. Figure 5–5 shows two intersecting cables (A-B and C-D) which intersect at the cable junction, point X. You know the two cables intersect because both the top and front views of point X are on the same projection line. Point X in one view (top view) is directly above point X

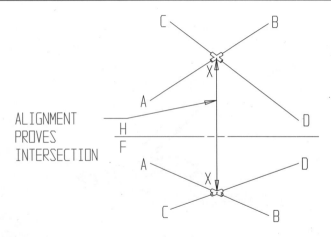

Figure 5–5 Two intersecting cables.

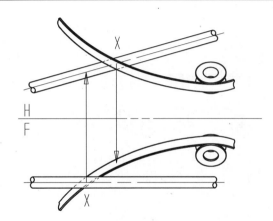

Figure 5–6 The pipe and belt do not intersect.

in the adjacent view (front view), proving that point X is a point on both cables and that they intersect.

Conversely, when lines are nonintersecting, the apparent point of intersection does not align between adjacent views, which proves that they do not intersect. Nonintersecting lines, or skewed lines, are not parallel and do not touch. Figure 5–6 shows a pipe and a belt. Do they intersect? Point X in the top view in Figure 5–6 does not align with a point of intersection in the front view, and point X in the front view does not align with a point of intersection in the top view. The pipe is below and in front of the belt. (The next section, Visibility of a Point and a Line will explain how to test for the visibility.)

Figure 5–7 illustrates the special case of a profile alignment. One is two intersecting cables and the other is two skewed (nonintersecting) cables. Can you use the projection line method to prove which two cables intersect?

No, an additional view is necessary, because both endpoints of one cable and the crossing point of the other cable use the same projection line. Study the additional views (right-side views) in Figure 5–8. The projection line method works between the front and right-side views. Which cables intersect?

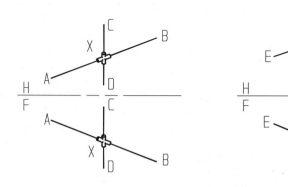

Figure 5–7 Special case of projection line method.

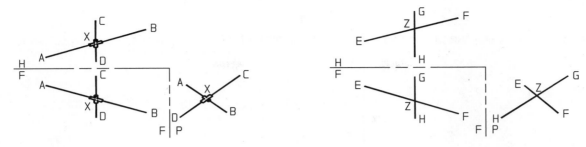

Figure 5–8 Which lines intersect?

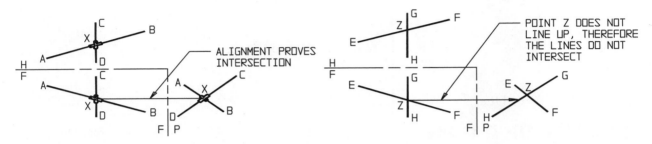

Figure 5–9 Lines A-B and C-D intersect. Lines E-F and G-H do not intersect.

When both endpoints of a line and the crossing point with another line use the same projection line to project to an adjacent view, a third view is needed to test for intersection. With the right-side view, it can be seen in Figure 5–9 that X in the front view and X in the right-side view align with each other, thus proving that cables A-B and C-D intersect at point X. With the right-side view, it can be seen in Figure 5–9 that Z in the front view and Z in the right-side view do not align with each other, thus proving that cables E-F and G-H are skewed (nonintersecting).

EXERCISE

Problem: The center line of two struts are shown in Figure 5–10. Do they intersect?

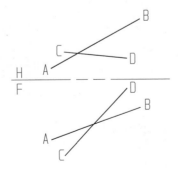

Figure 5–10 Do lines A-B and C-D intersect?

Answer

Figure 5–11 uses the projection line method to prove that the two struts do not intersect.

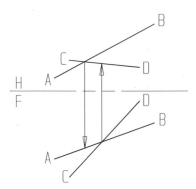

Figure 5–11 No, lines A-B and C-D do not intersect.

▼ VISIBILITY OF A POINT AND A LINE

When a point is not located on a line, it is important to know the relationship between the two. For example, is the point in front of the line or behind? Is the point above or below the line? When you are looking down on the top view, whichever is highest in elevation will be visible. Looking toward the front view, whichever is closer to the frontal projection plane will be visible.

Two adjacent views are necessary for testing for visibility.

Back to the gas line and end of hopper shoot studied in Figure 5–4. Is the end of the hopper shoot in front or behind of the gas line? Is the end of the hopper above or below the gas line?

Figure 5–12 shows how the height (above or below) was found. The top view shows width and depth, but not

DETERMINE HEIGHT:
WHICH IS HIGHER,
POINT F OR LINE DE?

ANSWER:
YOU SEE LINE DE BEFORE
POINT F; THEREFORE, LINE DE
IS HIGHER THAN POINT F.

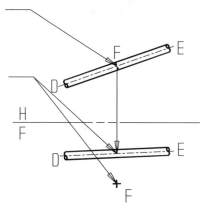

Figure 5–12 Visibility in top view.

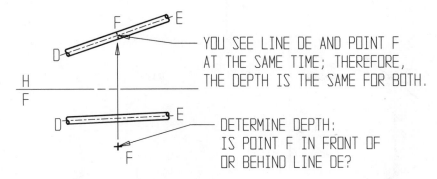

Figure 5–13   Visibility in front view.

height; therefore, you must look at the adjacent view (front view) to find the height.

Figure 5–13 checks visibility in the front view. The front view shows width and height, but not depth. Therefore, look at the adjacent view (top view) for depth.

▼ VISIBILITY OF LINES

When two lines do not intersect, it is important to know which line is visible and which is hidden. When you are looking down on the top view, whichever line is highest in elevation will be visible. Looking toward the front view, whichever line is closer to the frontal projection plane will be visible. Two views are necessary for testing for visibility. Each view shows only two dimensions; the adjacent view shows the missing

dimension. Remember to show the third dimension in the appropriate view.

Example

A new press was installed out on the shop floor of a sister company in another state. A conveyor line was in the way and had to be moved. While the new conveyor system was added to the floor print, the old system was never erased (Figure 5–14). Because all original floor plans are kept in your department, you have been asked to revise the floor plan. The only information you have been given is that the new conveyor line is higher and in front of the old conveyor line. Which line represents the new conveyor line?

The new conveyor line is 11-12. Figure 5–15 shows the final results. Figures 5–16 and 5–17 show each step separately.

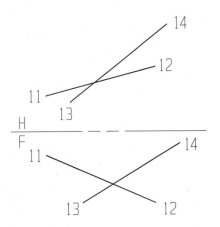

Figure 5–14   Which conveyor line is higher and in front?

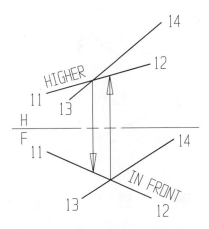

Figure 5–15   Line 11-12 is both higher and is in front of line 13-14.

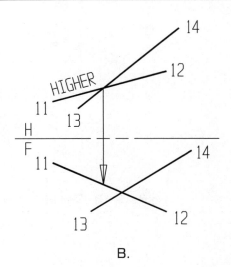

Figure 5–16 Top view visibility.

Figure 5–16 shows how the top view visibility was found. The top view shows width and depth, but not height (Figure 5–16(A)). Look at the front view for height, but *remember* to show it in the top view by labeling the higher line (Figure 5–16(B)).

Figure 5–17 checks visibility in the front view. The front view shows width and height, but not depth. Look at the top view for depth, but *remember* to show it in the front view by labeling the line closest to the frontal plane.

This method is a general one and is applied not simply to line problems, but to all situations where visibility is in question (Figure 5–18). Earlier in the chapter, we decided the pipe and belt did not intersect (Figure 5–6), therefore the visibility must be determined. Sketch the visibility to the Figure 5–18.

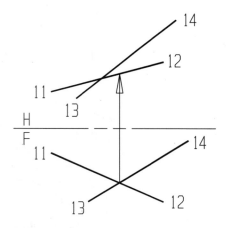

Figure 5–17 Front view visibility.

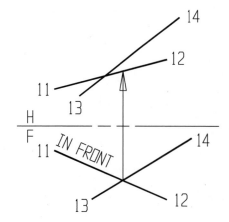

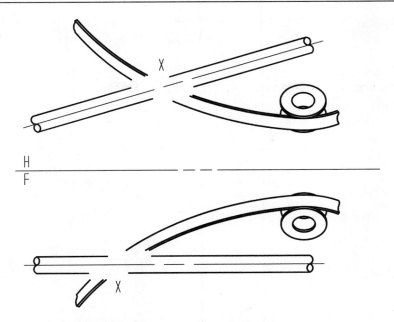

Figure 5-18 Determine visibility between pipe and belt.

In order to determine whether the pipe or the belt is visible in the top view of Figure 5–18, a projection line is dropped down from the place in the top view where the pipe and the belt cross. The belt is closer to the top (higher) than the pipe; therefore, the belt is drawn visible and the pipe is drawn hidden in the top view. To determine the visibility in the front view, a projection line is drawn upward from the crossing of the belt and the pipe in the front view. The pipe in the top view is closer to the frontal plane (closer to the observer) than the belt; therefore, the pipe is drawn visible and the belt is drawn hidden in the front view (Figure 5–19).

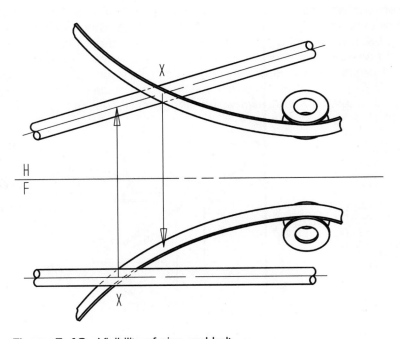

Figure 5-19 Visibility of pipe and belt.

EXERCISE

Problem: Let's try a practice problem. Complete the views in Figure 5–20 by sketching in the visibility for the angle-iron W-X and channel Z-Y.

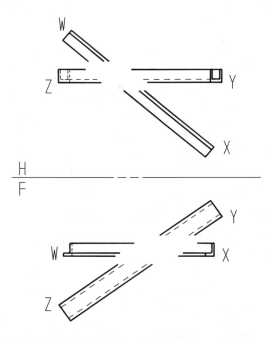

Figure 5–20 Determine visibility between angle iron and channel.

Answer

In order to determine whether angle iron W-X or channel Z-Y is visible in the top view of Figure 5–20, a projection line is dropped down from the place where W-X and Z-Y cross in the top view. Line W-X is closer to the top (higher) than Z-Y; therefore, the angle iron W-X is drawn visible and channel Z-Y is drawn hidden in the top view. To determine the visibility in the front view, a projection line is drawn upward from the crossing of W-X and Z-Y in the front view. Line W-X in the top view is closer to the frontal plane (closer to the observer) than Z-Y; therefore, the angle W-X is drawn visible and the channel, Z-Y, is drawn hidden in the front view (Figure 5–21).

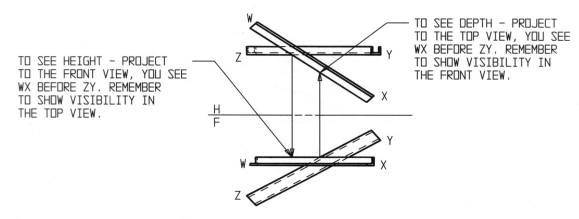

Figure 5–21 Visibility determined.

▼ PARALLEL LINES

When lines are parallel, they are an equal distance from each other throughout their length. Three views are needed to prove if in fact the lines are parallel or skewed (Figure 5–22). If lines are parallel to each other, they will appear parallel in at least three views, with two exceptions. One exception is when the lines appear as end views or points (Figure 5–23). The second exception is when the lines appear one behind the other (Fig-

ure 5–25). Figure 5–22 illustrates parallel lines A-B and C-D on the guide block. Notice they appear parallel to each other in all three views.

The first exception is demonstrated in Figure 5–23. Lines are parallel even when lines appear as end views or points (Figure 5–23). Sketch in the right-side view, to prove this.

Figure 5–24 shows the right-side view. The right-side view also demonstrates the second exception.

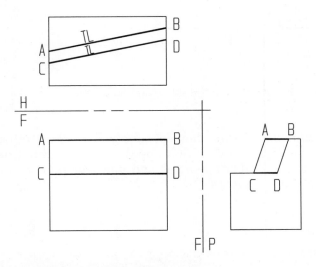

Figure 5–22 Parallel lines.

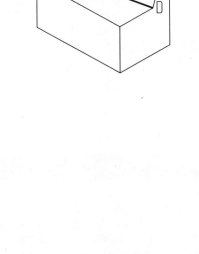

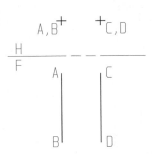

Figure 5–23 Complete right-side view to prove lines are parallel.

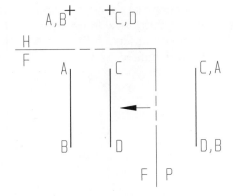

Figure 5–24 Lines A-B and C-D are parallel.

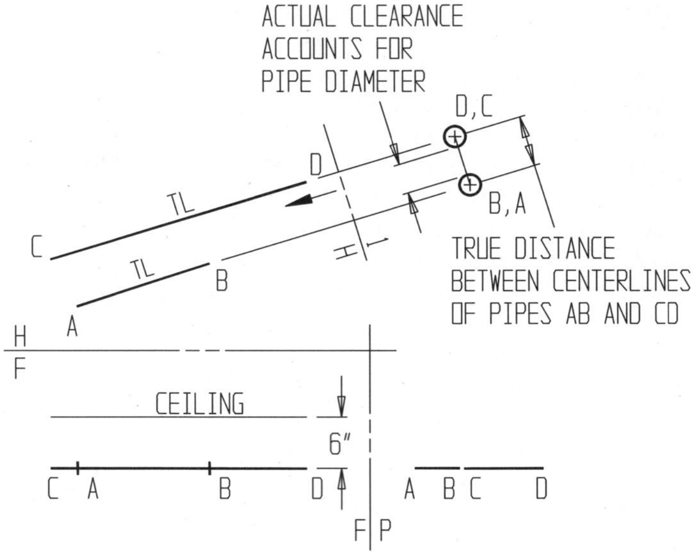

ACTUAL CLEARANCE
ACCOUNTS FOR
PIPE DIAMETER
D,C
D
TL
B,A
C
TL
B
TRUE DISTANCE
BETWEEN CENTERLINES
OF PIPES AB AND CD
H
A
F
CEILING
6"
C A B D
A B C D
F P

Figure 5–25 Parallel lines: One behind the other and as points.

Two pipes, A-B and C-D, are hung equidistant from a ceiling. The two exceptions to the rule "lines are parallel if they appear parallel in at least three views" are demonstrated in Figure 5–25. In the front and right-side views, they appear one behind the other, while in the top view they appear parallel and in true length. Lines that appear as end views or points, are illustrated in view 1.

EXERCISE

Problem: While the lines may appear parallel in two views, do not assume the third view will show the lines parallel or one of the two exceptions (Figures 5–23 through 5–25). Take the time to examine the third view. Figure 5–26 shows two different problems. One problem contains parallel lines and one problem contains skewed lines. Because the lines do not appear to cross, the test for intersecting and nonintersecting lines will not work; therefore, a third view is necessary to test for parallel lines. Sketch the right-side views to the problems in Figure 5–26. Which problem shows parallel lines and which problem shows skewed lines?

4
6
3
5
H
F
3
5
4
6
F P

4
5
3
6
H
F
3
5
4
6
F P

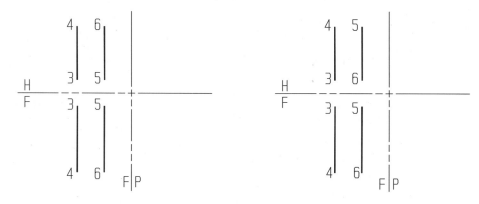

Figure 5–26 Complete the right-side views.

Answer

The right-side views should look like the ones in Figure 5–27. Figure 5–27(A) lines are parallel. Figure 5–27(B) lines are skewed. To help visualize, try holding two pencils in the proper positions.

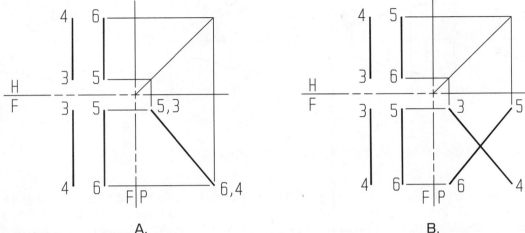

A.　　　　　　　　　　　　B.

Figure 5–27　Notice the transferred distance (depth) for point 6. (A) Lines are parallel. (B) Lines are not parallel.

Example

Let's examine one more example that demonstrates the principle that parallel lines will appear parallel in at least three views. In Figure 5–28 you are given four views of two parallel shafts. In the front and top views the shafts appear parallel. The true lengths of the shafts are shown in view 1. Notice that in view 1, only the center lines (shaft axes) were transferred. It is not necessary to transfer the entire shaft to find the information needed. In view 2, the shaft axes are seen as points, which also allows you to see the true distance between the shaft center lines. By adding the diameter of the shafts, the actual clearance between the pipes is found.

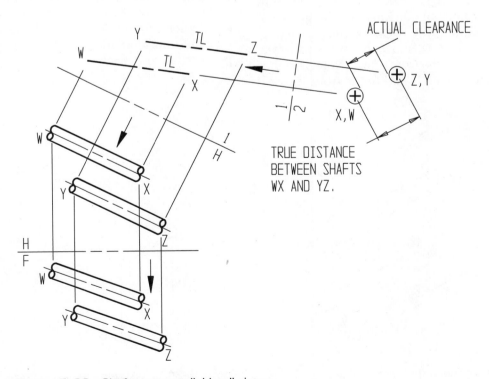

Figure 5–28　Shafts are parallel in all views.

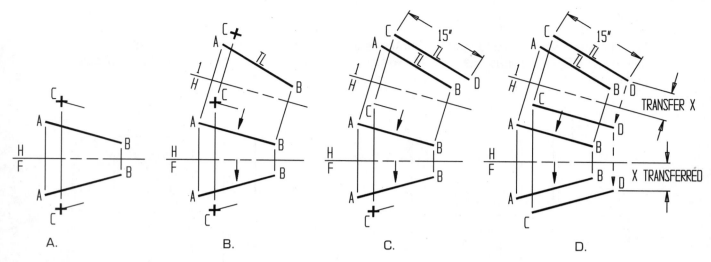

Figure 5–29 Constructing a line parallel and equal length to an existing line.

On occasion it is necessary to construct a line parallel and equal in length to an existing line and through a given point. Figure 5–29 illustrates this process. In part (A), you are given shaft A-B and one end of a 15–inch shaft, C-D, that runs parallel to shaft A-B. Because we know shaft C-D is parallel to A-B, we can draw a construction line parallel to A-B. However, since C-D is oblique in the principal views, we cannot locate point D without doing a primary auxiliary view. A top-adjacent or front-adjacent view may be completed. Remember the line of sight must be perpendicular to the oblique line in order to view the line true length (Figure 5–29(B)).

Because A-B and C-D are parallel, the line of sight will be perpendicular to both oblique lines. The length of 15" is measured in the view 1, where C-D is true length (Figure 5–29(C)). Figure 5–29(D) shows point D's measurement transferred back to the front view. (**Note**: Distance X must be the same in view 1 and in the front view.) If

you can transfer height from front view to first auxiliary view, you can transfer height from first auxiliary view to front view. Projection lines are drawn as hidden lines with arrowheads to show direction of endpoints being transferred. (Review Locating Principal Views of an Object Using Auxiliary Views in Chapter 3.)

▼ PERPENDICULAR LINES

Although two lines may make a variety of angles with each other, perpendicular lines occur often and deserve separate study. Lines are perpendicular when there is a 90° angle between them. Lines may be perpendicular whether they are intersecting or nonintersecting. Figure 5–30 shows two pieces of duct that intersect at 90° to each other, while Figure 5–31 illustrates the same two types of duct that are nonintersecting, yet still appear perpendicular to each other.

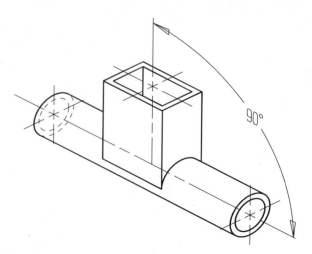

Figure 5–30 Intersecting and perpendicular ducts.

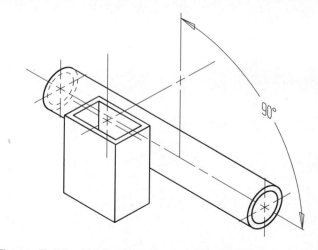

Figure 5–31 Nonintersecting and perpendicular ducts.

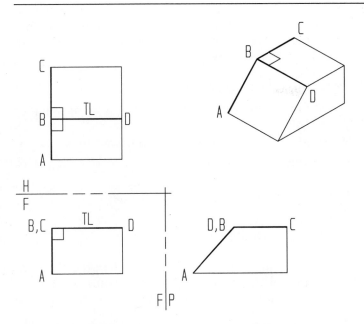

Figure 5-32 Perpendicular lines on a wedge.

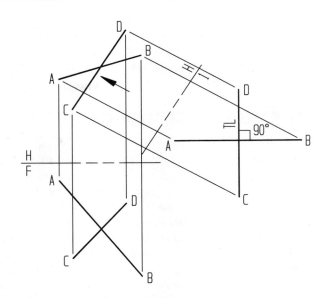

Figure 5-33 Establishing perpendicularity: Intersecting lines.

Whether they are intersecting or nonintersecting, lines that are perpendicular to each other in space will appear at right angles to each other in any view that shows either or both lines in true length. The only exception is when one line appears as a point. Figure 5-32 shows a wedge with intersecting lines A-B, B-C, and B-D labeled. The fact that lines B-C and B-D are perpendicular to each other is seen in the top view where both lines appear in true length and the right angle is evident. Lines B-D and A-B are perpendicular, as seen in the top and front views. In both views, line B-D is drawn true length, and so the right angle is seen. Notice that line A-B is not true length in either view. What other observations can you make about perpendicular lines on this object?

A more general situation illustrating two intersecting perpendicular pipes (A-B and C-D) is shown in Figure 5-33. The angular relationship between the two pipes cannot be seen in the top or front views. This is because neither pipe appears true length in either view. In auxiliary view 1, which has been selected so that the line of sight is perpendicular to one of the pipes (C-D) in the adjacent view, pipe C-D appears true length and the right angle between the pipes is seen.

The same principles and procedures apply with nonintersecting lines. Figure 5-34 shows two views of nonintersecting perpendicular shafts, A-B and C-D. To see their perpendicularity, the line of sight is drawn perpendicular to either line C-D or A-B. Figure 5-34 demonstrates the line of sight perpendicular to line C-D in the top view, causing C-D to appear true length in view 1. A-B and C-D appear at right angles to each other in view 1, demonstrating that they are perpendicular in space. View 2 shows both the true length of shaft A-B

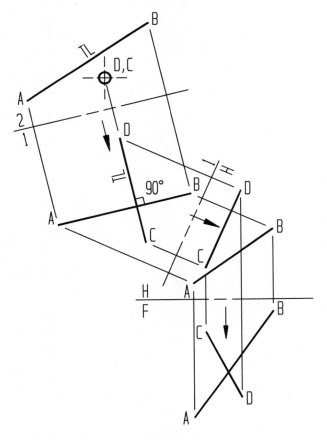

Figure 5-34 Shafts A-B and C-D are proven to be perpendicular where one shaft (A-B) appears true length (view 1).

and the end view of shaft C-D. This only occurs when the shafts are perpendicular to each other. Note that view 2 shows the true distance between the shafts. Shortest distances will be covered in more detail in the next section.

▼ SHORTEST DISTANCES

Sometimes specific information, such as clearance or shortest distance between two pipes, is required. Other examples of when to solve for shortest distance are: finding the true length of a new connecting strut, the shortest tunnel to be dug, the shortest possible (perpendicular) pipe to be installed in an existing pipe, location of the fitting for the shortest pipe to be installed, the pitch diameter of gears on parallel shafts, clearance between two power lines, etc. These problems are solved using information learned in Chapter 4. Simplify the problems first by reducing the objects to only

the necessary lines and points needed to solve the problem. This process enables you to categorize problems in one of the following ways:

- Shortest distance between a point and a line
- Shortest distance between two non-intersecting perpendicular lines
- Shortest distance between two skewed lines
- Shortest level (horizontal) line connecting two skewed lines
- Clearance between two parallel lines
- Shortest line at a given grade
- Slope connecting two skewed lines

No matter which type of problem in the preceding list you must solve for, the steps of procedure are all the same. The *shortest distance* is always perpendicular to

HELPFUL HINTS

Study Figure 5–35 and try to visualize these two shafts in space as you study these four views. Copy or trace the figure, cut it out, and fold 90° on each reference line to create a box. Study the different lines of sight. Repeat this process for Figure 5–36.

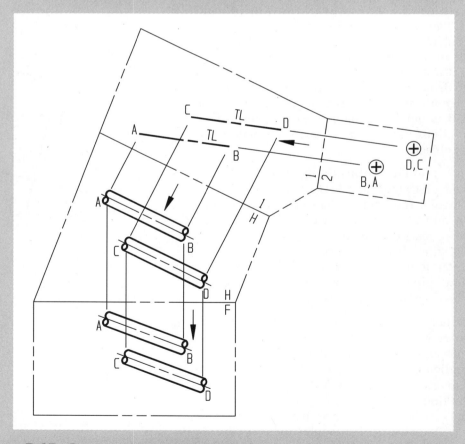

Figure 5–35 Cut-out to help visualize parallel lines.

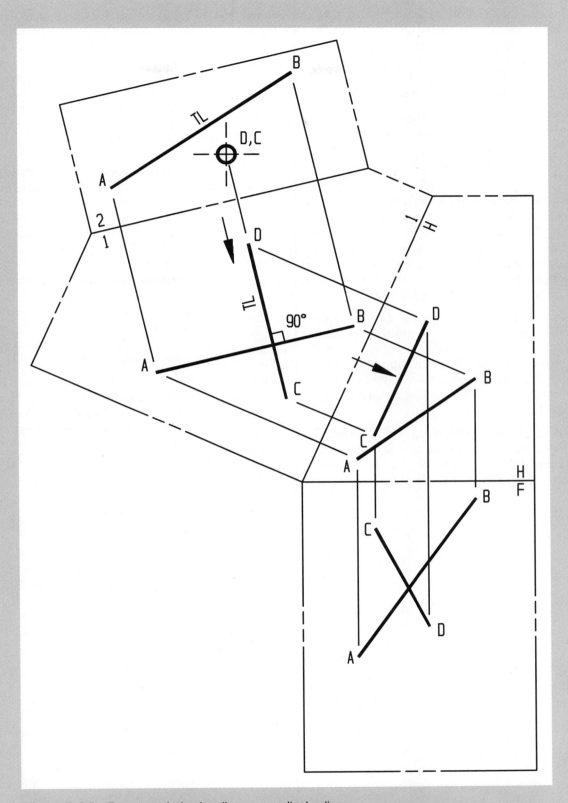

Figure 5-36 Cut-out to help visualize perpendicular lines.

the line(s) and is only found where at least one line appears as a point. To accomplish this your line of sight must be parallel to the true length line. Remember to always check the principal views first to see if auxiliary views are in fact needed. *In other words we haven't learned any new steps, just when to apply the true length and point view lines of sight!*

Study the following figures and note that each problem was solved using the same procedure. For space saving and comparison purposes, each problem (whether it be a cable, water pipe, strut, etc.,) will be represented by a line locating the center of each element.

Shortest Distance Between a Point and a Line. A new injection molding press has been ordered for your shop. The planned location of the new machine, does not allow for easy access to a water line. With E-F being the existing water line, and Y being the water connection location on the new machine, how much connecting pipe must be ordered (Figure 5–37)?

Shortest Distance Between Parallel Lines. A new hoist is needed out in the mold department. The location of the new hoist is complicated by the fact that there are two water lines running 6" below ceiling level. There must be at least a 12" clearance between the pipes in order for the hoist to run properly (Figure 5–38). Is there enough room?

Shortest Distance Between Two Skewed Lines. You are part of a team designing a new aircraft. Two cables, R-S and T-U, on an aircraft must be connected with the shortest possible stabilizing cable, X-Y (Figure 5–39).

Either line may be taken to a point view and the other one will be carried along and remain oblique. The shortest distance will be perpendicular to the oblique line through the point view of the other line (view 2).

Shortest Horizontal Distance Between Two Skewed Lines. The shortest horizontal distance is only found in an elevation view, therefore the plane method must be used. This method is discussed in Chapter 7, after the lines of sight for planes are covered.

Shortest Distance at a Given Grade or Slope Between a Line and a Point. The shortest distance and slope must be found in the same elevation view, therefore the plane method must be used. This method is discussed in Chapter 7, after the lines of sight for planes are covered.

Transferring New Location to Principal Views. If the location of the shortest distance needs to be drawn in the principal views the new location must be transferred back to the front view. Always draw transferring projection lines parallel to the line of sight for the adjacent view (Chapter 3, page 46–49).

We will review this process one view at a time, using the same examples used to solve shortest distance problems (Figures 5–37 and 5–39). Figures 5–40 through 5–45 are designed to prove that the steps of procedure are the same for projecting the shortest distance back to the front view no matter which problem is given.

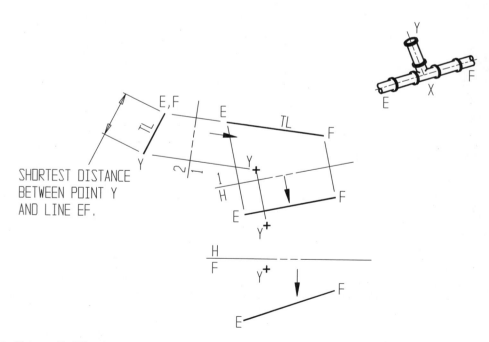

Figure 5-37 Shortest distance is found in the view were the line appears as a point.

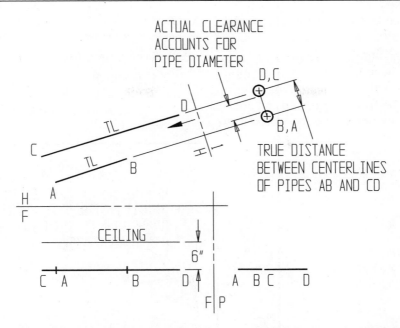

Figure 5–38 Clearance is found in view where lines A-B and C-D appear as points.

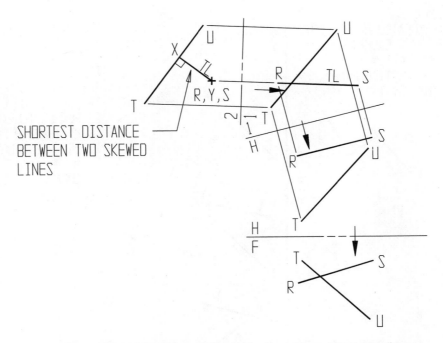

Figure 5–39 Shortest distance found in the view where one line appears as a point and the other line is oblique. The shortest distance (connecting line) will be perpendicular to both lines.

Project the shortest distance line from view 2 to view 1 (Figures 5–40 and 5–41).

Study the view showing the shortest distance for each problem and how the line was located in the existing adjacent view. Remember, if the shortest distance can be measured, then it must be true length, and the adjacent view must show the line perpendicular to the line of sight.

Review

A. True-length line

1. If the line of sight is perpendicular to the object line in one view, it will appear true length in the adjacent view and can be measured.

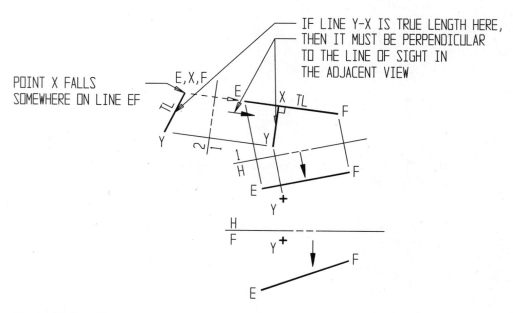

Figure 5–40 Shortest distance between a point and a line projected from view 2 to view 1.

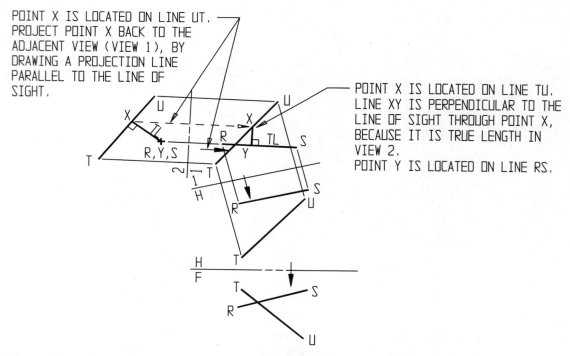

Figure 5–41 Shortest distance between two skewed lines projected from view 2 to view 1.

The shortest distance between a line and a point was found by drawing a connecting line from the point view of the line (E-F) to the point (X) (Figure 5–40, view 2). The connecting line (Y-X) is true length. To see the connecting line (Y-X) in the adjacent view (Figure 5–40, view 1), it must be drawn perpendicular to the line of sight through point Y, locating point X on line E-F.

The shortest distance between two skewed lines was found by drawing a connecting line from the point view of the line (R-S) perpendicular to line (T-U) (Figure 5–41, view 2). The connecting line (X-Y) is true length. To see the connecting line (X-Y) in the adjacent view, point X must be projected to line T-U in the adjacent view (Figure 5–41, view 1). Line X-Y must be drawn perpendicular to the line of sight through point X locating endpoint Y on line R-S.

Project the shortest distance line from view 1 to the top view (Figures 5–42 and 5–43).

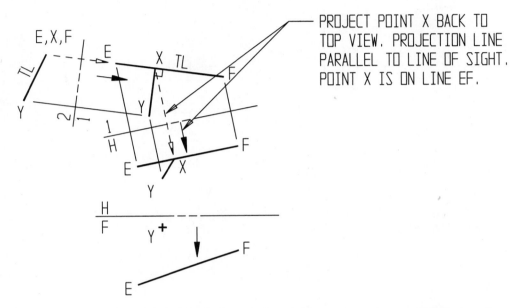

Figure 5–42 Shortest distance between a point and a line projected from view 1 to the top view.

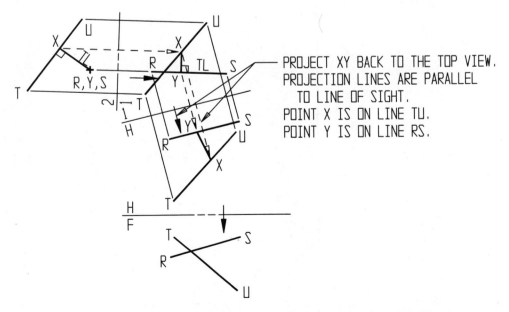

Figure 5–43 Shortest distance between two skewed lines projected from view 1 to the top view.

Project the shortest distance line from the top view to the front view (Figures 5–44 and 5–45).

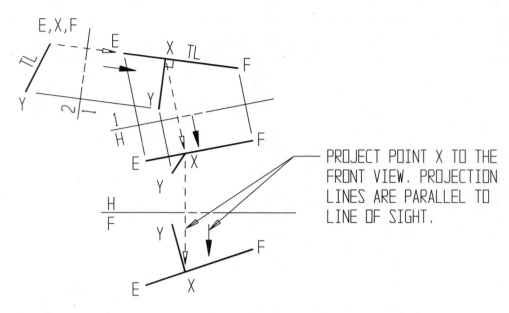

Figure 5–44   Shortest distance between a point and a line from the top view to the front view.

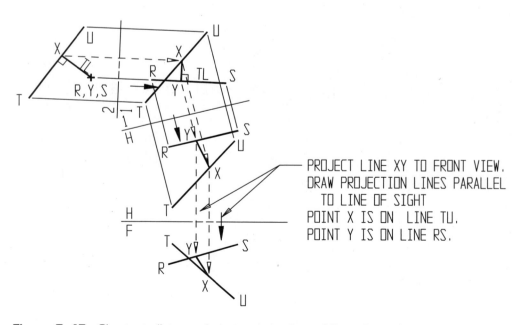

Figure 5–45   Shortest distance between two skewed lines from the top view to the front view.

EXERCISE

Problem: Sketch the shortest distance for the following problem right on the figure. Figure 5–46 shows the center lines of two mine shafts, K-L and M-N. A new tunnel, S-T, will be located the shortest distance possible between the two existing tunnels. Project the shortest tunnel, S-T, to the front view. (The shortest distance can be found by finding the view where line M-N or K-L is represented as a point. The answer in Figure 5–47 locates the point view of K-L.)

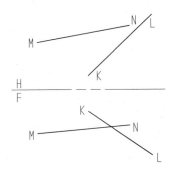

Figure 5–46   Locate the shortest distance between mine shafts M-N and K-L.

Answer

Figure 5–47 shows one of the solutions, finding the point view of line K-L, for Figure 5–46. The process of locating the shortest distance between two lines was used. Find the true length of one of the lines and carry the other line along (view 1). Locate the point view of the true length line and carry the other line along (view 2).

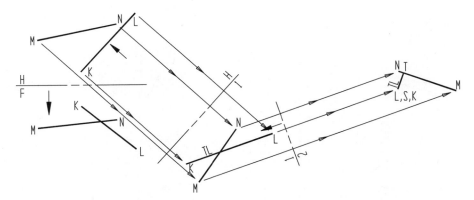

Figure 5–47   Shortest distance (S-T) between two tunnels.

Figure 5–48 shows the shortest distance tunnel, S-T, projected back to the front view.

Problem: Sketch the construction lines necessary to find the slope angle and compass bearing of the shortest distance line, S-T (Figure 5–48).

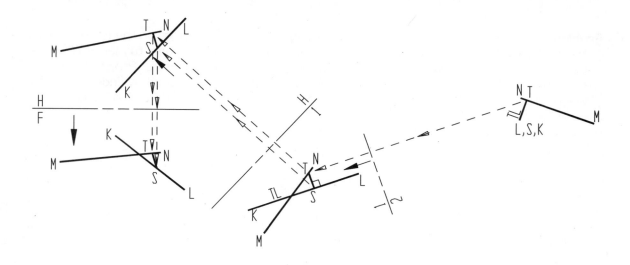

Figure 5–48 Project shortest distance between skewed lines back to the front view.

Answer

Figure 5–49 shows the correct solution to solve for bearing and slope of the shortest distance line. Bearing is read in the top view and the shortest distance line may be any line type. Slope is found where the line is true length in an elevation view (H/? reference line). Is line S-T true length in an elevation view? No (1/2 reference line). The elevation views are front (H/F reference line) or top-adjacent views (H/1 reference line), but the line is not TL in either view. What should you do? Yes, draw a new top-adjacent auxiliary view of line S-T only (H/3 reference line).

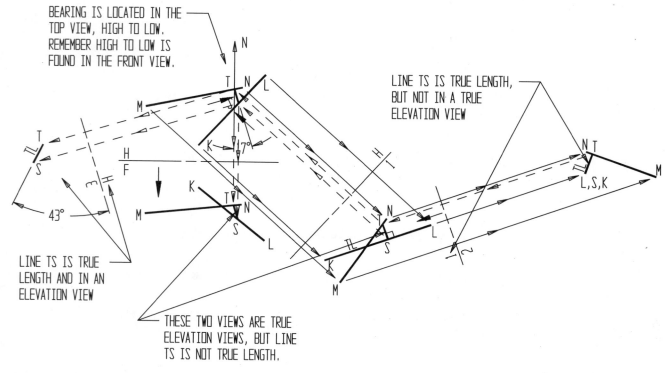

Figure 5–49 Slope of line T-S is 43°. Compass bearing of line T-S is S17°E.

HELPFUL HINTS

The following shows step by step how to locate the shortest distance between a line and a point.

I. Determine **line of sight**.

   A. Locate **true-length line**.

      1. Is the true-length line **in** one of the **principal planes**?

         a. If the line of sight is perpendicular to the object line in one view, it will appear true length in the adjacent view and can be measured.

      2. If the above rule does not apply, it is an oblique line and cannot be measured as is.

         a. Draw a **primary auxiliary** view **to find the true length**.

            (1) Draw the line of sight perpendicular to the oblique line in either principal view.

   B. Draw the **point view of a true-length line**.

      1. The line of sight must be drawn parallel to the true-length line.

         Line of sight locates:

         Shortest distance between a line and a point.

         Shortest distance between two parallel lines.

         A perpendicular line.

         Shortest distance between two skewed lines.

         Shortest level (horizontal) distance between two lines.

         Shortest distance between two lines at a given slope.

II. Draw the **reference line** perpendicular to the line of sight and label.

III. Draw **projection lines** parallel to the line of sight.

IV. Transfer **points of measurement** to the new view from the related view and label. Always go back to the last line of sight to take the measurement.

V. Connect the points and **label** true-length lines, edge views, and/or true-shape planes.

All problems should be analyzed first. Ask yourself, what am I **solving for**, what is already **given**, and what are the necessary **conditions needed to solve** the problem?

▼ SHORTEST DISTANCE BETWEEN A LINE AND A POINT

The shortest distance between a point and a line is shown where the line appears as a point (Figure 5–50).

EXERCISE

Problem

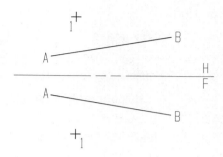

Figure 5–50 Locate the shortest distance between the point and line.

When solving a problem organize yourself first by asking the following questions. Write the answers next to the questions.

What am I **solving for**?

What is **given?**

    What views are given?

    What type of line do I have?

    What other information do I know?

What do I **need to solve** this problem?

Answer

What am I **solving for**?

    ■ Shortest distance between a line and a point

What is **given**?

    What views are given?

    ■ Top and front

    What type of line do I have?

    ■ The answer is oblique, because:

(Use Steps of Procedure I.A.1.a and I.A.2. to answer this question.)

I. Determine **line of sight**

  A. Locate **true-length line**

    1. Is the true-length line **in** one of the **principal planes**?

      a. If the line of sight is perpendicular to the object line in one view, it will appear true length in the adjacent view and can be measured.

    2. If the above rule does not apply, it is an oblique line and cannot be measured as is.

Problem: Sketch the solution on the figure. The correct answer will be shown on the next figure.

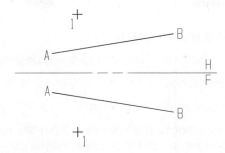

Figure 5–51 Line A-B is oblique.

What other information do I know?

■ Line A-B and point 1

What do I **need to solve** this problem?

■ Auxiliary view is needed because:

(Use Steps of Procedure I.2.a.1, II, III, IV, and V to answer this question.)

I. Determine **line of sight.**

    2. If the above rule does not apply, it is an oblique line and cannot be measured as is.

        a. Draw a **primary auxiliary** view **to find the true length.**

            (1) Draw the line of sight perpendicular to the oblique line in either principal view.

II. Draw the **reference line** perpendicular to the line of sight and label.

III. Draw **projection lines** parallel to the line of sight.

IV. Transfer **points of measurement** to the new view from the related view and label. Always go back to the last line of sight to take the measurement.

V. Connect the points and **label** true-length lines, edge views, and/or true-shape planes (Figure 5–52).

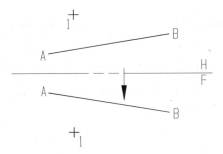

Figure 5–52 Sketch the true length of line A-B.

■ Successive auxiliary view is needed to locate point view of true-length line.

I. Determine **line of sight**.

   B. Draw the **point view of a true-length line**.

      1. The line of sight must be drawn parallel to the true-length line.

II. Draw the **reference line** perpendicular to the line of sight and label.

III. Draw **projection lines** parallel to the line of sight.

IV. Transfer **points of measurement** to the new view from the related view and label. Always go back to the last line of sight to take the measurement.

V. Connect the points and **label** true-length lines, edge views, and/or true-shape planes (Figure 5–53).

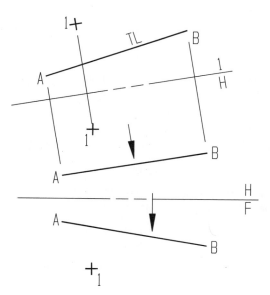

Figure 5–53 Locate the point view line A-B.

Locate the shortest distance line 1-2 in view 2. Project line 1-2 to view 1. Label line 1-2 where it appears true length.

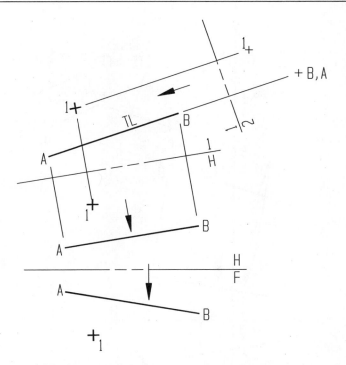

Figure 5–54 Sketch in the shortest distance between line A-B and point 1 and project back to view 1.

Project line 1-2 back to the top and front views.

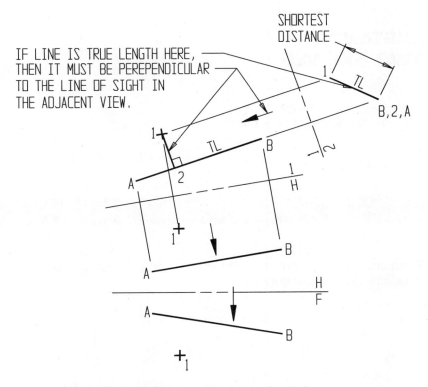

Figure 5–55 Project shortest distance to front view.

Answer

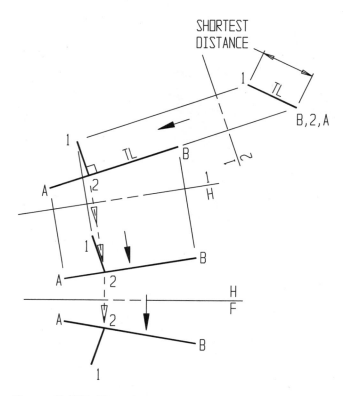

Figure 5–56 The shortest line is perpendicular between the line and point.

▼ SHORTEST DISTANCE BETWEEN TWO SKEWED LINES

The shortest distance between two skewed lines is shown in the view where one line appears as a point (Figure 5–57).

EXERCISE

Problem: When solving a problem organize yourself first by asking the following questions. Write the answers next to the questions.

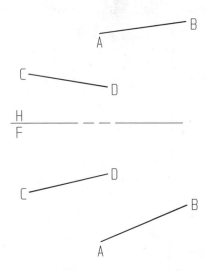

Figure 5–57 Locate the shortest distance between the two skewed lines.

What am I **solving for**?

What is **given**?

What views are given?

What type of line do I have?

What other information do I know?

What do I **need to solve** this problem?

Answer

What am I **solving for**?

■ Shortest distance between two skewed lines

What is **given?**

What views are given?

■ Top and front

What type of lines do I have?

■ The answer is oblique, because:

(Use Steps of Procedure I.A.1.a and I.A.2. to answer this question.)

I. Determine **line of sight**.

  A. Locate **true-length line**.

    1. Is the true-length line **in** one of the **principal planes**?

      a. If the line of sight is perpendicular to the object line in one view, it will appear true length in the adjacent view and can be measured.

    2. If the above rule does not apply, it is an oblique line and cannot be measured as is (Figure 5–58).

Sketch the solution on the figure. The correct answer will be shown on the next figure.

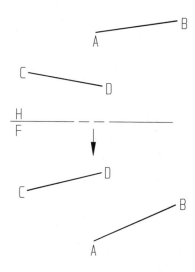

Figure 5-58 Both lines are oblique.

What other information do I know?

■ Lines A-B and C-D

What do I **need to solve** this problem?

■ Auxiliary view is needed because:

(Use Steps of Procedure I.2.a.1 to answer this question.)

---

2. If the above rule does not apply, it is an oblique line and cannot be measured as is.

a. Draw a **primary auxiliary** view **to find the true length**.

(1) Draw the line of sight perpendicular to the oblique line in either principal view (Figure 5–59).

Line of sight may be perpendicular to either line.

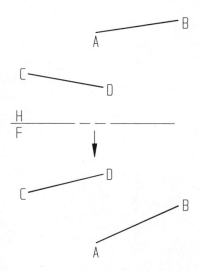

Figure 5–59 Sketch the true length of line C-D and carry line A-B along.

▪ Successive auxiliary view is needed to locate point view of true-length line (Figure 5–60).

  I. Determine **line of sight**.

    B. Draw the **point view of a true-length line**.

      1. The line of sight must be drawn parallel to the true-length line.

  II. Draw the **reference line** perpendicular to the line of sight and label.

  III. Draw **projection lines** parallel to the line of sight.

  IV. Transfer **points of measurement** to the new view from the related view and label. Always go back to the last line of sight to take the measurement.

  V. Connect the points and **label** true-length lines, edge views, and/or true-shape planes.

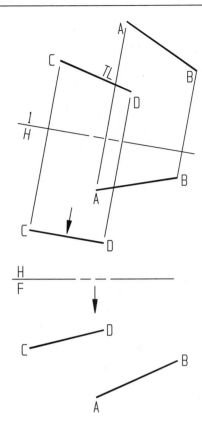

Figure 5–60 Sketch the shortest distance between lines A-B and C-D.

Locate the shortest distance line X-Y (Figure 5–61).

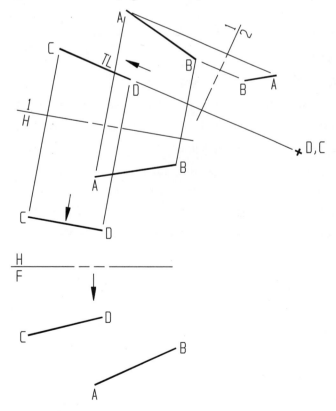

Figure 5–61 Locate the shortest distance, line X-Y, between lines A-B and C-D.

Project line X-Y back to view 1. Line X-Y is true length in view 2; therefore, it will be perpendicular to the line of sight in the adjacent view (view 1) (Figure 5–62).

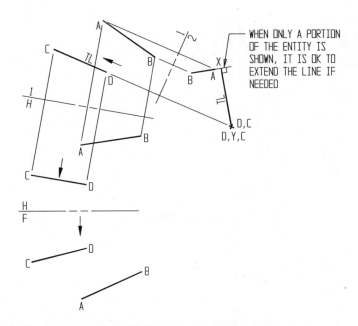

Figure 5–62 Project shortest distance X-Y to view 1.

Project line X-Y to the top view. Draw projection lines parallel to the line of sight. Transfer the measurements from the related view (view 2) (Figure 5–63).

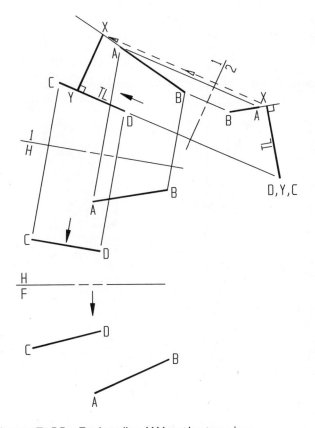

Figure 5–63 Project line X-Y to the top view.

Project line X-Y to the front view. Draw projection lines parallel to the line of sight. Transfer the measurements from the related view (view 1) (Figure 5–64).

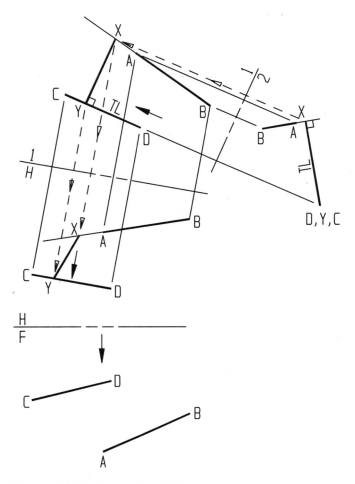

Figure 5–64 Project line X-Y to the front view.

Answer

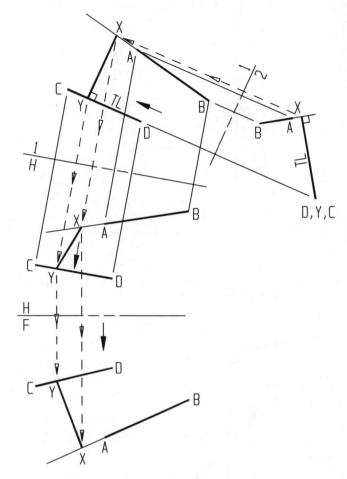

Figure 5-65 Locate the bearing of the shortest distance line, X-Y.

Problem: Locate the bearing of line X-Y on Figure 5–65.

Answer

The bearing is located by first determining elevation of line X-Y. The front view shows point Y is higher than point X. The top view shows the direction of line X-Y is NE. The bearing is N30°E.

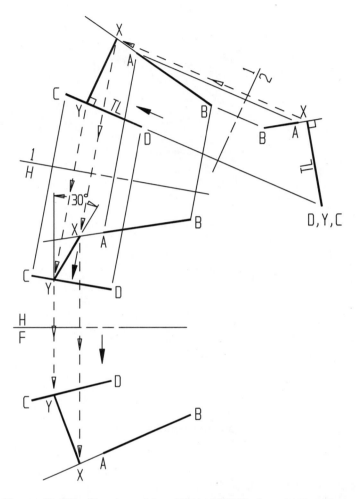

Figure 5–66   Bearing of line X-Y is N30°E. Locate the slope angle for line X-Y.

Problem: Locate the slope of line X-Y on Figure 5–66.

Answer

Line X-Y is TL in view 2, but this is not a true elevation view. The front view and view 1 show elevation, but line X-Y is not TL in either view; therefore, a new view must be created (view 3) that shows line X-Y TL in an elevation view. The slope angle is 54°.

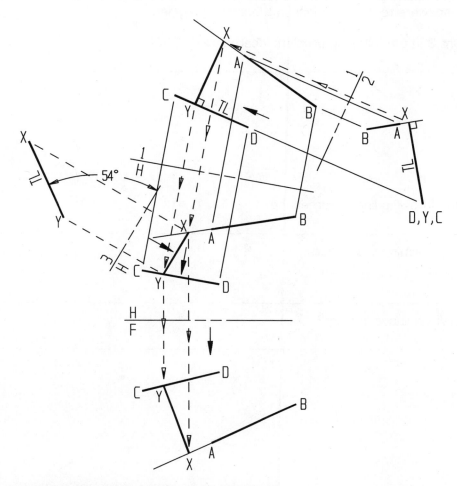

Figure 5–67   Slope angle for line X-Y is 54°.

Name _____

Date _____

Course _____

◄ CHAPTER 5 STUDY QUESTIONS ►

The Study Questions are intended to assess your comprehension of chapter material. Please write your answers to the questions in the space provided.

1. Is point 3 in the following problem located on the 1-2?

_____

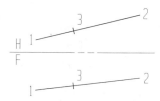

2. Are the following lines intersecting?

_____

If not, determine the visibility.

    Which line is on top?

_____

    Which line is in front?

_____

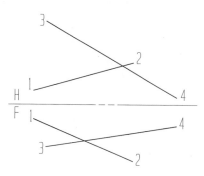

3. Indicate whether the following statements are true or false.

    True   False   If lines appear parallel in any two adjacent views, they are parallel in reality.

    True   False   Perpendicular lines do not have to intersect.

    True   False   If two lines are perpendicular in actuality, they will appear perpendicular in any orthographic view.

4. If lines are parallel in space, they will appear _____ in at least three views, except two. List the two exceptions.

_____

_____

5. How must parallel lines appear in order to measure the shortest distance between them?

    _____

    _____

6. How must the line of sight be placed in order to see the true distance between parallel lines?

    _____

    _____

7. If two lines are perpendicular and nonintersecting, how must they appear in order to measure the shortest distance between them?

    _____

    _____

8. How must the line appear in order to measure the shortest distance between a point and a line?

    _____

    _____

9. If two lines are skewed, how must they appear in order to see the shortest distance between them?

    _____

    _____

10. When finding the shortest distance between skewed lines, does it matter which line remains oblique and which one is seen as a point?

    _____

    _____

Notice that with each problem in this chapter, you are learning new applications of several of the fundamental views. You are moving into more varied and advanced aspects of descriptive geometry.

The following problems may be drawn using instruments on the page provided, created using a 2D or 3D system, or a combination of drawing board and CAD. Refer to Appendix A for additional dimensions needed to solve problems three-dimensionally.

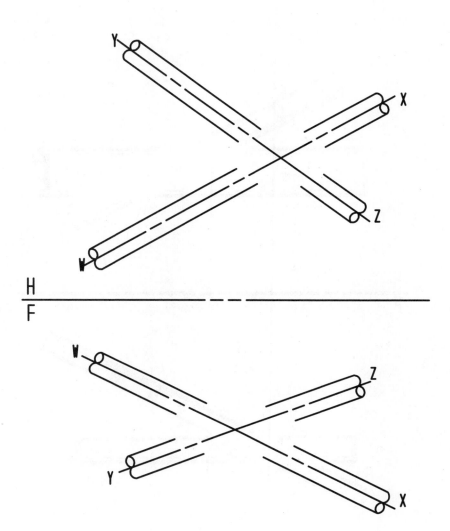

H
—————————————————
F

		ADG INC.			
		VISIBLITY OF SHAFTS			
	SIZE	DATE	DWG NO.	SCALE 1" = 1"	REV
DRW. BY			SECTION #	SHEET	

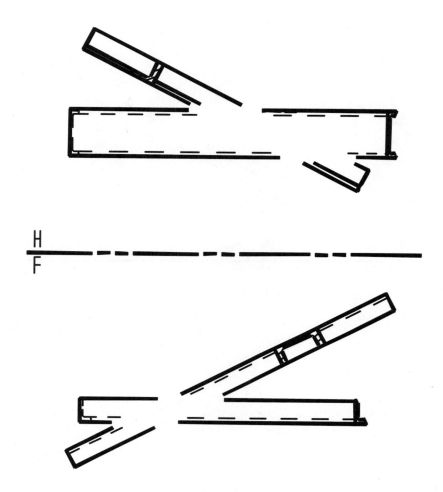

H
F

		ADG INC.			
		INTERSECTION			
	SIZE	DATE	DWG NO.	SCALE	REV
				1" = 10"	
DRW. BY			SECTION #	SHEET	

Chapter 5, Problem 3: THREE .25 INCH DIAMETER RODS ARE SHOWN. CONSTRUCT A PROFILE VIEW. SHOW THE CORRECT VISIBLITY IN ALL VIEWS.

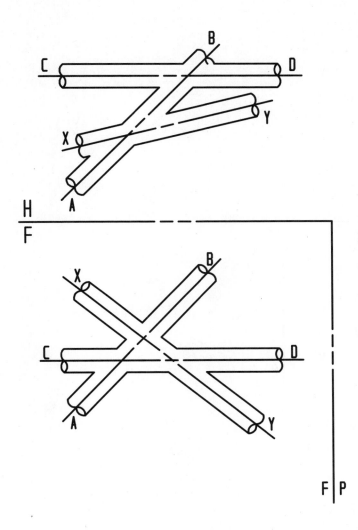

						ADG INC.		
						RODS		
	SIZE	DATE		DWG NO.		SCALE		REV
						1" = 1"		
DRW. BY					SECTION #		SHEET	

Chapter 5, Problem 4: GIVEN TWO VIEWS OF PIPELINES AB AND CD, DETERMINE IF THEY ARE PARALLEL IN SPACE.

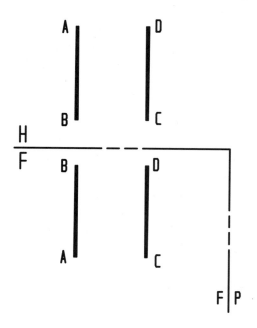

ARE THE LINES PARALLEL?	ADG INC.			
	PIPELINES PARALLEL?			

	SIZE	DATE	DWG NO.	SCALE 1" = 1"	REV
DRW. BY			SECTION #	SHEET	

Chapter 5, Problem 5: YOU ARE ASKED TO DRAW A PIPELINE, WZ, PARALLEL TO THE EXISTING PIPELINE, XY. THEY ARE EQUAL IN LENGTH AND ARE BOTH 2 INCHES IN DIAMETER. WHAT IS THE CLEARANCE (DISTANCE) BETWEEN THEM?

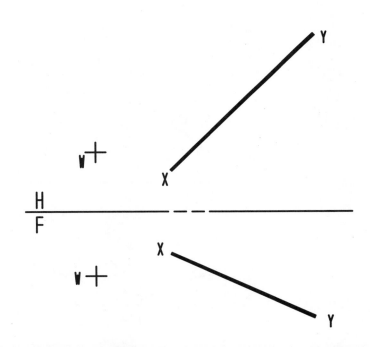

CLEARANCE =	ADG INC.			
	PIPELINE CLEARANCE			
	SIZE DATE	DWG NO.	SCALE 1" = 10"	REV
DRW. BY		SECTION #	SHEET	

Chapter 5, Problem 6: PINION GEAR X DRIVES SPUR GEAR Y. LINES AB AND CD REPRESENT THE
CENTER LINES OF THE GEAR SHAFTS. SHAFT CD IS THE SAME LENGTH AND PARALLEL TO SHAFT AB.
LOCATE THE CENTER LINE OF SHAFT CD IN THE FRONT VIEW.

PINION GEAR X IS LOCATED ON SHAFT AB WITH THE PITCH DIAMETER THROUGH POINT E. GEAR Y IS
IS LOCATED ON SHAFT CD WITH THE PITCH DIAMETER TANGENT TO GEAR X PITCH DIAMETER.
SHOW THE PITCH DIAMETERS AS PHANTOM LINES IN ALL VIEWS. WHAT IS THE PITCH
DIAMETER OF PINON GEAR X? WHAT IS THE PITCH DIAMETER OF SPUR GEAR Y?

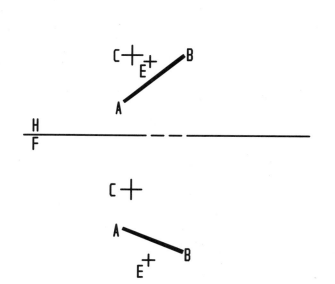

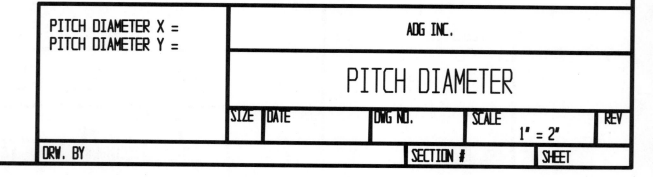

PITCH DIAMETER X =	ADG INC.			
PITCH DIAMETER Y =				
	PITCH DIAMETER			
	SIZE	DATE	DWG NO.	SCALE	REV
				1" = 2"	
DRW. BY			SECTION #		SHEET

Chapter 5, Problem 7: A HOIST IS SUPPORTED BY AN A-FRAME, WHOSE TWO LEGS ARE PERPENDICULAR TO THE LOAD-BEARING BEAM, XY, AND TO EACH OTHER. DRAW THE CENTER LINES OF THE A-FRAME LEGS FROM POINT Z TO THE FLOOR.

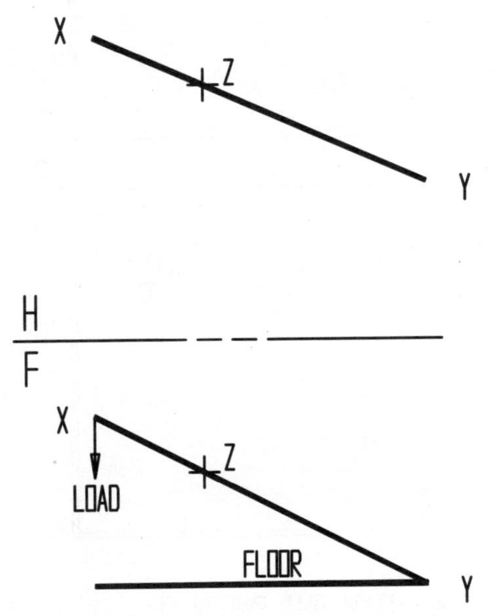

Y

Z

X

H
—
F

Y

Z

FLOOR

X

LOAD

Chapter 5, Problem 8: A LENGTH OF 4-INCH SQUARE TUBING IS USED AS A BRACE AND IS ANCHORED TO THE FLOOR AT POINT A, 6 FEET, 9 INCHES IN FRONT OF THE BACK WALL. THE TUBING IS ATTACHED TO THE BACK WALL, (POINT B) 7 FEET ABOVE THE FLOOR AND DIRECTLY BEHIND THE FLOOR ANCHOR (POINT A). ANOTHER LENGTH OF 4-INCH SQUARE TUBING IS USED AS A STIFFENING BRACE. IT IS PERPENDICULAR TO THE MAIN BRACE, WITH ONE END LOCATED 6 FEET, 3 INCHES TO THE RIGHT OF POINT B AND 11 INCHES ABOVE THE FLOOR. DETERMINE THE TRUE LENGTH OF THE CENTER LINE OF THE MAIN BRACE, AND THE TRUE LENGTH OF THE CENTER LINE OF THE STIFFENING BRACE FROM THE WALL TO THE CENTER LINE OF THE MAIN BRACE. HOW FAR FROM THE FLOOR, MEASURED ALONG THE MAIN BRACE, IS THE CONNECTION?

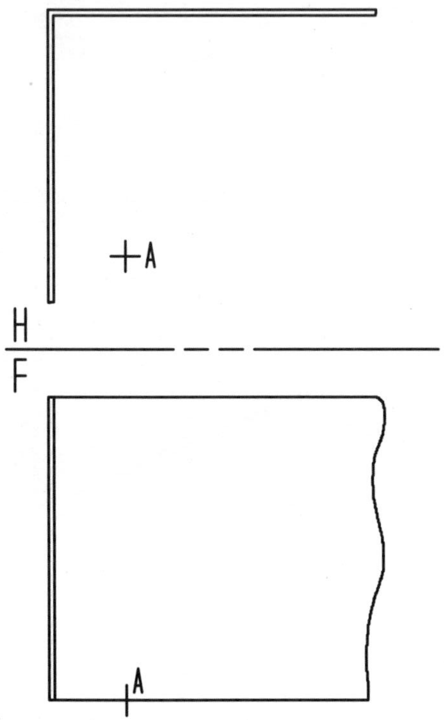

+A

H
―
F

A

ADG INC.

BRACE

MAIN BRACE =
STIFFENING BRACE =
INTERSECTION POINT =

SIZE	DATE	DWG NO.	SCALE	REV
			3/16" = 1'0"	

DRW. BY		SECTION #	SHEET

Chapter 5, Problem 9: DETERMINE THE SHORTEST VENTILATION SHAFT TO BE DUG FROM A POINT X ON THE EARTH'S SURFACE
TO THE CENTER LINE OF A TUNNEL, YZ. FIND THE TRUE LENGTH, GRADE, AND BEARING OF THE VENTILATION SHAFT.

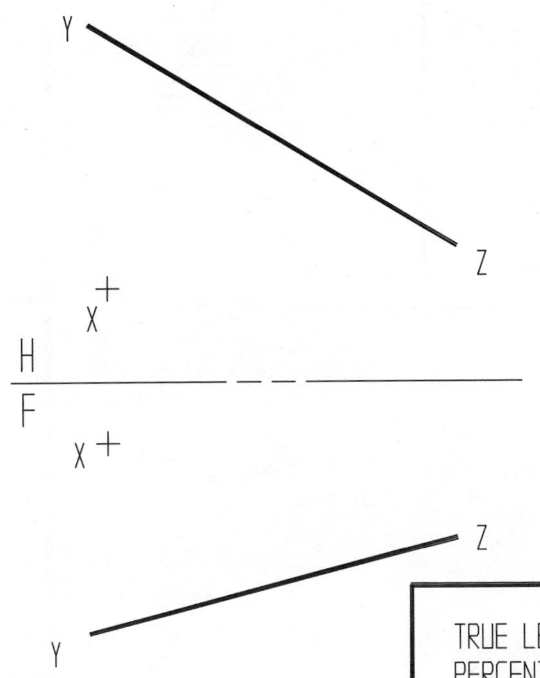

Z

Y

+
X

H
———
F

+
X

Z

Y

ADG INC.

SHORTEST AIRWAY TO TUNNEL

SIZE	DATE		DWG NO.		SCALE	REV
					1" = 50'	
			SECTION #		SHEET	

TRUE LENGTH =
PERCENT GRADE =
BEARING =

DRW. BY

Chapter 5, Problem 10: LINE AB IS THE CENTER LINE OF A GAS PIPE RUNNING THROUGH A MECHANICAL ROOM OF A BUILDING. A TEE FITTING IS TO BE INSERTED ALONG THIS PIPE TO ALLOW FOR A STRAIGHT CONNECTING PIPE TO THE BOILER AT C. DETERMINE THE TRUE LENGTH AND SLOPE ANGLE OF THE SHORTEST CONNECTING PIPE FROM C TO AB. SHOW THE POSITION OF THE CONNECTOR IN ALL THE VIEWS.

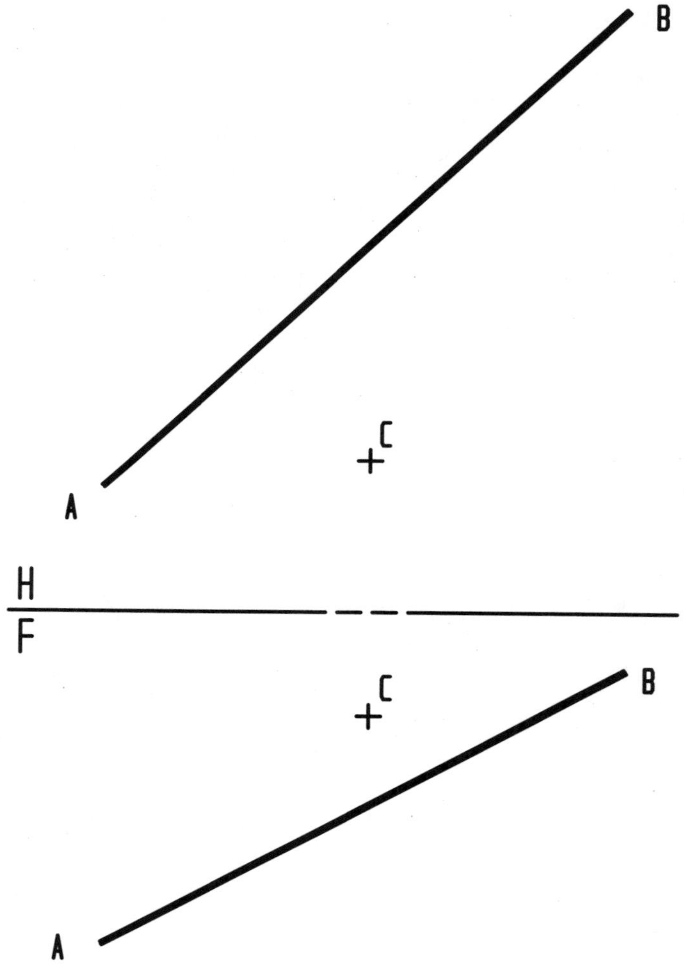

B

+ C

A

H
―
F

+ C

B

A

TRUE LENGTH =
SLOPE ANGLE =

ADG INC.

GAS LINE

SIZE	DATE	DWG NO.	SCALE	REV
			1" = 20"	

DRW. BY

SECTION #

SHEET

Chapter 5, Problem 11: YOU HAVE A PLOT PLAN AND FRONT VIEW OF A CORNER LOT IN A RESIDENTIAL AREA. AB AND BC REPRESENT THE CENTER LINES OF THE EXISTING WATER MAIN. IF THE WATER METER AT THE HOUSE IS LOCATED AT D, WHAT WOULD BE THE LENGTH AND SLOPE OF THE SHORTEST PIPE TO THE EXISTING WATER LINE? SHOW THE PIPE IN ALL VIEWS.

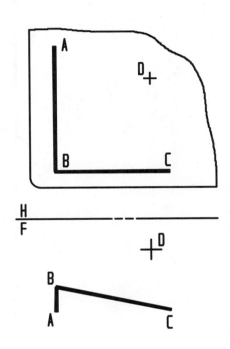

SHORTEST PIPE?
   TRUE LENGTH =
   SLOPE ANGLE =
	ADG INC.			

	WATER LINE			
	SIZE	DATE	DWG NO.	SCALE 1" = 100'	REV
DRW. BY		SECTION #		SHEET

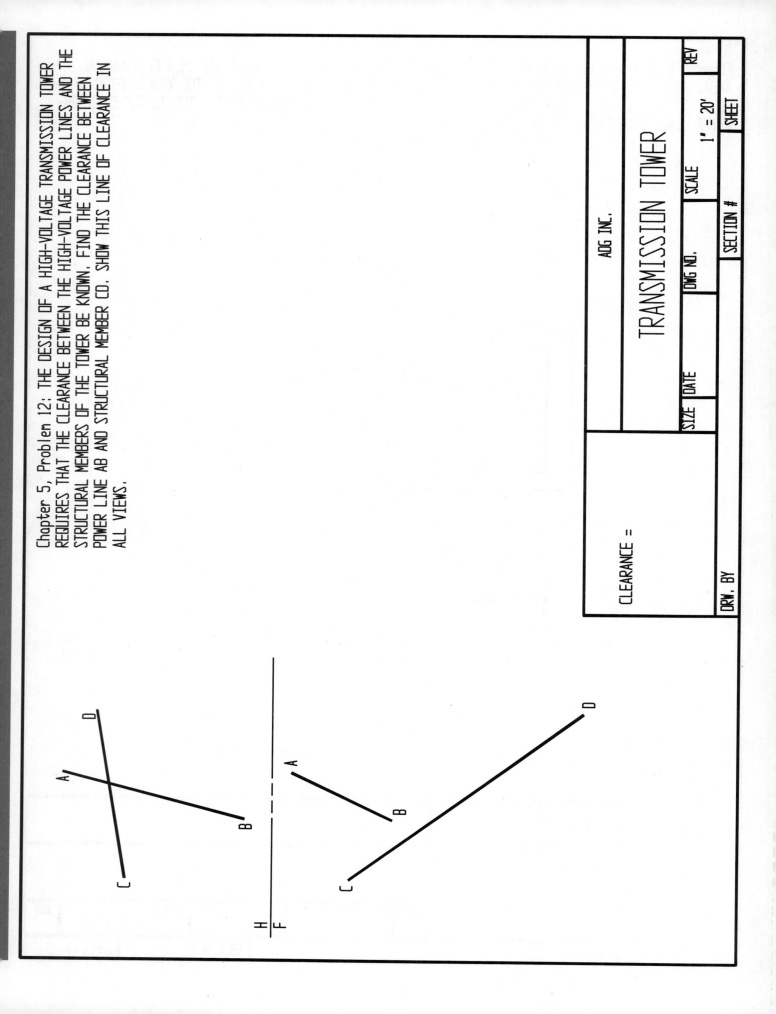

Chapter 5, Problem 12: THE DESIGN OF A HIGH-VOLTAGE TRANSMISSION TOWER
REQUIRES THAT THE CLEARANCE BETWEEN THE HIGH-VOLTAGE POWER LINES AND THE
STRUCTURAL MEMBERS OF THE TOWER BE KNOWN. FIND THE CLEARANCE BETWEEN
POWER LINE AB AND STRUCTURAL MEMBER CD. SHOW THIS LINE OF CLEARANCE IN
ALL VIEWS.

ADG INC.

TRANSMISSION TOWER

SIZE	DATE	DWG NO.	SCALE 1" = 20'	REV
		SECTION #		SHEET

CLEARANCE =

DRW. BY

Chapter 5, Problem 13: A BRACE NEEDS TO BE CONSTRUCTED TO SEPARATE TWO PIPES AT THEIR CLOSEST POINT. FIND THE SHORTEST DISTANCE BETWEEN THE PIPES AB AND CD. SHOW THE BRACE IN ALL VIEWS.

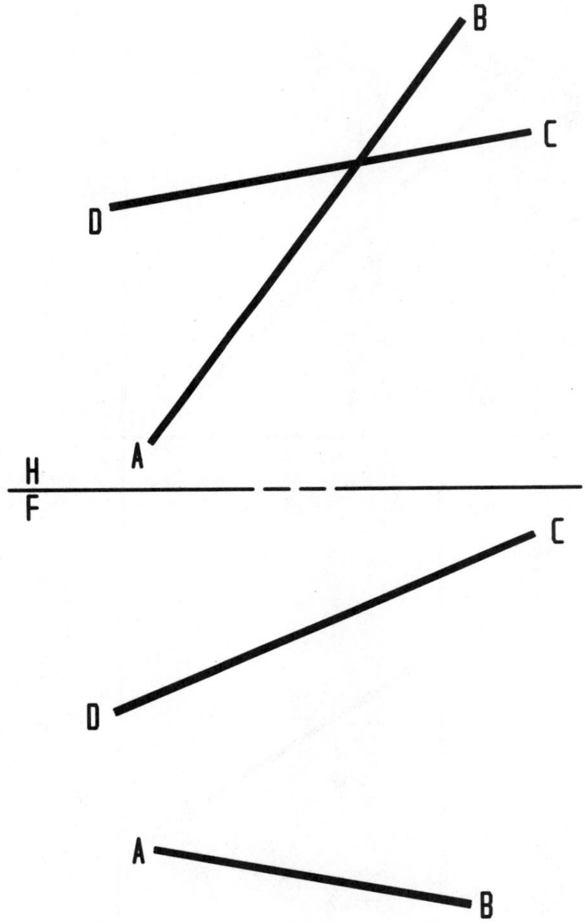

TRUE LENGTH =

ADG INC.

BRACE

SIZE	DATE	DWG NO.	SCALE	REV
			1" = 2'	

DRW. BY		SECTION #	SHEET

Chapter 5, Problem 14: A VERTICAL MAST, VM, 18 FEET HIGH AND A WIRE, AB, ARE LOCATED AS SHOWN. A GUY WIRE FROM POINT X IS TO BE FASTENED AS HIGH AS POSSIBLE TO THE MAST AT POINT Y, YET IT MUST CLEAR WIRE AB BY 2 FEET. FIND THE HIGHEST POINT ON THE VERTICAL MAST AT WHICH THE GUY WIRE CAN BE FASTENED. SHOW THE GUY WIRE, XY, IN ALL VIEWS.

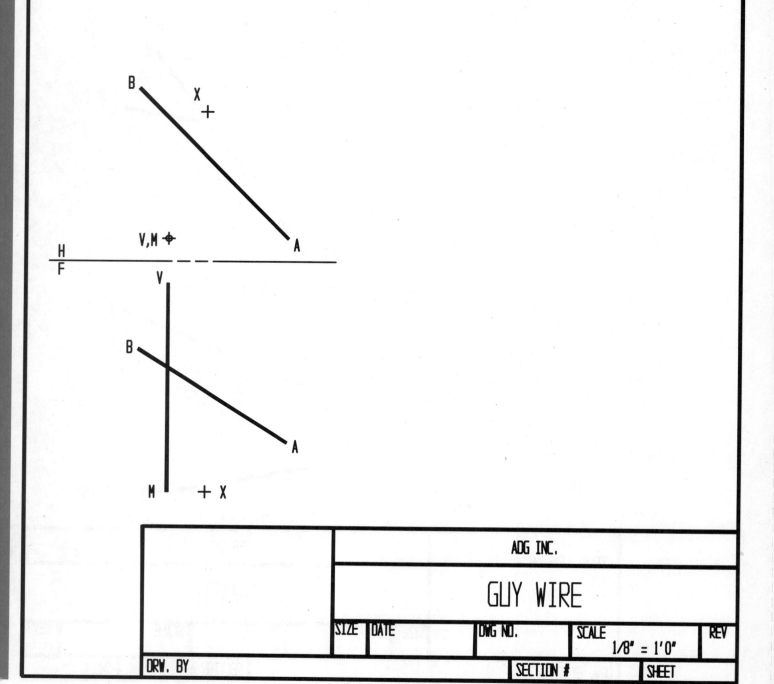

	ADG INC.			
	GUY WIRE			
SIZE DATE	DWG NO.	SCALE 1/8" = 1'0"		REV
DRW. BY		SECTION #		SHEET

Planes

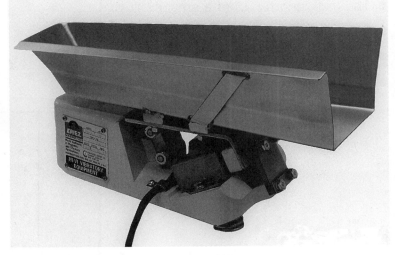

The downslope of this feeder may be set 0° to 10°.

The first part of this chapter deals with the representation of planes in space and their projections onto standard projection planes. The rest of the chapter will encompass when, where, why, and how to determine the edge view of a plane and true shape of a plane displayed in a principal view and an auxiliary view.

After completing this chapter, you will be able to:

▸ *Explain what a plane surface is.*

▸ *Describe four different methods by which a plane may be formed.*

▸ *Explain the difference between normal and oblique planes.*

▸ *Explain how to determine the type of plane.*

▸ *Demonstrate how to find the edge view of a plane surface.*

▸ *Define and demonstrate how to find the bearing (strike) of a plane.*

▸ *Define and demonstrate how to find the slope (dip) of a plane.*

▸ *Define and demonstrate how to find the dip direction of a plane.*

▸ *Explain and demonstrate how to find the true shape of a plane.*

Almost all objects and many engineering problems consist of planes. The basic principles involving planes are applicable in most industrial fields. A **plane** is a surface which is not curved or warped. It is a surface in which any two points may be connected by a straight line, and the straight line will always lie completely within the surface. In other words, every point on that line is also on the plane.

Figure 6–1 is a pictorial drawing of an object composed of a series of planes. Surfaces A, B, C, and D each satisfy the definition of a plane. You could place a pencil on any one of these surfaces, in any position, and it would lie flat on the surface of the plane.

As mentioned in Chapter 3, at times specific lines of sight are needed in order to view:

■ The true length (TL) of a straight line

■ Point view of a true length line

■ Edge view (EV) of a plane

■ A plane in its true size (TS) and shape

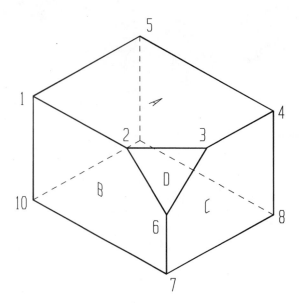

Figure 6–1 Planes.

▼ THE FORMATION OF PLANES

A plane may be constructed in four different ways, each of which is commonly used in engineering drawings. Figure 6–2 shows the four methods of representing a plane.

Two parallel lines form plane A-B-C-D in Figure 6–2(A). In conjunction with this figure, examine Figure 6–1, and notice that parallel lines 1-2 and 7-10 could be said to form plane B. Two intersecting lines, E-F and G-H, form plane E-H-F-G in Figure 6–2(B). Any three points, not in a straight line, form plane L-M-N in Figure 6–1 form plane D in the same manner.

Point Z, and line, X-Y, form plane X-Y-Z in Figure 6–2(D). Plane D in Figure 6–1 could also be a plane formed by line 2-3 and point 6.

A plane may represent only a portion of a surface. For example, why draw an entire wall, when only a small portion of it needs to be viewed. A plane may extend indefinitely beyond its established limits in order to solve for specific information. For example, if the solution for a problem does not fall on the part of the surface you choose to draw, just extend the plane. (This is illustrated in Figure 6–12, page 195.)

▼ PLANE TYPES

A plane may appear as an edge view (EV), true shape (TS), or foreshortened (plane appears smaller than it really is) (Figure 6–3).

A plane that appears as an edge view in either the front or side views, and is not perpendicular to a principal view line of sight, is called an **inclined plane** (Figure 6–3(A)). A plane that is parallel to any one of the primary projection planes, such as the horizontal or frontal

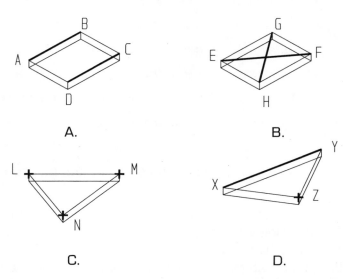

Figure 6–2 Formation of planes. (A) Two parallel lines. (B) Two intersection lines. (C) Three points. (D) A line and a point.

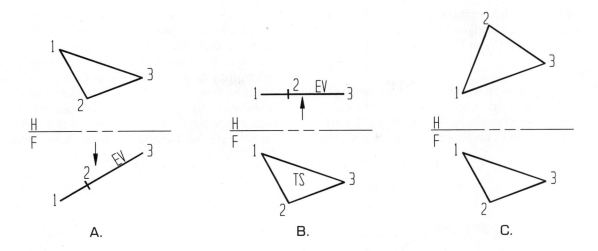

Figure 6–3   (A) Inclined plane. (B) True shape plane. (C) Oblique plane.

planes, will be seen true shape and size is called a **normal plane.** This is because the line of sight is perpendicular to the plane (Figure 6–3(B)). These true shape planes will be either vertical or horizontal (Figure 6–4(A), (B) and (C)). A plane that does not appear as an

edge view in a principal view, is called an **oblique plane.** (Figure 6–3C)

The **type of plane** is determined by the principal plane in which the plane is true shape. (Figure 6–4). If the line

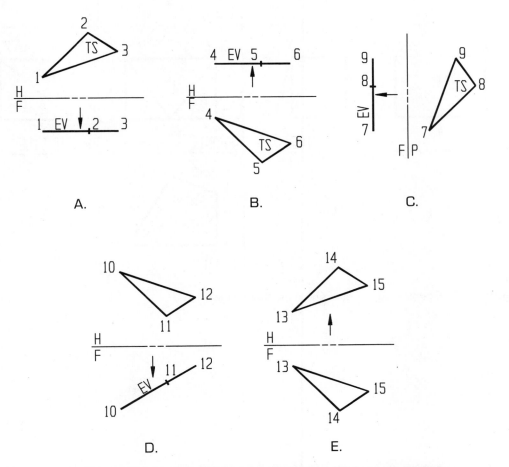

Figure 6–4   Plane types. (A) Horizontal. (B) Frontal. (C) Profile. (D) Inclined. (E) Oblique.

of sight is perpendicular to the edge view, the plane will be true shape in the adjacent view. (See the next section, Edge View of a Plane.)

Example

Figure 6–5 shows a pictorial and an orthographic drawing of an object showing all six primary views. Examine this object carefully. Planes A, G, E, and F are shown true shape in orthographic views because they are each perpendicular to a principal line of sight. For example, imagine yourself looking at the front of this object. In the front view, all you see of plane A is its edge. It looks like a straight line. The edge view of plane A is perpendicular to the line of sight to view the top view, which means the plane is parallel to the horizontal projection plane. Because plane A is parallel to the horizontal projection plane its true size and shape can be seen in the top view.

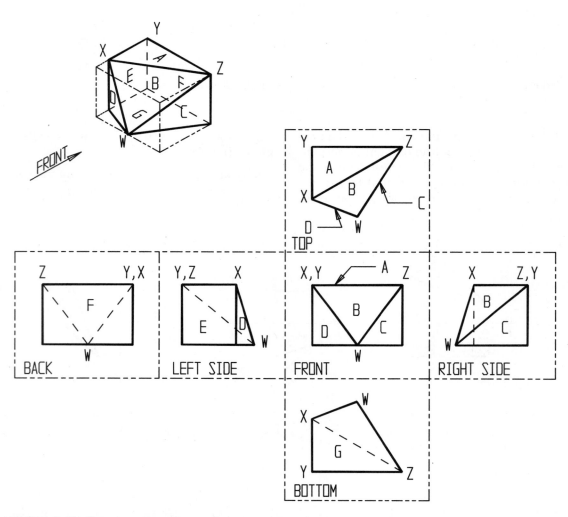

Figure 6–5 Examine edge views and true shape views.

EXERCISE

Problem: Ask yourself in which view you see plane E as an edge view, and in which view you see its true size and shape? Do the same with planes G and F (Figure 6–5).

Answer

Plane E is seen as an edge view in the front view and true shape in the left side view. Plane G is seen as an edge view in the front view and true shape in the bottom view. Plane F is seen as an edge view in the side views and true shape in the back view.

Planes C and D are inclined because they appear as an edge view in one view, but are not perpendicular to the line of sight for the adjacent view. Plane B is oblique because it does not appear as an edge view in any view; therefore, it is inclined to each of the primary planes. Because of their position, the true shape and size of these planes is not seen in any of the primary views.

▼ EDGE VIEW OF A PLANE

The *edge view* (EV) of a plane may be used to determine slope of a roof, highway, or ski slope. It may also be used to find the true shape of a particular plane on an A-frame structure or a dovetail clip. The intersection of two planes, such as two roofs on a multilevel house, may be found using the edge view method. Edge views of planes are needed to find the dihedral angle, the angle between two planes, the angle between two control tower walls or two steel plates to be welded together.

Any plane will be seen as an edge (a straight line) in a view in which *any* line on the plane appears as a point. Sometimes the edge view may be found in a principal view. However, sometimes auxiliary views are essential to find the missing information. This is the **third fundamental view**.

You have already learned how to view a line as a point. (Remember, the point view is found when the line of sight is parallel to a true-length line in the adjacent view. Which means the third fundamental view, edge view of a plane, can only be completed after the first two fundamental views have been located. In other words, we cannot disregard what we learned in the previous chapters, we must always locate the true length line first, then the point view of the true length line.)

I. Determine **line of sight**.

   A. Locate **true-length line**.

   B. Locate the **point view of the true-length line**.

   C. Locate an **edge view** of a plane (all points fall in a line).

      1. The line of sight must be drawn parallel to a true-length line that lies in the plane.

II. Draw the **reference line** perpendicular to the line of sight and label.

III. Draw **projection lines** parallel to the line of sight.

IV. Transfer **points of measurement** to the new view from the related view and label.

V. Connect the points and **label** true-length lines, edge views, and/or the true-shape planes.

EXERCISE

Problem: Go back to Figure 6–5 and locate the edge view of plane A.

Answer

Figure 6–6 shows the answer is already in a principal view, the front view.

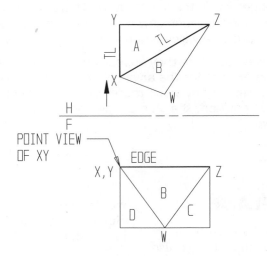

Figure 6–6 First three fundamental views are demonstrated in plane A.

Problem: Circle, on Figure 6–6, the line in plane A that demonstrates all three fundamental views (true length of any line in plane A, point view of the true-length line in plane A, edge of plane A).

Answer

Line X-Y is **true length** in the top view. The line of sight for the front view is parallel to line X-Y in the top view: therefore, line XY is seen as a **point view** in the front view. In the view where a line is seen as a point, the rest of the points in plane A will fall in a straight line forming an **edge view** of the plane in the front view (Figure 6–6).

Problem: Study Figure 6–6, and circle the line that demonstrates all three fundamental views (true length of any line in plane B, point view of the true length line in plane B, edge of plane B).

Answer

The edge view is not found in a principal plane, but the true length of line X-Z is found in the top view. Figure 6–7, demonstrates an auxiliary view is necessary to locate the point view of line X-Z. Transfer the rest of the points of plane B and an edge view of plane B is formed. (To aid you in visualizing plane W-X-Z, the object lines in Figure 6–7 that are not necessary to solve for the edge view of plane W-X-Z are left light. The projection lines are drawn as hidden lines, so they will not be confused with the light object lines.)

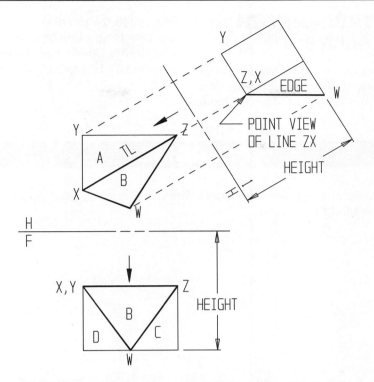

Figure 6–7 Auxiliary view completed to find edge view of plane B.

For the next problem, the height of point Z in Figure 6–5 will be lowered (Figure 6–8(A)). Notice line X-Z is no longer true length in the top view. Without locating the true length of any line in plane B, the edge view of plane B cannot be found (Figure 6–8(B)). The edge view of plane B can be located by creating a true-length line of any line in plane B in a primary auxiliary view followed by a secondary auxiliary view locating the point view of the true-length line and an edge view of plane B. However, if the slope of the plane is needed, the edge view of the plane will have to be in an elevation (front, side, or top-adjacent) view. The following section will discuss how to create a true-length line in a principal view, so that the edge view can be found in a top-adjacent view.

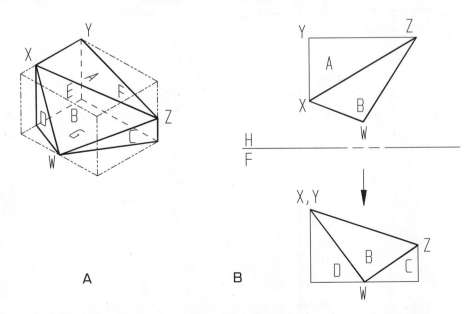

Figure 6–8 Plane B does not show a true-length line in the principal views.

▼ LOCATING A TRUE-LENGTH LINE IN THE PLANE

The edge view of the plane can only be found in the view where a true-length line in the plane appears as a point. In other words, we can't disregard what we have learned in the previous chapters, we must always locate the true-length line first, then the point view of the true-length line. Sometimes the entire solution, or part of the solution, is in a principal plane(s).

EXERCISE

Problem: Locate and label the true-length lines in Figure 6–9(A) and (B). Sketch in the appropriate line of sight.

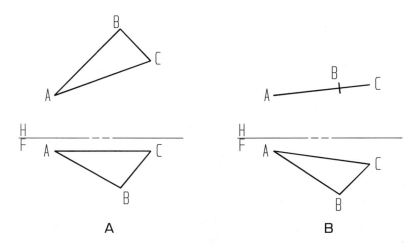

A   B

Figure 6–9 Locate the true-length line in each problem.

Answer

The line of sight to view the top view is drawn in the front view. It is perpendicular to line A-C in the front view; therefore, line A-C in the adjacent view (top view) is true length. (Figure 6–10(A)) In order for a plane to appear as an edge view, it must include a point view of a true length line. This means the line of sight will be parallel to the true length line in the adjacent view (Figure 6–10(B)).

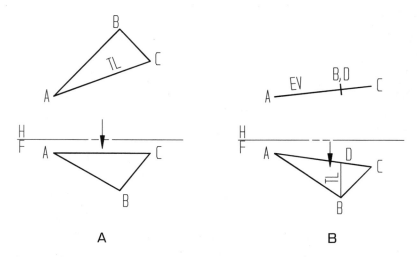

A   B

Figure 6–10 (A) True-length line already given. (B) True-length line created.

If a true-length line is not present in a principal view, one can be created anywhere in the oblique plane. This is done by creating a construction line perpendicular to the line of sight for the adjacent view (Figure 6–11). The construction line may be located anywhere in the plane, but it is easiest to use an existing point. Make sure the points are projected to the correct line, for example, point 5 is on line D-E in the front view. Make sure you project it to line D-E in the top view and not line D-F.

If a plane includes a vertical line, the plane may have to be extended in order to locate the other end of the line (Figure 6-12).

1. Extend one line of the plane in any direction, any distance, in the adjacent views. In Figure 6–12(C) line G-I was extended in the adjacent views to the right. How far the line is extended does not matter. This creates new endpoint W in the top and front views.

2. Draw a new line from the new endpoint, W, to the endpoint of the other line in the plane, H.

3. Create a true-length line, G-Z, in the new plane, G-W-H.

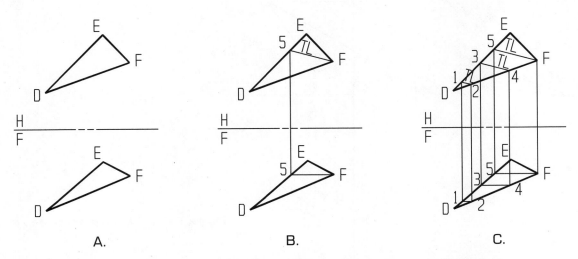

Figure 6–11 (A) No TL. (B) TL created using existing point F. (C) Multiple TLs created anywhere inside the plane.

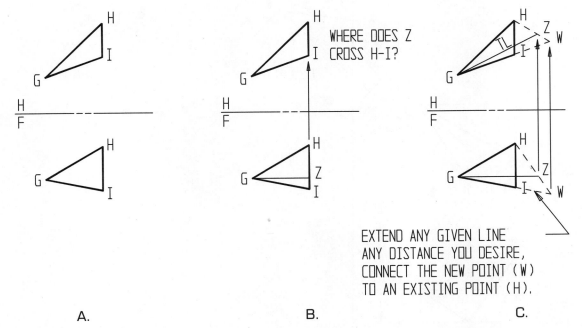

Figure 6–12 (A) No TL. (B) Cannot project Z. (C) Extend plane in order to project Z.

EXERCISE

Problem: On Figure 6–13, sketch the edge view for plane B in a top-adjacent view.

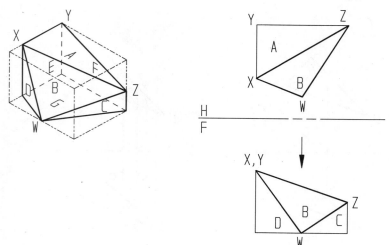

Figure 6–13 Sketch the edge view for plane B.

Answer

Figure 6–14 shows the edge view of plane B.

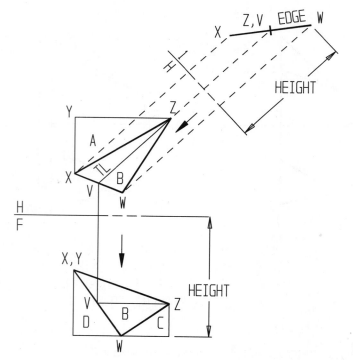

Figure 6–14 Edge view for plane B.

▼ BEARING (STRIKE) OF THE PLANE

The bearing (compass reading) of a horizontal (level) line within a plane is called **strike**. Strike is a geological term used to describe orientation of various layers of the earth's crust.

Compass bearing of a horizontal line is measured where any line in the plane is true length in the top view; therefore, the angle that the true-length line makes in the top view is read using a protractor from a north line. Because the horizontal line does not have a low end, it is general practice to read the acute angle from north; therefore, the orientation will be due north, northeast, or northwest. N 70° W is an example of compass bearing.

The following elements are identical to finding the bearing of a line. The bearing is used solely for map direction; therefore, it is read using a navigational compass (not a drawing compass). The compass is held horizontally; therefore, the bearing is found (read) only in the top orthographic view. North is assumed to be directed toward the top of the page, unless otherwise specified. The bearing is measured independently of the slope.

Figure 6–15 shows the strike of plane B

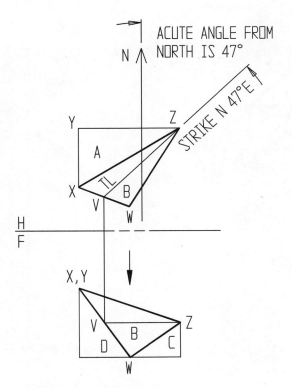

Figure 6–15 Strike of plane B is N47°E.

EXERCISE

Problem: What is the strike for plane A-B-C in Figure 6–16.

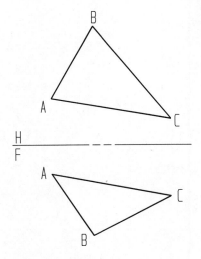

Figure 6–16 Strike of plane A-B-C.

Answer

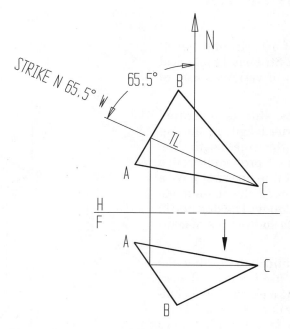

Figure 6–17   The strike is N 65.5° W.

▼ SLOPE (DIP) OF THE PLANE

The **slope** of a plane is the angle the edge view of the plane makes with the edge view of the horizontal plane. The geological term for slope is **dip** and is used in conjunction with strike.

Remember the building with a wheelchair ramp (Figure 6–18)? The angle between the ground (edge view of horizontal plane) and the ramp is the slope. If a wheelchair

ramp's slope is too steep it is difficult to wheel or push a wheelchair up the ramp or to control it coming down the ramp.

In orthographic drawings the slope can only be measured when the horizontal plane appears as an edge view (reference line H/?, such as H/F or H/1) and the plane appears as an edge view. The case of an inclined plane, such as plane 1-2-3-4 in Figure 6–18, the true slope is seen in the front view. But in the case of oblique

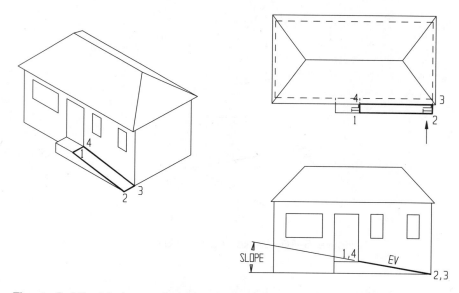

Figure 6–18   Maximum slope for a wheelchair ramp, plane 1-2-3-4 is 1:12.

planes (Figure 6–19), the true slope of the plane can be seen only in an elevation view that shows the edge view of the plane.

An edge view of a plane in an auxiliary view, other than a top-adjacent view, will not demonstrate true slope. This is because the view does not show true height (or have a horizontal line of sight). To prove this to yourself, review the cutout figures in the Helpful Hints section of Chapter 3.

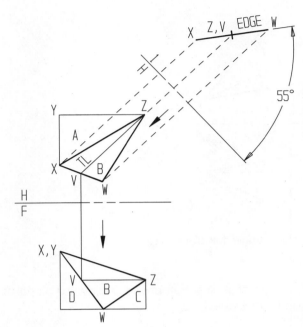

Figure 6–19 The slope angle of plane B is 55°.

EXERCISE

Problem: What is the slope angle for plane A-B-C in Figure 6–20?

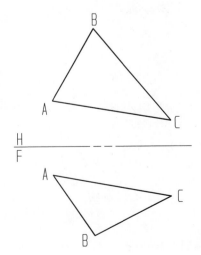

Figure 6–20 What is the slope angle for plane A-B-C?

Answer

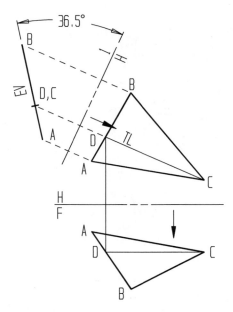

Figure 6-21 The slope angle for plane A-B-C is 36.5°.

The dip is generally given in conjunction with the general direction of the downward slope of the plane (See the next section, Dip Direction of the Plane.)

▼ DIP DIRECTION OF THE PLANE

Dip **direction of slope** is the direction a ball will roll when placed on a plane. It will roll perpendicular to the true-length line in top view toward the low side (found in elevation view) of the plane (Figure 6–22 and 6–23).

Because the dip direction is always 90° to the bearing the degrees are not given, only direction— NE, NW, SE, or SW. This direction is given in conjunction with dip. For example, for a dip of 24° and a NE dip direction, the plane would be described as having a dip of 24° downward in a northeasterly direction.

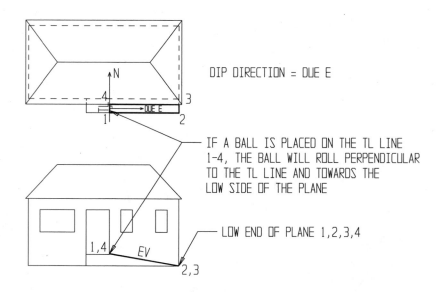

DIP DIRECTION = DUE E

IF A BALL IS PLACED ON THE TL LINE 1-4, THE BALL WILL ROLL PERPENDICULAR TO THE TL LINE AND TOWARDS THE LOW SIDE OF THE PLANE

LOW END OF PLANE 1,2,3,4

Figure 6-22 Dip direction of plane 1-2-3-4 is due east.

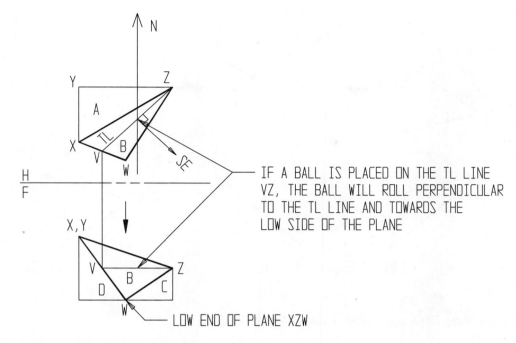

IF A BALL IS PLACED ON THE TL LINE VZ, THE BALL WILL ROLL PERPENDICULAR TO THE TL LINE AND TOWARDS THE LOW SIDE OF THE PLANE

LOW END OF PLANE XZW

Figure 6–23   Dip direction of plane B is SE.

EXERCISE

Problem: Notice in Figure 6–24 the top views are all the same, but the front views are different. Label the low side of each plane.

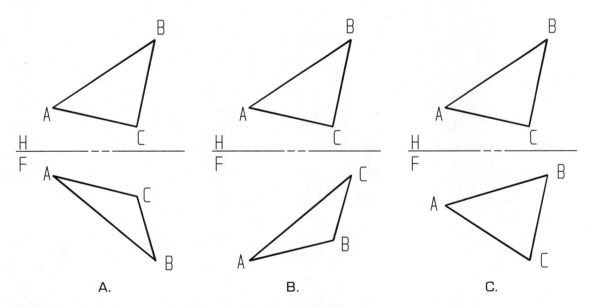

A.   B.   C.

Figure 6–24   Label the low side for each of the planes above.

Answer

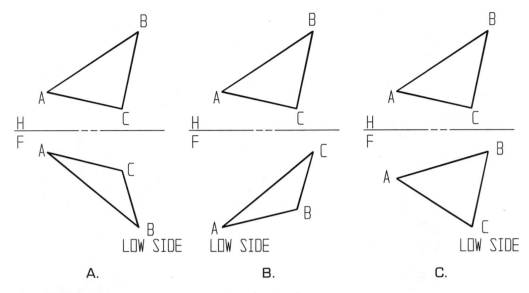

Figure 6–25 The low side is found in the front view.

Problem: For the purpose of practicing dip direction, list only the dip direction for each of the problems shown in Figure 6–24.

List the strike for each of the problems shown in Figure 6–24.

Answer

Dip direction:

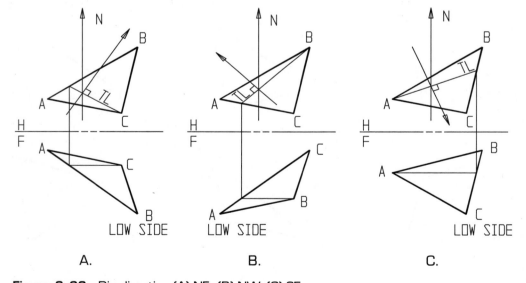

Figure 6–26 Dip direction (A) NE. (B) NW. (C) SE.

Strike:

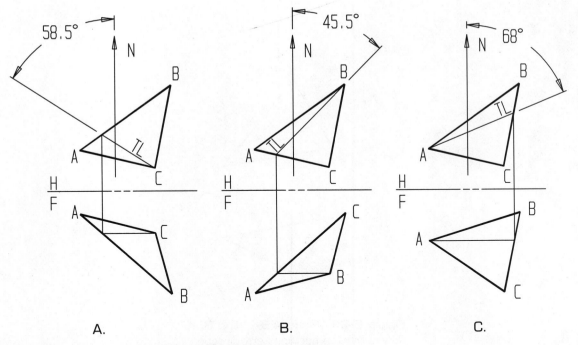

A.   B.   C.

Figure 6–27 Strike (A) N58.5°W. (B) N45.5°E. (C) N68°E.

Problem: The slope for plane A-B-C in Figure 6–28 is 36.5°. What is the dip direction?

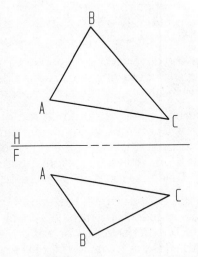

Figure 6–28 What is the dip
direction for plane A-B-C?

Answer

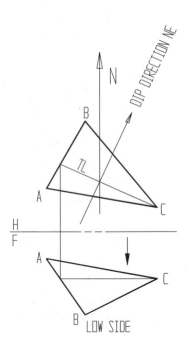

Figure 6–29 Plane A-B-C dips 36.5° downward in a northeasterly direction.

HELPFUL HINTS

The compass bearing (or strike) of a plane gives the orientation of a plane. Because the line used to measure the bearing is level, the direction the plane's slope must also be given. It is called dip direction. It is 90° to the bearing toward the low side of the plane, and is written in conjunction with dip.

To help visualize strike and dip direction find something stiff to draw on, like a piece of cardboard. Draw a straight line at any angle from one side of the cardboard to the other. Label the line TL, to help you remember that it is horizontal and true length.

Pick up the cardboard and place a thumb on top of each end of the line. Hold the cardboard level and rotate your thumbs so the line changes direction. This is *compass bearing*—the direction a horizontal line is traveling. Move your right thumb until the line is traveling from your left thumb to your right thumb approximately N45°E. At this point, if a ball was placed on the TL line it would not roll because the TL line and the plane are level.

Holding the TL line at the endpoints, tip the end of the plane closest to you down. (This is dip (slope), and for now it doesn't matter how much you move it as long as you move it.) Remember to keep the TL line horizontal and at a N45°E direction. This is dip direction. If a ball was placed on the TL line it would roll perpendicular to the true-length line toward the low end of the plane. In this case, SE. If tipped the other way the ball would roll in a NW direction.

To see the slope, or dip, of the plane you must move around the plane until you are looking parallel to the true length line and the plane appears as an edge view. (As you are moving about the plane, remember the true-length line must stay horizontal and at a N45°E direction.) This is slope and it tells you how much the plane was tipped. To prove the true-length line may be anywhere in the plane, draw a line parallel to your true-length line and repeat the above steps.

EXERCISE

Problem: The following sections contain two problems in each figure. For practice purpose only, each topic, strike, dip, and dip direction will be examined individually. Read the following steps and sketch the correct solutions on the figure provided. The correct answers will be shown in the next figure.

Strike

Make a note card of the following information to use while completing homework assignments.

> **Strike (compass bearing of a horizontal line.)**
>
> **1.** Find a true length line in
> **2.** the top view and
> **3.** measure the acute angle
> **4.** from North.
> **5.** North is assumed to be directed toward the top of the page, unless otherwise specified.
> **6.** The strike is measured independently of the slope (dip).

Steps 1 and 2. *Find a true length line in top view.*

Figure 6–30 Sketch in the true-length lines and lines of sight for planes A-B-C and D-E-F.

Steps 3 and 4. *Measure the acute angle from North.*

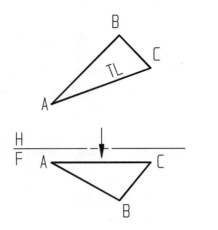

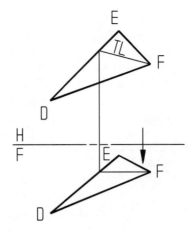

Figure 6–31 Measure the strike for planes A-B-C and D-E-F.

Answer

The correct solution for strike of a plane (Figure 6–32).

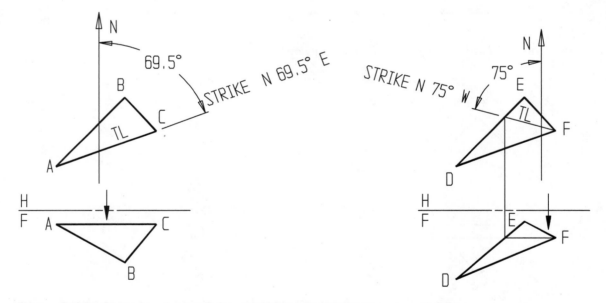

Figure 6–32 Strike for plane A-B-C is N69.5°E and N75°W for plane D-E-F.

Dip (Slope) of a Plane

Problem: Make a note card for the following information to use while completing homework assignments.

Dip of a plane

1. Angle the

2. edge view of the plane makes

3. with the edge view of the horizontal plane (H/1 reference line).

Steps 1,2, and 3. *The angle the edge view of the plane makes with the edge view of the horizontal plane (H/1 reference line).*

Figure 6–33 Measure the dips for planes A-B-C and D-E-F.

Answer

The correct solution for dip of a plane is seen in Figure 6–34. The same answer would have resulted from projecting the top-adjacent view from the other end of the true-length line, as long as the line of sight remains parallel to the true-length line.

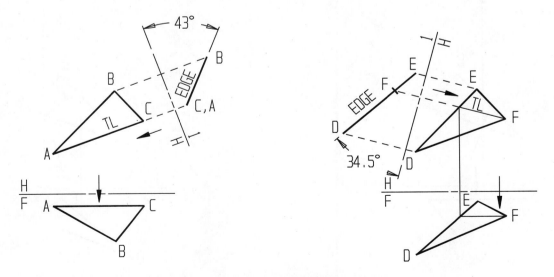

Figure 6–34 Dip for plane A-B-C is 43° and 34.5° for plane D-E-F.

Dip Direction

Problem: Make a note card for the following information to use while completing homework assignments.

Dip direction

1. Draw a line perpendicular to
2. true length line in
3. top view towards
4. the low side (found in elevation view).

Steps 1, 2 and 3. *Draw a line perpendicular to the true length line in the top view.*

Figure 6–35 Locate the true-length line for planes A-B-C and D-E-F.

Step 4. *Toward the low side.*

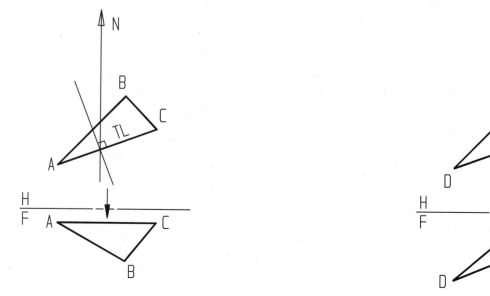

Figure 6–36 Determine the low side of the planes A-B-C and D-E-F.

Answer

The correct solution for dip direction of a plane (Figure 6–35).

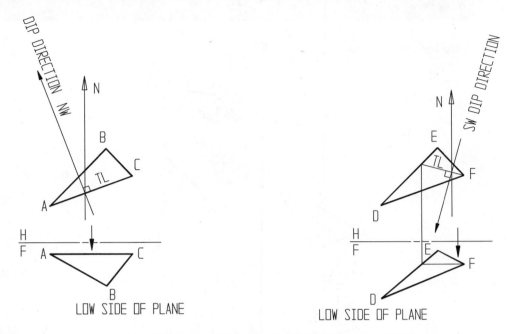

Figure 6–37 Dip direction for plane A-B-C is NW and SW for plane D-E-F.

Problem: The dip and dip direction are read in conjunction with one another. How should the answers for planes A-B-C and D-E-F in Figures 6–34 and 6–37 be read?

Answer

Plane A-B-C dips 43° downward in a northwesterly direction. Plane D-E-F dips 34.5° downward in a southeasterly direction.

▼ TRUE SHAPE OF THE PLANE

The **true shape** (TS) of a plane may be used to determine the area of a surface in order to calculate how many shingles, ceramic tiles, gallons of paint, and so on, may be required. Runners in an injection mold must be milled in specific locations. These locations may only be dimensioned on true shape views.

To solve for the true shape of a plane, the line of sight must be perpendicular to the edge view of the plane. This is the **fourth** and final **fundamental view** and builds on the first three fundamental views. In other words, we can't disregard what we learned in the previous chapters, we must always locate a true-length line first, the point view of the true-length line second, and the edge view of the plane third, before the true shape of the plane can be found.

I. Determine line of sight.

    A. Locate **true-length line.**

    B. Locate the **point view of the true-length line.**

    C. Locate an **edge view of a plane** (all points fall in a line).

    D. Draw the **true shape of a plane.**

        1. The line of sight must be drawn perpendicular to the edge view of a plane.

II. Draw the **reference line** perpendicular to the line of sight and label.

III. Draw **projection lines** parallel to the line of sight.

IV. Transfer **points of measurement** to the new view from the related view and label.

V. Connect the points and **label** true-length line, edge views, and/or true-shape plane.

We already found the edge view to Figure 6–13 in Figure 6–14.

EXERCISE

Problem: Solve for true shape (Figure 6–38).

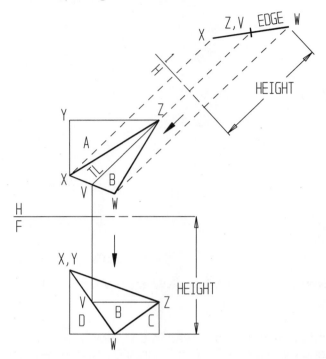

Figure 6–38 Draw the true shape of plane B.

Answer

True shape solution (Figure 6–39).

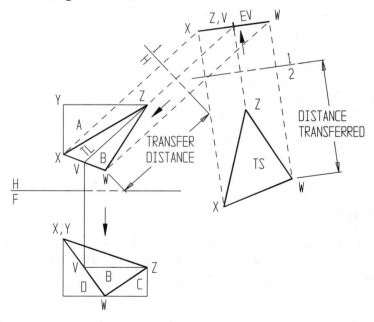

Figure 6–39 True shape of plane B.

HELPFUL HINTS

You may want to ask the following questions before you start to draw.

When solving a problem organize yourself first by asking the following questions.

What am I **solving for**?

What is **given**?

What views are given?

What type of plane do I have?

What other information do I know?

What do I **need to solve** this problem?

EXERCISE

Problem: To find the true shape of plane A-B-C-D, all four fundamental views (lines of sight) must be completed (Figure 6–40). Highlighting the plane with colored pencils helps to focus only on the plane you are working with.

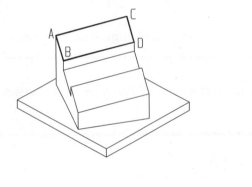
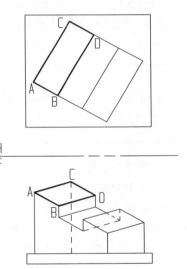

Figure 6–40 Isometric and orthographic of a guide block.

If you are not sure where to start, answer the following questions first.

What am I **solving for**?

What is **given**?

What views are given?

What type of plane do I have?

What other information do I know?

What do I **need to solve** this problem?

Answer

What am I **solving for**?

■ True shape of plane A-B-C-D

What is **given**?

What views are given?

■ Top and front

What type of plane do I have?

■ The answer is oblique, because a principal line of sight is not perpendicular to an edge view of the plane.

What other information do I know?

■ *None—I don't even have a true-length line.*

What do I **need to solve** this problem?

■ Auxiliary views are needed, because all four fundamental views must be found.

I. **Determine line of sight**.

A. Locate **true-length line**.

B. Locate the **point view of the true-length line**.

C. Locate an **edge view of a plane** (all points fall in a line).

D. Draw the **true shape of a plane**.

1. The line of sight must be drawn perpendicular to the edge view of a plane.

Problem: Read the following steps and sketch in the correct solution. The correct answer will be shown in the next figure.

I. **Determine line of sight.**

A. Locate **true-length line**.

1. Is the true length line **in** one of the **principal planes**?

a. If the line of sight is perpendicular to the object line in one view, it will appear true length in the adjacent view and can be measured.

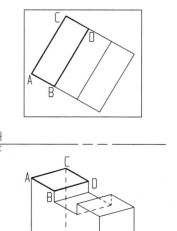

Figure 6–41 Locate a true-length line in plane A-B-C-D.

Locate the point view of a true-length line.

  I. Determine line of sight.

    B. Draw the **point view of a true-length line**.

      1. The line of sight must be drawn parallel to the true-length line.

  II. Draw the **reference line** perpendicular to the line of sight and label.

  III. Draw **projection lines** parallel to the line of sight.

  IV. Transfer **points of measurement** to the new view from the related view and label. Always go back to the last line of sight to take the measurement.

  V. Connect the points and **label** true-length lines, edge views, and/or true-shape planes.

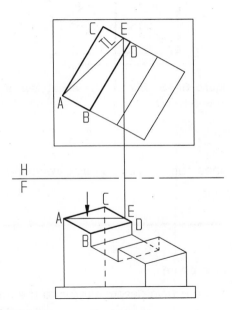

Figure 6–42   Sketch the point view for line A-E.

Locate the edge view of a plane.

  I. Determine line of sight.

    C. Draw an **edge view of a plane** (all points fall in a line).

      1. The line of sight must be drawn parallel to a true-length line that lies in the plane.

  II. Draw the **reference line** perpendicular to the line of sight and label.

  III. Draw **projection lines** parallel to the line of sight.

  IV. Transfer **points of measurement** to the new view from the related view and label. Always go back to the last line of sight to take the measurement.

  V. Connect the points and **label** true-length lines, edge views, and/or true-shape planes.

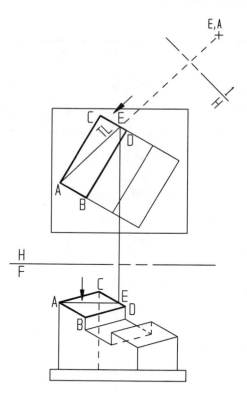

Figure 6–43 Sketch the edge view of plane A-B-C-D.

(Remember only true-length lines can appear as point views in adjacent views.)
Locate the true shape of a plane.

I. Determine line of sight.

    D. Draw the **true shape of a plane**.

        1. The line of sight must be drawn perpendicular to the edge view of a plane.

II. Draw the **reference line** perpendicular to the line of sight and label.

III. Draw **projection lines** parallel to the line of sight.

IV. Transfer **points of measurement** to the new view from the related view and label. Always go back to the last line of sight to take the measurement.

V. Connect the points and **label** true-length lines, edge views, and/or true-shape planes.

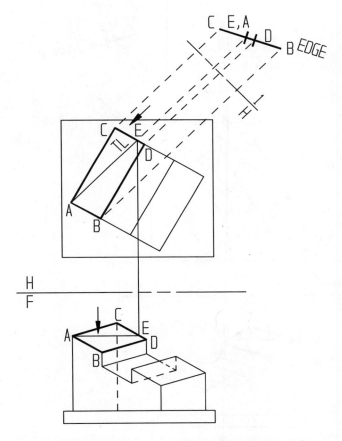

Figure 6–44 Sketch the true shape to plane A-B-C-D.

Figure 6–45 shows actual true shape of plane A-B-C-D.

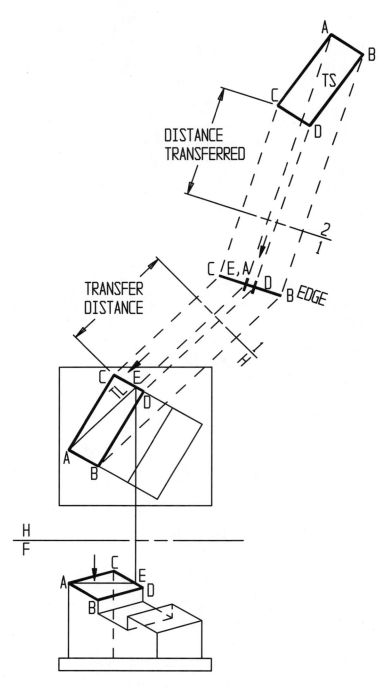

Figure 6–45 True shape of plane A-B-C-D.

Name _____

Date _____

Course _____

CHAPTER 6 STUDY QUESTIONS

The Study Questions are intended to assess your comprehension of chapter material. Please write your answers to the questions in the space provided.

1.   Define the term **plane** as used in descriptive geometry.

   _____

   _____

2.   Identify the following types of planes.

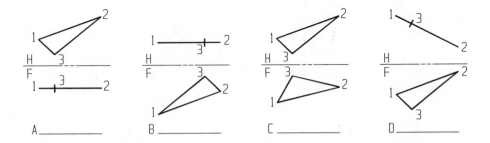

A_____   B_____   C_____   D_____

3.   Create a true-length line in plane 1-2-3.

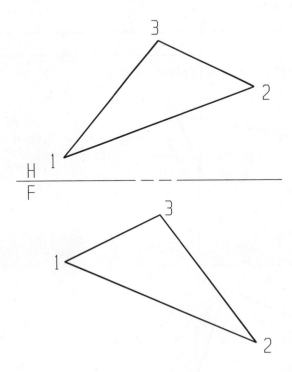

4. Define strike of a plane.

_____

_____

5. Determine the strike of the oblique plane below.

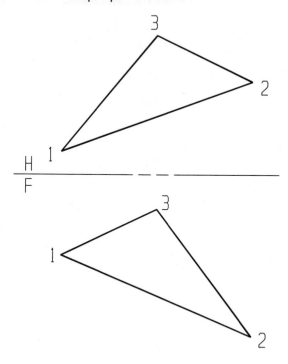

6. Define slope of a plane.

_____

_____

7. Determine the dip of the following plane.

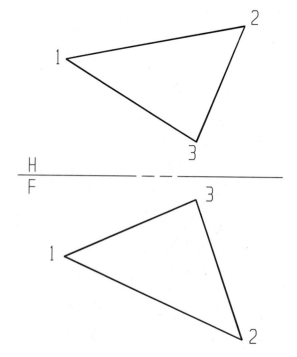

8.   Determine the dip direction of the plane in number 7.

_____

_____

9.   Sketch the true shape for plane.

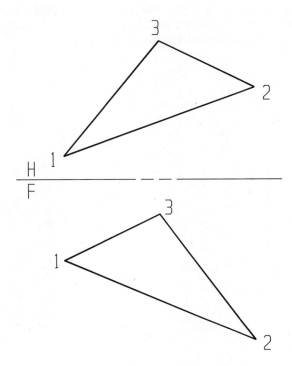

10.   Describe the relationship between the four fundamental views.

_____

_____

_____

_____

The problems in this chapter are based on a variety of industrial applications. In solving these problems you should try to utilize direct approaches that get to the heart of the problem. At this analysis stage of the design process, many problems use center lines and single lines to represent the problem situation. Yet, when necessary, relevant sizes are given.

The following problems may be drawn using instruments on the page provided, created using a 2D or 3D CAD system, or a combination of drawing board and CAD. Refer to Appendix A for additional dimensions needed to solve problems three-dimensionally.

Chapter 6, Problem 1: A TRANSISTION PIECE FOR A FUME EXHAUST DUCT SYSTEM
IS SHOWN. FIND THE TRUE SHAPE AND SIZE OF PLANES A, B, C, AND D. THE
INTERIOR SURFACES OF THE DUCT ARE TO RECEIVE A PROTECTIVE COATING OF
EPOXY RESIN PAINT. HOW MANY SQUARE FEET OF METAL WILL NEED PAINTED?
(DISREGARD METAL THICKNESS) WHAT TYPE OF PLANE IS PLANE B?

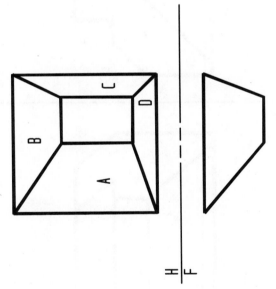

H
F

ADG INC.

TRANSISTION PIECE

SIZE	DATE	DWG NO.	SCALE	REV
			1/4" = 1'0"	
		SECTION #	SHEET	

TOTAL SQUARE FOOTAGE TO
BE PAINTED =

PLANE TYPE OF PLANE B?

DRW. BY

Z

H
F

Z

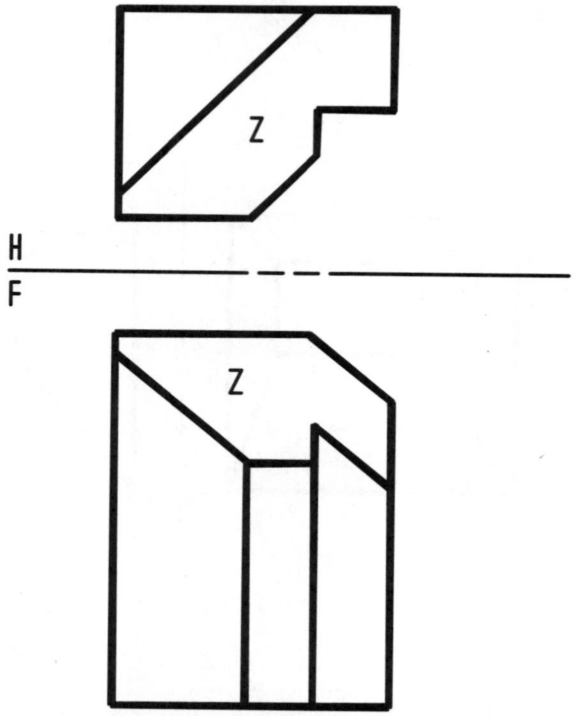

SLOPE =

ADG INC.

CLUTCH STOP

SIZE	DATE	DWG NO.	SCALE		REV
			FULL		

DRW. BY

SECTION #

SHEET 1 OF 1

Chapter 6, Problem 3: A RAFTER FOR A DECORATIVE GAZEBO IS SHOWN BELOW. SHOW ALL NECESSARY DIMENSIONS NEEDED TO CUT THE RAFTER TO SIZE. WHAT IS THE NOMINAL SIZE OF THE RAFTER?

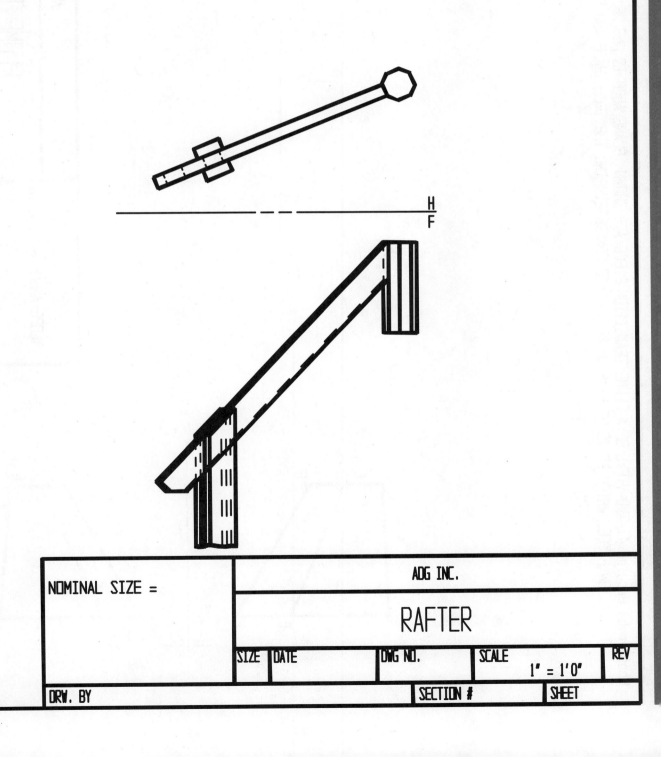

H
F

NOMINAL SIZE =		ADG INC.			
		RAFTER			
	SIZE	DATE	DWG NO.	SCALE 1" = 1'0"	REV
DRW. BY			SECTION #	SHEET	

Chapter 6, Problem 4: TWO VIEWS OF A FLUME TRANSITION PIECE ARE SHOWN. PLANE ABCD IS ONE OF THE SIDE WALLS OF THIS TRANSITION PIECE, WHICH IS TO BE MADE OF REINFORCED CONCRETE. FIND THE TRUE SHAPE AND SLOPE ANGLE OF PLANE ABCD.

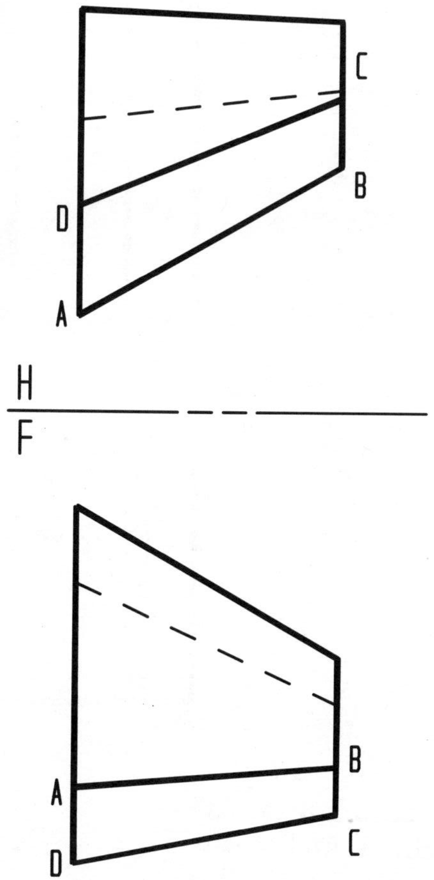

SLOPE ANGLE =

ADG INC.

FLUME TRANSITION

SIZE	DATE	DWG NO.	SCALE
			1/4" = 1'0"
		SECTION #	SHEET

DRW. BY

REV

Chapter 6, Problem 5: THE FRONT AND RIGHT SIDE VIEWS OF A
ANGLE PLATE ARE SHOWN. DRAW A VIEW THAT SHOWS THE INCLINED
SURFACE A IN TRUE SHAPE AND SIZE.

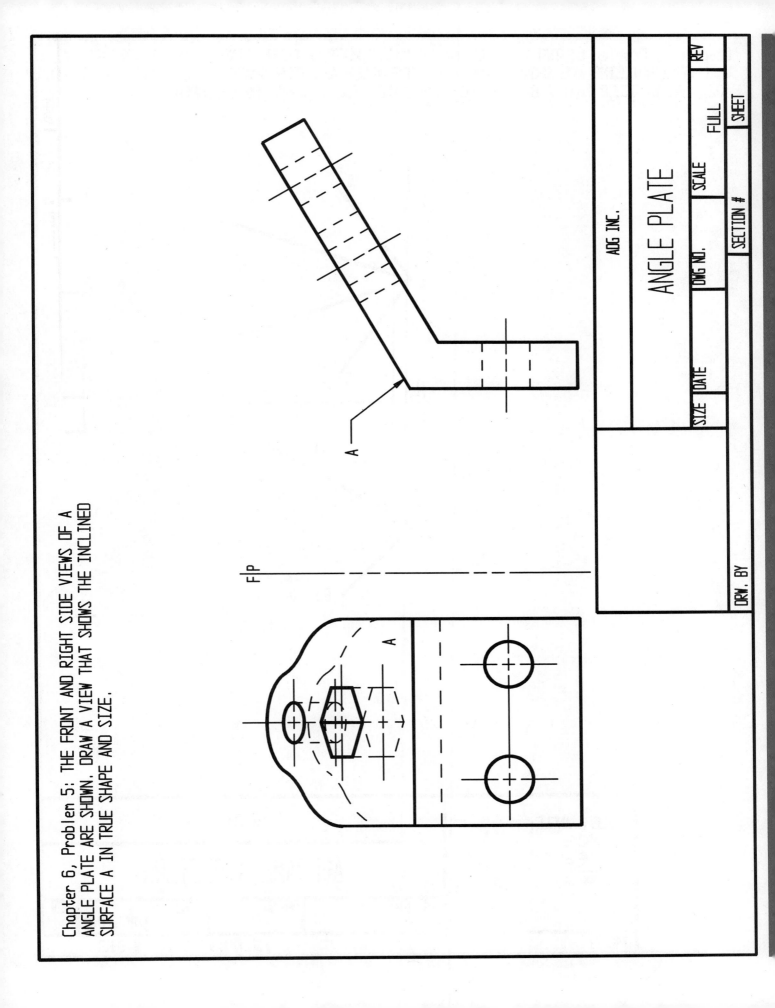

F/P

A

ADG INC.

ANGLE PLATE

DRW. BY

SIZE | DATE | DWG NO. | SCALE | FULL | REV
SECTION # | SHEET

Chapter 6, Problem 6: TWO VIEWS OF AN A-FRAME STRUCTURE CONNECTING TO THE FLOOR AND THE WALL OF A BUILDING ARE SHOWN. FIND THE SLOPE ANGLE AND TRUE SHAPE AND SIZE OF PLANES ABD, ADE, ABC IN ORDER THAT A BRACKET FOR THE CONNECTION AT A MAY BE DESIGNED.

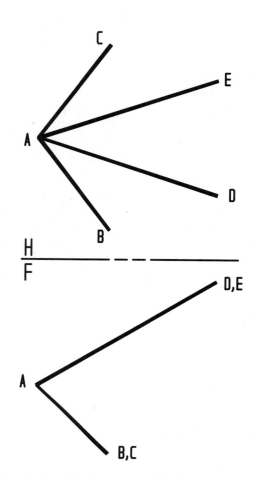

SLOPE ANGLE:	ADG INC.				
ABD =	A-FRAME STRUCTURE				
ADE =					
ABC =					
	SIZE	DATE	DWG NO.	SCALE 1/8" = 1'0"	REV
DRW. BY			SECTION #		SHEET

Chapter 6, Problem 7: POINTS R, S, AND T ARE ON THE UPPER BEDDING PLANE OF A STRATUM OF COAL. FIND THE STRIKE, DIP, AND DIP DIRECTION OF THIS STRATUM.

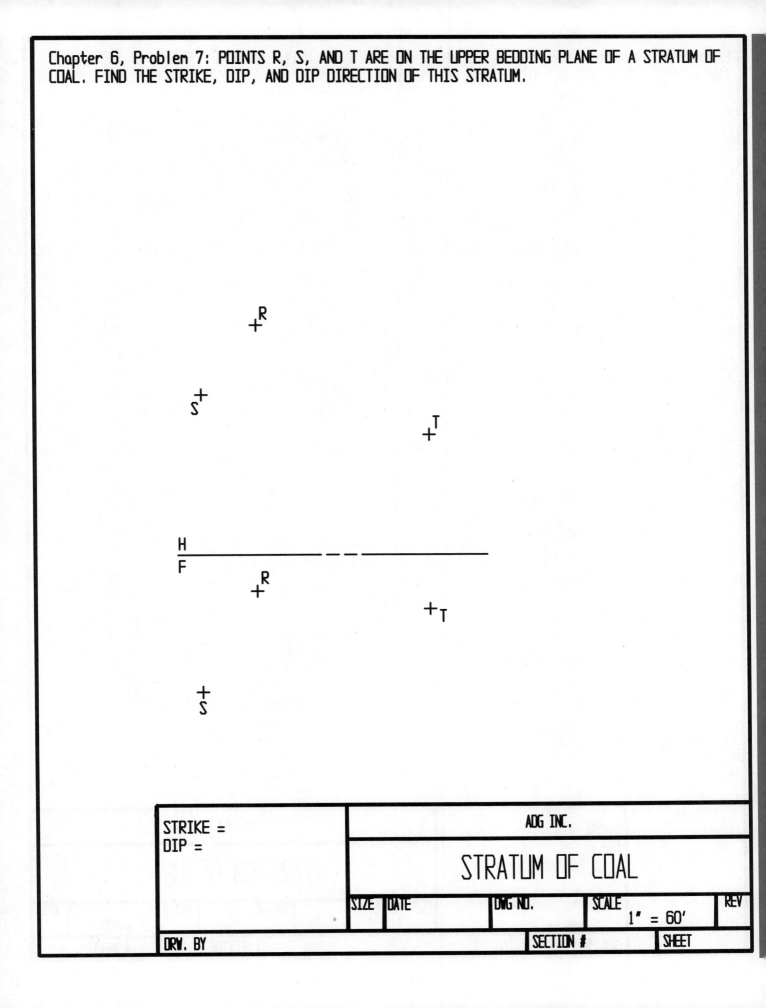

STRIKE =
DIP =

ADG INC.

STRATUM OF COAL

SIZE	DATE		DWG NO.		SCALE 1" = 60'	REV
DRW. BY				SECTION #		SHEET

Y
+

+Z

X+

$\dfrac{H}{F}$ ——————— – – ———————

+Z

X+

+Y

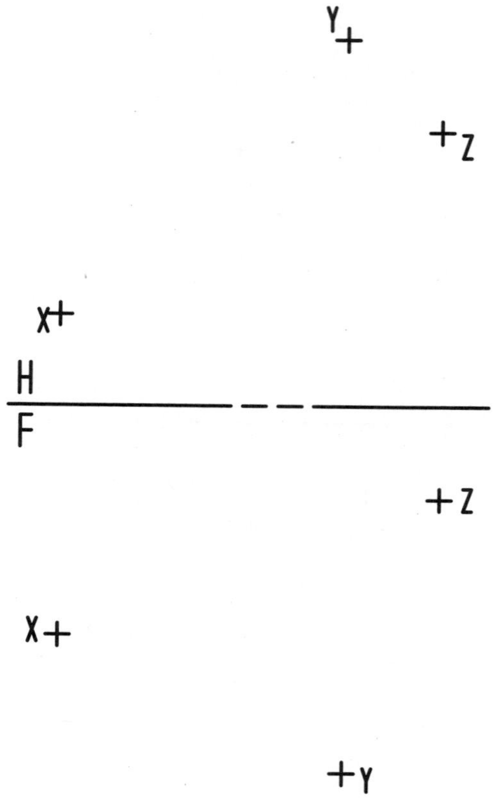

STRIKE =
DIP =

ADG INC.

STRATUM OF ORE

SIZE	DATE		DWG NO.	SCALE 1" = 400'	REV
DRW. BY				SECTION #	SHEET

Chapter 6, Problem 9: IN THE CONSTRUCTION OF PIPELINES IT IS OFTEN NECESSARY TO DETERMINE THE TRUE ANGLE OF THE BEND IN THE PIPE. FIND THE TRUE ANGLE BETWEEN THE CENTER LINES OF PIPES AB AND BC. CONSTRUCT A BEND WITH A 36-INCH RADIUS BETWEEN THE PIPE SECTIONS. PIPE SECTIONS AB WILL HAVE A 90° CONNECTION (TEE) INSTALLED FOUR FEET FROM A. SHOW THE POSITION OF THE CONNECTION IN ALL VIEWS.

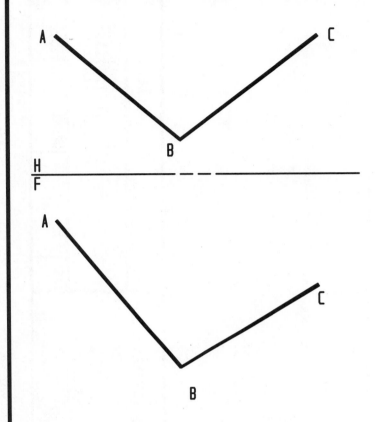

ANGLE BETWEEN AB AND BC?		ADG INC.			
		PIPELINES			
	SIZE	DATE	DWG NO.	SCALE 1" = 4'	REV
DRW. BY			SECTION #		SHEET

Chapter 6, Problem 10: THE TOP AND FRONT VIEWS OF A SECTION OF RIGID TUBING USED FOR OIL IN A GAS ENGINE ARE SHOWN. WHAT ARE THE TRUE ANGLES OF THE BENDS AT B AND C?

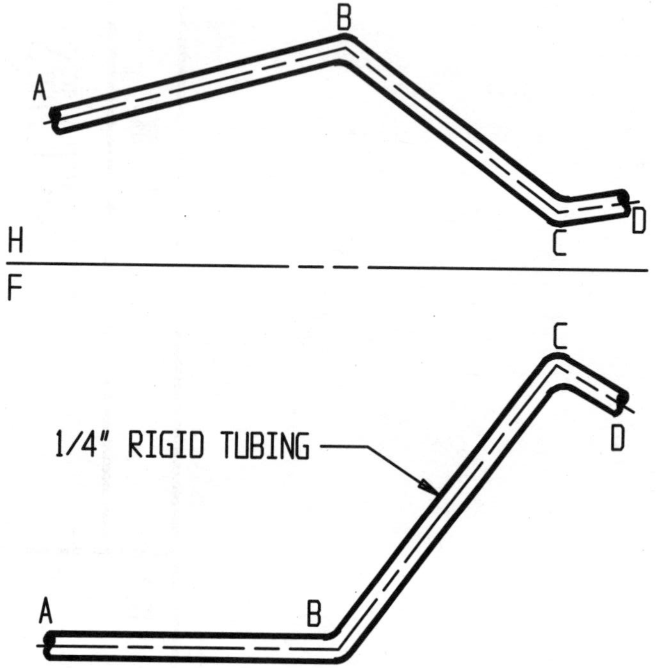

1/4" RIGID TUBING

H
F

A
B
C
D

ANGLE AT BEND
B=
C=

ADG INC.

RIGID TUBING

SIZE | DATE | DWG NO. | SCALE 1" = 2" | REV

DRW. BY | SECTION # | SHEET

Chapter 6, Problem 11: THREE POINTS (A, B, AND C) LOCATED ON THE CENTER LINE OF A 24-INCH DIAMETER CONDUIT FOR CONVEYING WATER TO A POWER PLANT ARE SHOWN. THE CONDUIT DESIGN CALLS FOR A LONG RADIUS ELBOW TO CONNECT THE STRAIGHT SECTIONS OF PIPE. THE RADIUS IS TO BE EIGHT TIMES THE DIAMETER OF THE PIPE. FIND THE TRUE SHAPE AND SIZE OF THE SWEEP ANGLE OF THE ELBOW, AND THE TRUE LENGTHS OF THE STRAIGHT PIPES FROM POINT A TO THE ELBOW AND FROM POINT C TO THE ELBOW. SHOW THE CENTER LINE OF THE ELBOW IN THE TOP AND FRONT VIEWS.

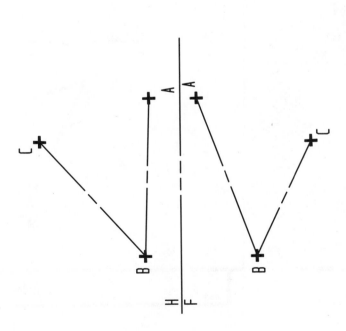

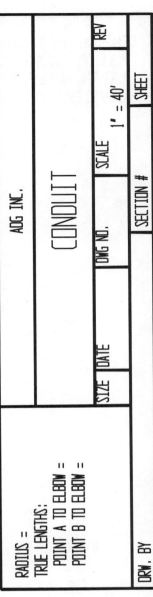

ADG INC.

CONDUIT

| SIZE | DATE | | DWG NO. | SCALE 1" = 40' | REV |

| | | | SECTION # | SHEET |

RADIUS =
TRUE LENGTHS:
    POINT A TO ELBOW =
    POINT B TO ELBOW =

DRW. BY

Chapter 6, Problem 12: A FLAT BELT TRAVELS THROUGH POINT M, AROUND A 4-INCH DIAMETER
PULLEY, LOCATED SOMEWHERE NEAR POINT P, AND THROUGH POINT N. SHOW THE PULLEY DIAMETER
AND THE ANGLE OF BELT CONTACT. IN ADDITION, SHOW THE PULLEY DIAMETER IN THE FRONT AND
TOP VIEWS, DISREGARDING ITS THICKNESS.

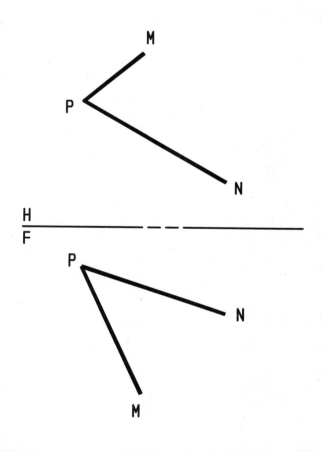

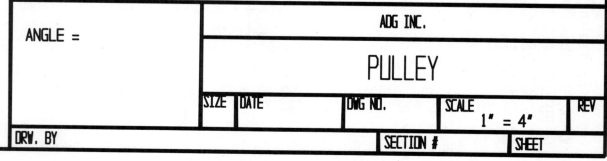

ANGLE =	ADG INC.				
	PULLEY				
	SIZE	DATE	DWG NO.	SCALE 1" = 4"	REV
DRW. BY			SECTION #		SHEET

Plane Relationships

Imagine building this house without knowing the necessary dimensions provided in auxiliary views.

After completing this chapter, you will be able to:

- *Determine the shortest horizontal, perpendicular, or grade line from a point to a plane.*

- *Explain why all shortest lines from a given point to a given plane have the same bearing.*

- *Explain why all shortest lines from a given point to a given plane appear true length in the view that shows the plane as an edge.*

- *Determine the angle between a line and a plane.*

- *Determine the angle between two intersecting planes.*

- *Determine the shortest horizontal, perpendicular, or grade line between two lines, using the plane method.*

Sometimes specific information about spatial relationships is needed, such as the shortest distance between a plane and a point. Examples include: the shortest distance for a guy wire between a roof and a pole; the shortest horizontal distance for a tunnel between a specific point and a vein of ore; the shortest distance for a chute between a conveyor line and a hopper; the angle between a billboard and the metal support frame; or the angle between two sloping sides of a concrete bridge footer. With the information learned in Chapter 6 you can solve for:

- Shortest (perpendicular) line from a point to a plane

- Shortest grade line from a point to a plane

■ Shortest horizontal line from a point to a plane

■ True angle between two planes (dihedral angle)

■ Angle between a line and a plane

Regardless of which of the problems listed above you must solve for, the Steps of Procedure are the same. The shortest distance between a plane and a line, or a point, is found in the view where one plane appears as an edge view. The angle between two planes (dihedral) is found in the view where both planes appear as edge views in the same view. To accomplish this, the line of sight must be parallel to the true-length view of the intersecting (common or connecting) line in both planes.

Remember to always check the principal views first to see if auxiliary views are in fact needed. *In other words we haven't learned any new steps, just when to apply the true length, point view, edge view, and true shape lines of sight!* Study the figures in this chapter and note that each problem was solved using the same procedure. For space saving and comparison purposes each problem (whether it be a cable, water pipe, strut, etc.,) will be represented by a line which is the center line of each element. (Review Chapter 6, page 194, Locating a True-Length Line Within the Plane.)

▼ SHORTEST DISTANCE FROM A POINT TO A PLANE

Figure 7–1 shows a tunnel entrance (point X) leading to a vein of ore (plane A-B-C-D). (For more mining terminology, see Chapter 11.) This figure will be referred to in the next three sections of this chapter. It is necessary to locate the shortest tunnel. To find the shortest distance to the stratum (upper plane of ore vein), the stratum must appear as an edge in an elevation view (view 1).

In auxiliary view 1, it is possible to locate three different shortest distances. As you will see later in this chapter; Figure 7–2 shows the **shortest perpendicular** tunnel, X-W; Figure 7–3 shows the shortest tunnel at a **specific percent grade,** X-Z; and Figure 7–4 shows the **shortest horizonal** tunnel, X-Y. Remember X-W, X-Y, and X-Z are all true length in view 1; therefore, they are perpendicular to the line of sight in the adjacent view. Because each tunnel (regardless of its grade) appears true length in view 1, they all have the same bearing

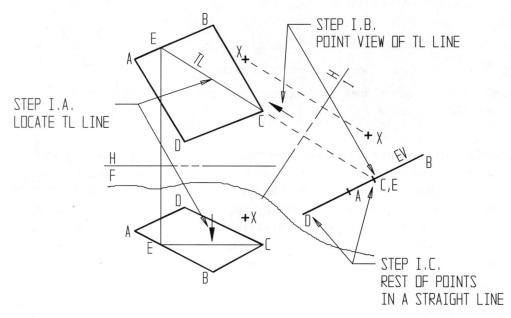

Figure 7–1 The first three fundamental lines of sight are used to locate the edge view of the plane.

and are perpendicular to the true-length line C-E. Try to visualize this circumstance, and refresh your memory on why these observations are true.

Shortest Distance (Perpendicular) from a Point to a Plane

Example
The shortest (perpendicular) distance from a point to a plane is seen in a view that shows the edge view of the plane.

In Figure 7–2, plane A-B-C-D represents the upper plane (surface) of an ore vein, and point X is a point representing the tunnel entrance. It is necessary to locate the shortest tunnel.

To find the shortest tunnel to the stratum, the stratum must appear as an edge (auxiliary view 1). Then, draw the shortest tunnel perpendicular to the edge view of the plane from point X, locating point W. Project the shortest tunnel's location to the top and front views. Remember, X-W is true length in view 1; therefore, it is perpendicular to the line of sight in the adjacent view.

Don't forget visibility. To determine visibility of the top view, imagine standing on the H/1 reference line looking down on the tunnel and stratum (view 1). Which do you see first—line X-W or plane A-B-C-D? View 1 (elevation view) shows line X-W is on top of plane A-B-C-D; therefore, line X-W will be visible in the top view.

To determine the visibility for the front view, imagine standing on the H/F reference line looking at the front of the tunnel and stratum (top view). Which line do you see first—X-W or C-D? The top view shows that line C-D is in front of line X-W; therefore, line X-W will be hidden in the front view from line C-D to W. (Review visibility of lines in Chapter 5.)

Shortest Specified Grade Distance From a Point to a Plane

You know that the shortest distance from a point to a plane is a line perpendicular to the plane, but at times the slope angle of the perpendicular is too steep for practical purposes. For example, in an ore mine (Refer back to Figure 7–1) the entrance tunnel leading to the vein of ore (plane) should be short for economic reasons, but it should also have the best slope for the transportation of ore. Such a passage can be designed as the shortest grade (or slope) line from the point of entrance to the plane of the stratum of ore.

Example
In Figure 7–3, plane A-B-C-D represents the upper plane (stratum) of an ore vein, and point X is a point representing the tunnel entrance. It is necessary to locate the shortest tunnel at a –20 percent grade from point X. To find the shortest grade line to the stratum, the stratum must appear as an edge in an elevation view (top-adjacent view—view 1).

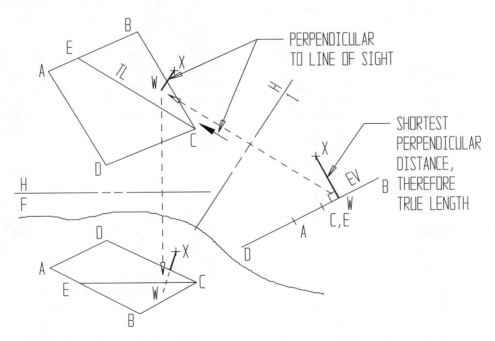

Figure 7–2   Shortest distance from a point to a plane is perpendicular.

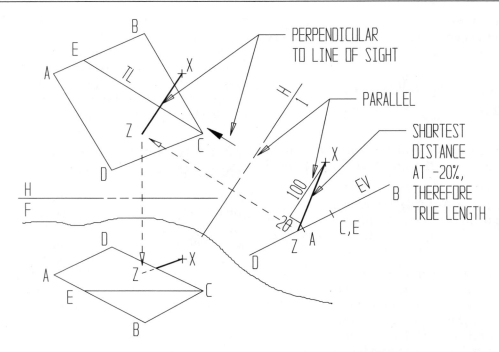

Figure 7–3 Shortest distance from a point to a plane at a specific grade.

Draw the shortest tunnel at a specific percent grade (–20%) from point X to the edge view of the plane, locating point Z. Remember, downhill is away from the top view. Project line X-Z to the top and front views. Remember, X-Z is true length in the top-adjacent view; therefore, it is perpendicular to the line of sight in the adjacent view.

Don't forget visibility. To determine the visibility for the top view, imagine standing on the H/1 reference line looking down on the top of the tunnel and stratum (view 1). Which do you see first—line X-Z or plane A-B-C-D. View 1 (elevation view) shows that line X-Z is on top of plane A-B-C-D; therefore, line X-Z will be visible in the top view. To determine the visibility for the front view, imagine standing on the H/F reference line looking at the front of the tunnel and stratum (top view). Which line do you see first— X-Z or C-D? The top view shows that line C-D is in front of line X-Z; therefore, line X-Z will be hidden in the front view from line C-D to Z.

Shortest Horizontal Distance Between a Plane and a Point

As stated previously at times the shortest perpendicular distance between a point and a plane is too steep for your purposes. For example, in an ore mine (refer back to Figure 7–1) the entrance tunnel leading to the vein of ore (plane) should be short for economic reasons, but it may need to be level. Such a passage can be designed as

the horizontal line from the point of entrance to the plane of the stratum of ore.

Example
In Figure 7–4, plane A-B-C-D represents the upper plane (stratum) of an ore vein, and point X is a point representing the tunnel entrance. It is necessary to locate the shortest tunnel at a horizontal distance from point X. To find the horizontal line to the stratum, the stratum must appear as an edge in an elevation view (top-adjacent view—view 1).

Draw the shortest horizontal distance parallel to the edge view of the horizontal plane (reference line H/F or H/1) from point X to the edge view of the plane to locate point Y. Project line X-Y to the top and front views. Remember X-Y is true length in the top-adjacent view; therefore, it is perpendicular to the line of sight in the adjacent view.

Don't forget visibility. To determine visibility for the top view, imagine standing on the H/1 reference line looking down on the tunnel and stratum (view 1). Which do you see first—line X-Y or plane A-B-C-D? View 1, elevation view, shows that line X-Y is on top of plane A-B-C-D; therefore, line X-Y will be visible in the top view. To determine visibility for the front view, imagine standing on the H/F reference line looking at the front of the tunnel and stratum (top view). Which line do you see first—X-Y or C-D? The top view shows that line X-Y is behind line C-D; therefore, line X-Y will be hidden in the front view from line C-D to Y.

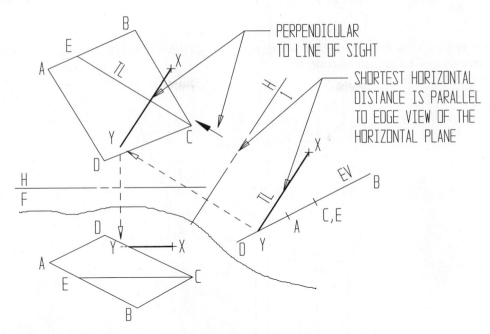

Figure 7–4   Shortest horizontal distance between a point and a plane.

EXERCISE

Problem: Locate the shortest perpendicular tunnel, shortest horizontal tunnel, and shortest tunnel at 35° slope angle from point S to plane W-X-Y-Z. To save space and time locate all three on Figure 7–5. In which view will the tunnels be true length?

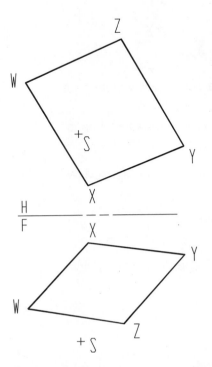

Figure 7–5   Locate   shortest   perpendicular   tunnel, shortest   horizontal tunnel, and shortest   tunnel at 35°.

Answer

All tunnels will be true length in view 1. Figure 7–6 shows the edge view of plane X-Y-W-Z, which is necessary to locate all tunnels.

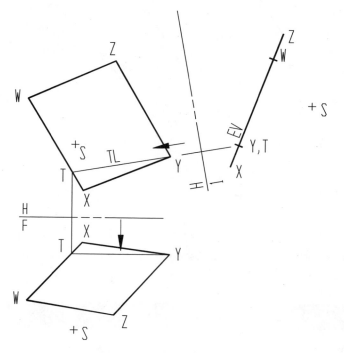

Figure 7–6 Shortest distances between a point and a plane.

I. Determine **line of sight.**

  A. Locate **true-length line.**

    1. Is the true-length line **in** one of the **principal planes?**

      a. If the line of sight is perpendicular to the object line in one view, it will appear true length in the adjacent view and can be measured.

    2. If the above rule does not apply, it is an oblique line and cannot be measured as is.

      a. Draw a **primary auxiliary** view **to find the true-length.**

        (1) Draw the line of sight perpendicular to the oblique line in either principal view.

  B. Draw the **point view of a true-length line.**

    1. The line of sight must be drawn parallel to the true length line.

  C. Draw an **edge view of a plane** (all points fall in a line).

    1. The line of sight must be drawn parallel to a true-length line that lies in the plane. The line of sight locates:

      ■ Shortest (perpendicular) line from a point to a plane

      ■ Shortest grade line from a point to a plane

      ■ Shortest horizontal line from a point to a plane

II. Draw the **reference line** perpendicular to the line of sight and label.

III. Draw **projection lines** parallel to the line of sight.

IV. Transfer **points of measurement** to the new view from the related view and label. Always go back to the last line of sight to take the measurement.

V. Connect the points and **label** true-length lines, edge views, and/or true-shape planes.

Figure 7–7 shows the shortest (perpendicular) tunnel.

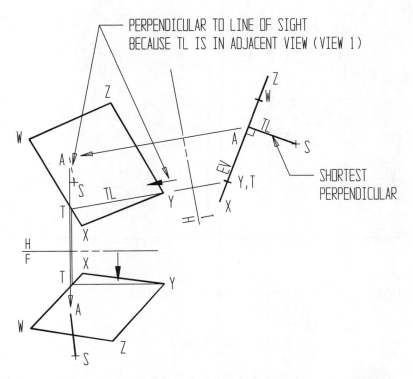

Figure 7–7 Shortest distance between point S and plane W-X-Y-Z is perpendicular.

Figure 7–8 demonstrates the shortest distance at a given slope angle.

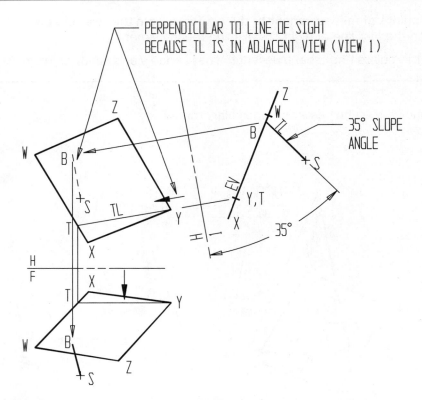

Figure 7–8 Shortest tunnel at a 35° slope angle from point S.

Figure 7–9 demonstrates the shortest horizontal distance.

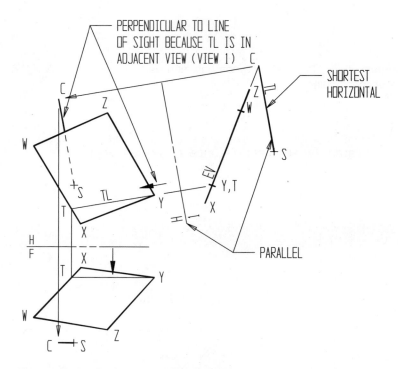

Figure 7–9 Shortest horizontal tunnel from point S to plane W-X-Y-Z.

Figure 7–10 compares all three tunnels.

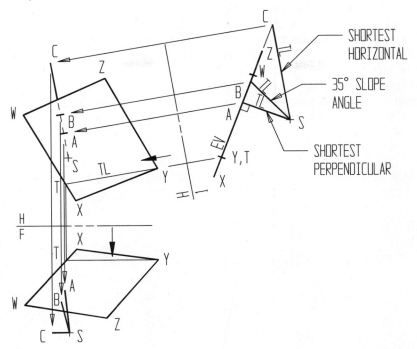

Figure 7–10 Shortest perpendicular tunnel, shortest tunnel at 35°, and shortest horizontal tunnel compared.

▼ THE TRUE ANGLE BETWEEN TWO PLANES (DIHEDRAL ANGLE)

The angle that is formed by two intersecting planes is called the **dihedral angle**. The dihedral angle is seen in its true size when both planes appear as edge views. Both planes appear as edge views when the line of intersection between the two planes is shown as a point (Figure 7–11).

Example

Figure 7–12 illustrates a hopper that feeds coal from an overhead storage compartment into a boiler stoker. The corners are to be reinforced with bent angles. It is necessary to find the true angle between planes A-E-H-D and A-B-F-E in order to design the bent angles properly. Carefully examine the procedure in the following paragraphs.

Locate the intersecting (connecting) line between the two planes, A-E-H-D and A-B-F-E. Yes, the connecting line between planes A-B-F-E and A-E-H-D is A-E.

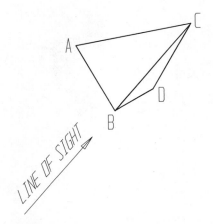

Figure 7–11 Angle between two planes. Line of sight parallel to the intersecting line between the two planes.

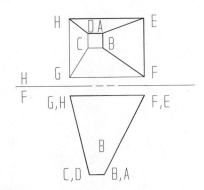

Figure 7–12 Find the dihedral angle between planes A-B-F-E and A-E-H-D.

Locate the true length of the intersecting line A-E (Figure 7–13, view 1). Next, draw point view of the intersecting line A-E. In view 2, where the line of intersection is seen as a point A-E, the two planes will be seen as edge views. The true size of the dihedral angle may be measured in view 2. As with all other examples in this text, try to visualize this problem by moving yourself through the four views.

The only new step is to identify the connecting line between the two planes. The rest of the steps are the first three fundamental views (lines of sight for true-length view of a line, point view of a line, and edge view of a plane) from previous chapters.

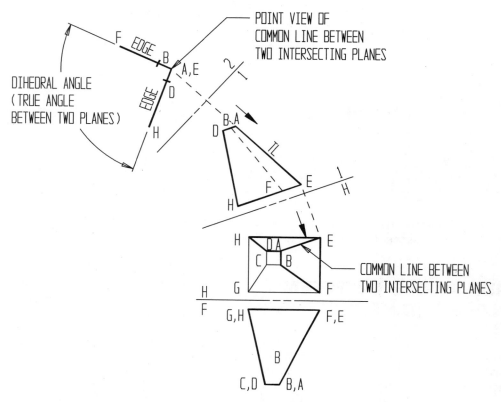

Figure 7–13 Dihedral angle between planes A-B-F-E and A-E-H-D.

HELPFUL HINTS

The following fundamental views will solve for the edge view of a plane, which will solve for:

■ Shortest (perpendicular) line from a point to a plane
■ Shortest grade line from a point to a plane
■ The angle between a line and a plane
■ The true angle between two planes (dihedral angle)

Review

   I. Determine **line of sight.**

     A. Locate **true-length line.**

       1. Is the true-length line **in** one of the **principal planes?**

      a. If the line of sight is perpendicular to the object line in one view, it will appear true length in the adjacent view and can be measured.

    2. If the above rule does not apply, it is an oblique line and cannot be measured as is.

      a. Draw a **primary auxiliary** view **to find the true length**.

        (1) Draw the line of sight perpendicular to the oblique line in either principal view.

  B. Draw the **point view of a true-length line**.

    1. The line of sight must be drawn parallel to the true-length line.

  C. Draw an **edge view of a plane** (all points fall in a line).

    1. The line of sight must be drawn parallel to a true-length line that lies in the plane.
The line of sight locates:

      ■ Shortest (perpendicular) line from a point to a plane

      ■ Shortest grade line from a point to a plane

      ■ The angle between a line and a plane

      ■ The true angle between two planes (dihedral angle)

II. Draw the **reference line** perpendicular to the line of sight and label.

III. Draw **projection lines** parallel to the line of sight.

IV. Transfer **points of measurement** to the new view from the related view and label. Always go back to the last line of sight to take the measurement.

V. Connect the points and **label** true-length lines, edge views, and/or true-shape planes.

EXERCISE

Problem: In Figure 7–14, a foundry hopper is fed with fresh sand through a pipe conveyor from point P to the rear panel of the hopper. Find the shortest pipe conveyor that can be used. Sketch the solution on Figure 7–14. Locate the shortest pipe conveyor in all views.

Answer

1. To begin the solution, find the edge view of the rear panel (A-B-C-D) (Figure 7–15). Lines A-D and B-C are true length in the front view, and point views in the right-side view; therefore, the edge view of plane A-B-C-D is also found in the right-side view.

2. Draw a perpendicular line from point P to the rear panel of the hopper, plane A-B-C-D (Figure 7–15). This perpendicular line represents the true length of the center line of the shortest pipe conveyor that can be used.

3. Project shortest distance (line P-R) to adjacent view (front view). The information in the side view determines the front view. If P-R is true length in the side view, it must appear perpendicular to the line of sight in the adjacent view (front view). Notice that P-R also appears perpendicular to the true-length lines, A-D and B-C, in the front view.

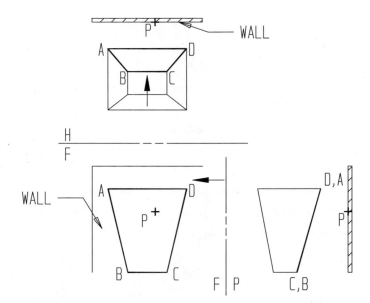

Figure 7–14 Locate the shortest distance between point P and plane A-B-C-D.

Remember visibility. Line P-R is connected to the rear panel of the hopper; therefore, it will be drawn hidden in the front view.

4. Project shortest distance (line P-R) to adjacent view (front view). Use dividers to transfer the distance for point R from the side view to the top view. Remember visibility. Line A-D of the panel is higher than line P-R; therefore, line P-R will be drawn hidden in the top view.

Problem: Find the location of the center of the hole in the rear panel for the pipe conveyor. Sketch the solution on Figure 7–15. (Review true shape in Chapter 6.)

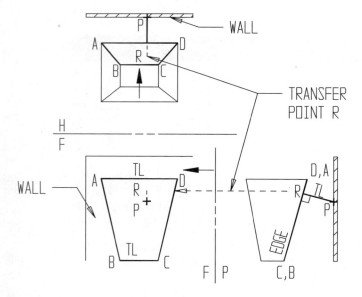

Figure 7–15 Shortest (perpendicular) distance between point P and plane A-B-C-D.

Answer

Figure 7–16 shows the location on plane A-B-C-D. Remember, the location dimensions will only show where the plane appears true shape.

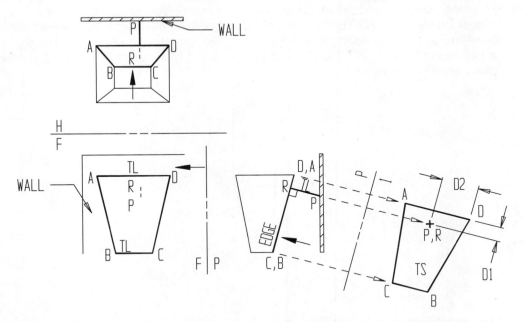

Figure 7–16 Dimensions to locate hole in hopper for pipe conveyor are shown in view 1, where the hopper plane is true shape and the pipe conveyor is shown as a point.

Problem: When turning a piece of stock on a lathe, a tool bit (Figure 7–17) is one choice of cutters used. The angle on the tool bit between the face (plane 1-2-3-4) and the flank (plane 1-4-5-6) is ground to a specific angle depending on the type of material to be cut. The angle for steel is 62° and cast iron is 71°. Is the tool bit in Figure 7–17 to be used when cutting steel or cast iron? **Hint:** Sometimes it helps to identify the planes you are working with by using a highlighting marker or colored pencils.

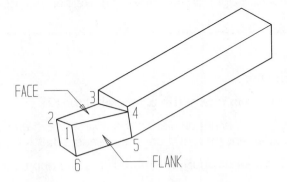

Figure 7–17   Is this tool bit ground to cut steel or cast iron?

Read the following steps and sketch in the correct solution. The correct answer will be shown in the next figure.

To locate the dihedral angle between two planes, both planes must be viewed as edge views in the same plane. In order to do this, you must use the first three fundamental views. You know the first step is to locate a true-length line. In order to see both planes as edge views in the same plane, does it matter which line is true-length? You are correct, yes it does. Which line must be true-length? You are correct again, the connecting line between the two planes. Which line is the connecting line in Figure 7–18? Correct again, line 1-4 is the connecting line between planes 1-2-3-4 and 1-4-5-6. Is line 1-4 true-length in a principal view or must you draw an auxiliary view?

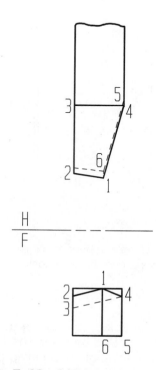

Figure 7–18 Orthographic of tool bit.

Correct, an auxiliary view is needed to find the true-length of line 1-4. Project the rest of the planes into view 1 also (sketch right on Figure 7–19). Remember visibility.

I. Determine **line of sight.**

   A. Locate **true-length line.**

      1. Is the true-length line **in** one of the **principal planes**?

         a. If the line of sight is perpendicular to the object line in one view, it will appear true length in the adjacent view and can be measured.

      2. If the above rule does not apply, it is an oblique line and cannot be measured as is.

         a. Draw a **primary auxiliary** view **to find the true length.**

            (1) Draw the line of sight perpendicular to the oblique line in either principal view.

II. Draw the **reference line** perpendicular to the line of sight and label.

III. Draw **projection lines** parallel to the line of sight.

IV. Transfer **points of measurement** to the new view from the related view and label. Always go back to the last line of sight to take the measurement.

V. Connect the points and **label** true-length lines, edge views, and/or true-shape planes.

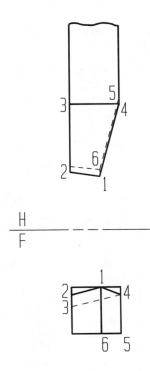

Figure 7–19 Locate the true length of the common line between planes 1-2-3-4 and 1-4-5-6.

The second fundamental view is to look parallel to the true-length line to see the point view of the true-length line.

I. Determine **line of sight.**

　　B. Draw the **point view of a true-length line.**

　　　　1. The line of sight must be drawn parallel to the true-length line.

II. Draw the **reference line** perpendicular to the line of sight and label.

III. Draw **projection lines** parallel to the line of sight.

IV. Transfer **points of measurement** to the new view from the related view and label. Always go back to the last line of sight to take the measurement.

V. Connect the points and **label** all true-length lines, edge view, and/or true-shape planes.

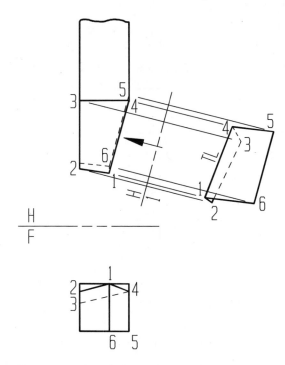

Figure 7–20 Locate the point view of connecting line 1-4.

The third fundamental view states that the edge view of the plane is found where a true-length line in the plane is seen as a point.

I. Determine **line of sight.**

    C. Draw an **edge view of a plane** (all points fall in a line).

        1. The line of sight must be drawn parallel to a true-length line that lies in the plane.

II. Draw the **reference line** perpendicular to the line of sight and label.

III. Draw **projection lines** parallel to the line of sight.

IV. Transfer **points of measurement** to the new view from the related view and label. Always go back to the last line of sight to take the measurement.

V. Connect the points and **label** all true-length lines, edge views, and/or true-shape planes.

Answer

The dihedral angle between planes 1-2-3-4 and 1-4-5-6 is 71°; therefore, this tool bit is used for cast iron (Figure 7–22).

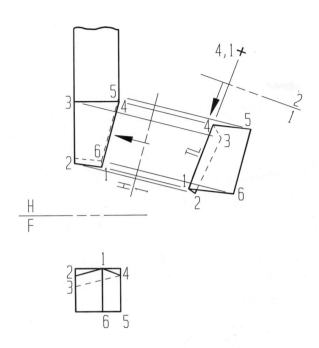

Figure 7–21 Locate the dihedral angle between planes 1-2-3-4 and 1-4-5-6.

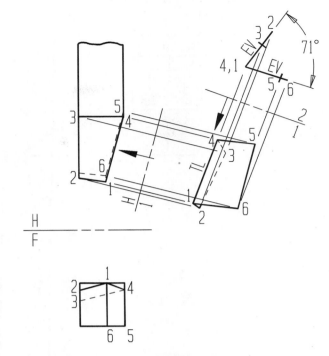

Figure 7–22 The dihedral angle between planes 1-2-3-4 and 1-4-5-6 is 71°.

▼ THE ANGLE BETWEEN A LINE AND A PLANE

You have already learned that the slope angle of a line is the angle between the line and the horizontal plane. Remember that slope angle is seen in a view that shows the true length of the line in an elevation view (horizontal plane is seen as an edge—H/F or H/? reference line).

In the same manner, the angle between a line and a plane is found where the line is viewed true length and the plane is seen as an edge. Such a view is obtained from the following order: first, show an edge view of the plane; second, show the plane in true shape; and finally, show the plane again as an edge view where the line appears true length.

Example

In this case, it is necessary to find the angle between the feeder pipe (line A-B) and the rear panel (plane W-X-Y-Z) of the sand hopper (Figure 7–23). In the right-side view, and edge view plane W-X-Y-Z is shown. Pipe A-B is shown, although it is not true length. In view 1 of Figure 7–24, the rear panel, plane W-X-Y-Z, is shown in true shape, and pipe A-B is simply projected to this view. Finally, in view 2, the true length of pipe A-B is seen along with the edge view of W-X-Y-Z. This condition has been accomplished by drawing the line of sight perpendicular to

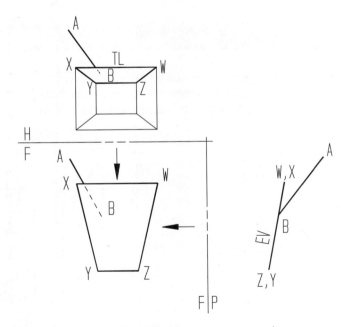

Figure 7–23 Plane W-X-Y-Z is in edge view in the side view.

A-B, causing the rear panel to appear as an edge view again, and showing A-B in its true length in the same view. It is here that you can measure the true angle between the pipe and the rear panel of the hopper.

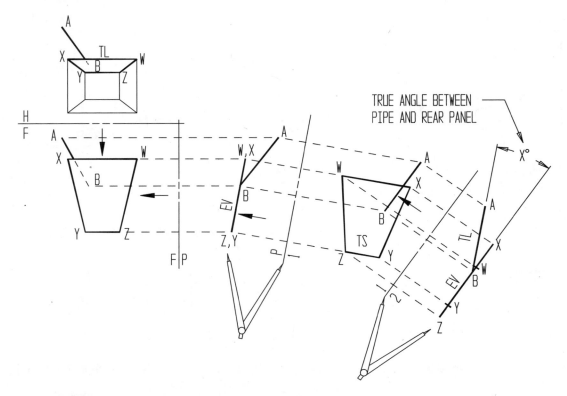

Figure 7–24 True angle between pipe and real panel.

EXERCISE

Problem: A clean room is a room that is dust free—for example, a paint room. An air line (line A-B) enters the clean room through the wall (plane C-D-E-F) (Figure 7–25). Locate the angle between the air line and the wall.

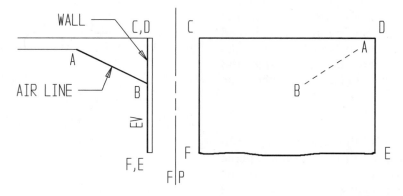

Figure 7–25 Locate the angle between line A-B and plane C-D-E-F.

Answer

To locate the angle between a line and a plane, the true-length view of the line and the edge view of the plane must be seen in the same view. The edge view of plane C-D-E-F already exists in the front view; however, line A-B is not true-length. Therefore, two more views are required—true shape of plane C-D-E-F (and then true-length of line A-B), and edge view of plane C-E-D-F. All of the fundamental lines of sight needed to locate the true shape of the plane (right-side view) have already been completed in the principal views.

Review

I. Determine **line of sight.**

  A. Locate **true-length line.**

   1. Is the true-length line **in** one of the **principal planes**?

     a. If the line of sight is perpendicular to the object line in one view, it will appear true length in the adjacent view and can be measured.

   2. If the above rule does not apply, it is an oblique line and cannot be measured as is.

     a. Draw a **primary auxiliary** view **to find the true length**.

       (1) Draw the line of sight perpendicular to the oblique line in either principal view.

  B. Draw the **point view of a true-length line**.

   1. The line of sight must be drawn parallel to the true-length line.

  C. Draw an **edge view of a plane** (all points fall in a line).

   1. The line of sight must be drawn parallel to a true-length line that lies in the plane.

  D. Draw the **true shape of a plane**.

   1. The line of sight must be drawn perpendicular to the edge view of a plane.

II. Draw the **reference line** perpendicular to the line of sight and label.

III. Draw **projection lines** parallel to the line of sight.

IV. Transfer **points of measurement** to the new view from the related view and label. Always go back to the last line of sight to take the measurement.

V. Connect the points and **label** all true-length lines, edge views, and/or true-shape planes.

Set the line of sight perpendicular to line A-B for a view adjacent to the true shape view (right-side view). This new view will show the line true length and the edge view of the plane (Figure 7-26).

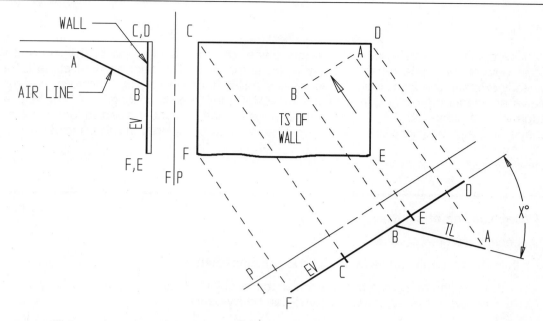

Figure 7-26 The angle between a line and a plane is found where the true-length line and the edge view of the plane is seen in the same view.

▼ PLANE METHOD FOR SOLVING PROBLEMS WITH POINTS AND LINES

You learned in Chapter 6, that planes can be created from parallel lines, intersecting lines, three points, and a point and a line (Review Figure 6–2). This information is helpful when trying to solve for the angle between two lines, shortest distance between two lines, shortest distance between a point and a line, and so on, as an alternative to the method learned in Chapter 5, Shortest Distances.

Shortest Distance Between Two Lines Using the Plane Method

Example

Figure 7–27 shows two cables, A-B and C-D. Find the shortest distance between lines A-B and C-D using the plane method. First, construct a plane containing cable C-D parallel to cable A-B or vice versa. To do this, draw line C-E any length parallel to A-B in the front view (Figure 7–28). Draw line E-D, in the front view, perpendicular to the line of sight for the top view. Point E is where lines C-E and E-D intersect. To find the top view of plane C-D-E, draw a line from C parallel to A-B until it intersects a line projected from point E in the adjacent view. This locates point E in the top view. Then connect E to D to complete

the plane. Remember construction line E-D is true length, because it was intentionally drawn perpendicular to the line of sight in the adjacent view (Fundamental rule I.A.i.a.).

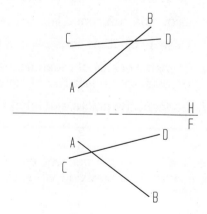

Figure 7-27 Locate shortest distance using the plane method.

Next, to find the point view of E-D create the line of sight parallel to line E-D (fundamental rule I.B.) (Figure 7–29). The rest of the plane will fall in line forming the edge view of plane C-D-E in the same view (fundamental rule I.C.). Project line A-B into the same view. Locate the true shape of plane C-D-E by creating the line of sight perpendicular to the edge

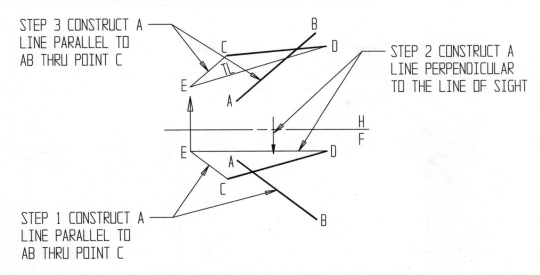

Figure 7–28   Construct a plane containing one line parallel to the other line.

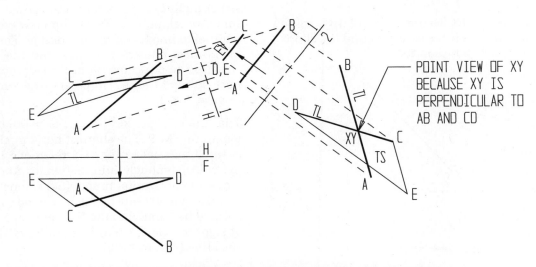

Figure 7–29   Shortest distance between two lines using the plane method.

view (fundamental rule I.D.). View 2 shows the true shape of plane C-D-E and the true-length of both cables A-B and C-D. In view 2, where the two lines appear to intersect, the common perpendicular (the stabilizing cable) appears as point X-Y.

The true length of X-Y is shown in view 1. The stabilizing cable, X-Y, may be projected back to the horizontal and front views by simple alignment (Figure 7–30).

Shortest Level (Horizontal) Line Connecting Two Nonintersecting Nonparallel Lines Using the Plane Method

A problem very similar to the one just described is one in which the required connecting line between two given lines is level, or horizontal, rather than perpendicular.

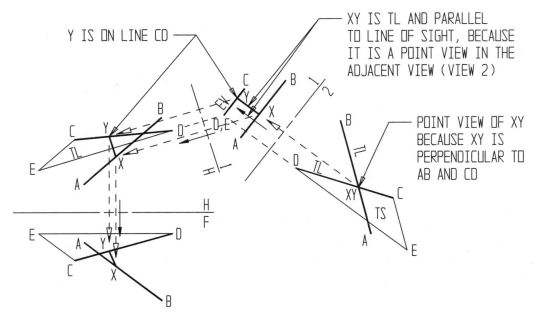

Figure 7–30 Shortest distance projected to the front view.

Example

Figure 7–31 shows the center lines of two mine shafts, K-L and M-N. It is necessary to find the shortest level tunnel, S-T, between them.

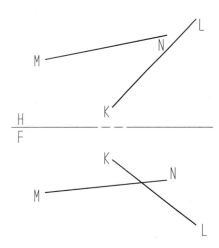

Figure 7–31 Locate the shortest horizontal tunnel between tunnels M-N and K-L.

First, create a plane M-N-P containing mine shaft M-N, with line N-P parallel to K-L until it intersects a line projected from P. To accomplish this, in the front view construct a line parallel to line K-L through point N. Construct a horizontal line through point P. (This will create a line perpendicular to the line of sight for the top view, making line M-P true length in the top view.) Point P is where the two con-

struction lines intersect. Draw a vertical construction line through point P. In the top view construct a line parallel to line K-L through point N. The intersection of the new construction line and the vertical construction line locates point P in the top view. Connect points M and P to create true-length line M-P (Figure 7–32).

The line of sight for view 1, is parallel to the true-length line M-P. The shortest horizontal tunnel (S-T) is to be parallel to the H/1 reference line. Because there isn't enough information to know where to locate the horizontal tunnel (S-T) along tunnels M-N and K-L, view 2 must be constructed. Since the tunnel will be parallel to the H/1 reference line, the line of sight for view 2 will be parallel to the H/1 reference line (Figure 7–33).

In view 2 the connecting tunnel (S-T) appears as a point where the two mine shafts appear to intersect. The position of the tunnel is now fixed and is located in the other views by projection. Note that S-T is parallel to the H/1 reference line, as is S-T in the front view, proving that tunnel S-T is level (Figure 7–34).

Shortest Line at a Given Grade or Slope, Connecting Nonintersecting, Nonparallel Lines

A third type of problem similar to the others discussed in this chapter is one in which the required line is not level, or perpendicular, but instead must have a specified grade or slope angle.

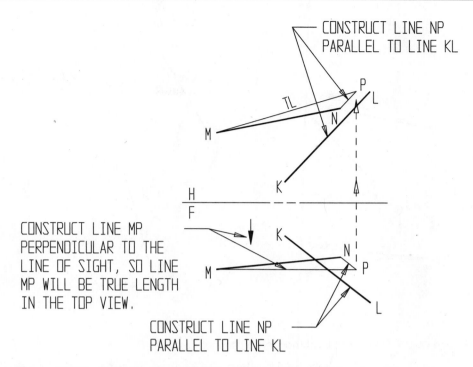

Figure 7–32   Creating plane M-N-P, with line N-P parallel to line K-L.

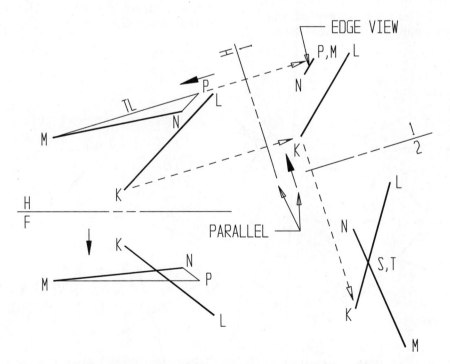

Figure 7–33   The shortest horizontal tunnel S-T appears as a point in view 2. Notice, to accomplish this the line of sight for view 2 must be parallel to the H/1 reference line.

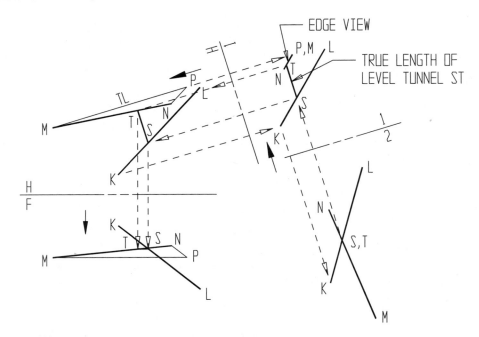

Figure 7–34 Shortest level (horizontal) tunnel.

Example

In Figure 7–35 two pipes, A-B and C-D, are to be connected with a pipe at a –40 percent grade from C-D to A-B.

As in the preceding two sections, draw a plane that contains one of the lines and is parallel to the other line. In auxiliary view 1, the pipes appear parallel. To position the shortest connecting pipe at a –40 per-

cent grade from C-D to A-B, you must draw the line of sight parallel to a line drawn at a –40 percent grade from reference line H/1. (Remember, away from the top view is downward.) In view 2, the pipes appear to intersect and the connector appears as a point. When projected back to view 1, the connecting pipe is shown in its true length. The true-length, center line to center line, is in view 1.

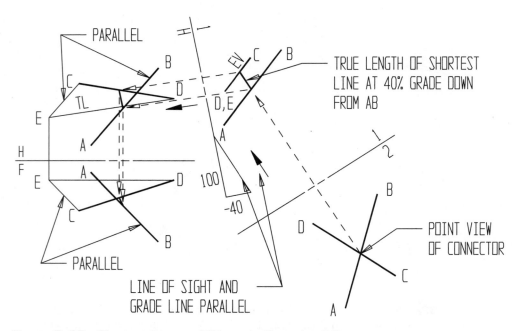

Figure 7–35 Shortest line at –40% grade from pipe C-D.

EXERCISE

Problem: In this case (Figure 7–36) pipes W-X and Y-Z are to be connected with a standard 60° Y fitting, upward from pipe Y-Z.

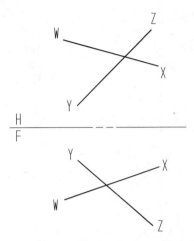

Figure 7–36 Locate the Y fitting at a standard 60° connecting between pipes W-X and Y-Z.

Read the following steps and sketch in the correct solution.

Create plane V-W-X parallel to line Y-Z, with V-X being true length in the top view (Figure 7–36). The answer will be given in the next figure.

Create the point view of line V-X. Also project the rest of the points into view 1 (Figure 7–37).

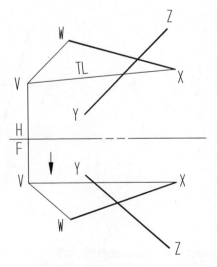

Figure 7–37   Locate the edge view of plane X-Y-Z.

Draw the line of sight for view 2 parallel 60° upward from the H/1 reference line and complete view 2.

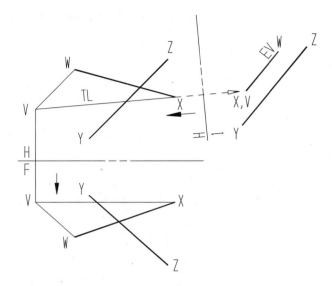

Figure 7–38 Create view 2 showing the location of the shortest connector at a 60° uphill slope from pipe Y-Z.

Project the shortest connector (S-T) back to the front view. In which view is S-T true length?

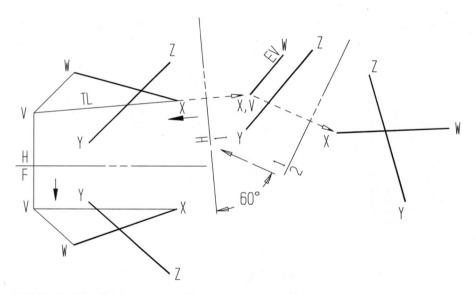

Figure 7–39 Project the shortest connector back to the front view.

Correct, S-T is true length in view 1.

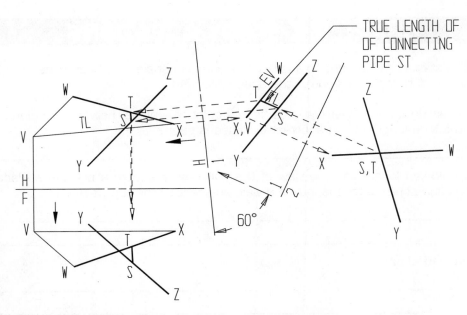

Figure 7–40 Shortest connector at a 60° uphill slope angle from pipe Y-Z.

Examine the front and top-adjacent views in Figures 7–35 and 7–40 carefully. Try to visualize the difference in orientation due to a 60° uphill slope from pipe Y-Z versus a 40% downward grade from pipe C-D. Toward the top view is upward and away from the top view is downward.

CHAPTER 7 STUDY QUESTIONS

The Study Questions are intended to assess your comprehension of chapter material. Please write you answers to the questions in the space provided.

1. In determining the shortest distance from a point to a plane, how will the plane appear when the true length of the shortest distance is seen?

    _____

2. How would you locate the top view of the shortest line from a point in a plane if the shortest line appears true length in a view adjacent to the top view?

    _____

    _____

3. Define dihedral angle.

    _____

    _____

4. Explain the procedure for finding the true dihedral angle between two planes.

    _____

    _____

5. Determine the dihedral angle between planes A-B-D and A-C-D.

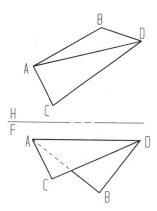

6. In order to find the angle between a line and a plane, how must the plane and the line appear?

    _____

    _____

7. Explain how you would solve the following problem: Lines M-N and R-S below represent two existing conveyor lines. They are to be connected by the shortest possible switching conveyor installed at a 30 percent grade downhill from R-S.

    _____

    _____

    _____

    _____

    _____

8. Illustrate your answer to question 7.

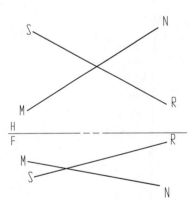

The problems in this chapter are based on a variety of industrial applications. In solving these problems you should try to utilize direct approaches that get to the heart of the problem. At this analysis stage of the design process, many problems use center lines and single lines to represent the problem situation. Yet, when necessary, relevant sizes are given.

The following problems may be drawn using drawing instruments on the page provided, created using a 2D or 3D CAD system, or a combination of drawing board and CAD. Refer to Appendix A for additional dimensions needed to solve problems three-dimensionally.

Chapter 7, Problem 1: A STORAGE HOPPER IS LOCATED AS SHOWN BELOW. SUPPORT BRACES MADE OF 4-INCH DIAMETER PIPE ARE TO BE FASTENED FROM POINTS X AND Y ON THE BUILDING WALLS TO THE NEAREST PLANES OF THE HOPPER. FIND THE TRUE LENGTHS OF THE TWO SHORTEST SUPPORT BRACES. THE BRACES WILL BE WELDED IN EACH POSITION.) DIMENSION THE PIPES FOR FABRICATION.

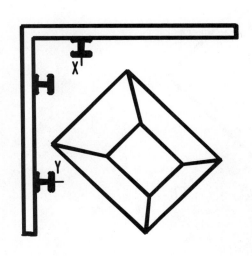

$$\frac{H}{F}$$

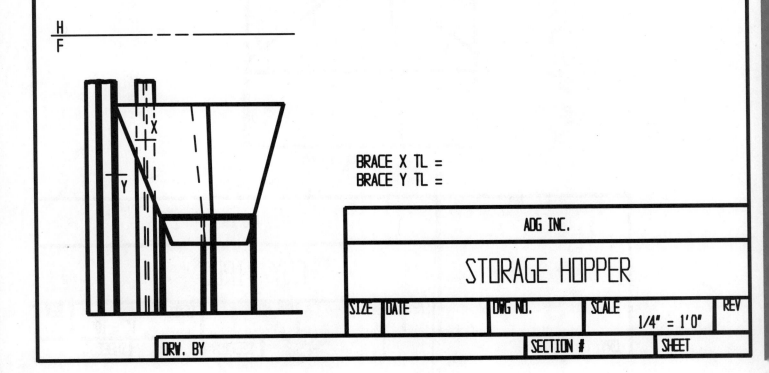

BRACE X TL =
BRACE Y TL =

ADG INC.

STORAGE HOPPER

SIZE	DATE		DWG NO.	SCALE		REV
				1/4" = 1'0"		

DRW. BY		SECTION #	SHEET

Chapter 7, Problem 2: TWO MASTS, A AND B, STAND NEXT TO A BUILDING. ONE GUY WIRE FROM THE TOP OF EACH MAST MUST BE ATTACHED TO THE NEAREST ROOF SURFACE AND SHOULD BE AS SHORT AS POSSIBLE. SHOW THE TWO GUY WIRES FROM A AND B TO THE ROOF PLANES, X AND Y, IN ALL VIEWS AND RECORD THE LENGTH OF EACH.

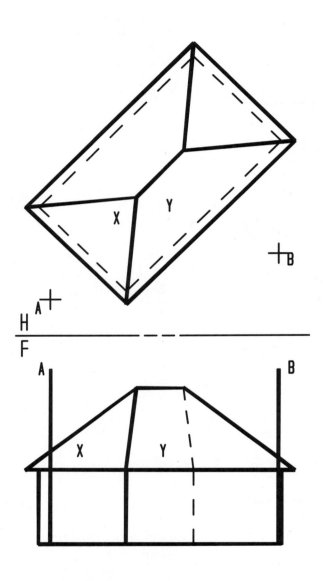

	ADG INC.			
X = Y =	GUY WIRES			
	SIZE	DATE	DWG NO.	SCALE 1" = 10'	REV
DRW. BY		SECTION #	SHEET	

Chapter 7, Problem 3: POINTS X, Y, AND Z HAVE BEEN LOCATED ON THE UPPER SURFACE (BEDDING PLANE) OF A VEIN OF ORE.
A NEW TUNNEL IS TO BE DUG FROM POINT A TO THE UPPER BEDDING SURFACE. FIND THE TRUE LENGTH OF THE SHORTEST TUNNEL AT
A -20 PERCENT GRADE FROM A TO THE UPPER BEDDING PLANE. SHOW THE TUNNEL IN ALL VIEWS.

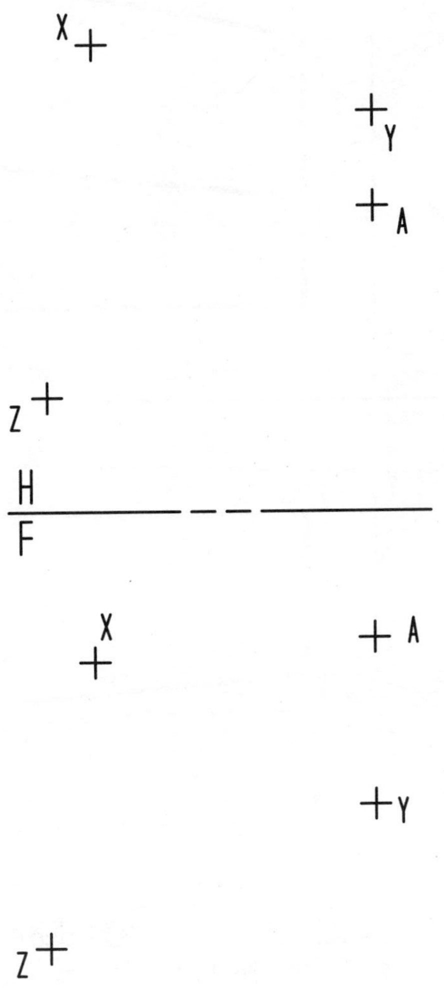

+Y

+A

+X

H
―――
F

+Z

+A

+X

+Y

+Z

AB =

ADG INC.

NEW TUNNEL

SIZE	DATE		DWG NO.		SCALE	REV
					1" = 400'	
		SECTION #		SHEET	

DRW. BY

Chapter 7, Problem 4: A METAL TANK IS TO FIT BELOW DECK IN THE FOWARD COMPARTMENT OF A SHIP. THE TANK HAS A LEVEL TOP, VERTICAL SIDES, AND A SLOPING BOTTOM, WXYZ. FROM POINT A AT THE NEARBY BULKHEAD, A PIPE MUST CONNECT TO PLANE WXYZ, AND THE PIPE MUST HAVE A 15° SLOPE AND A BEARING OF N45°W FROM A. FIND THE TRUE LENGTH OF THE CENTER LINE OF THE PIPE. SHOW THE PIPE IN ALL VIEWS.

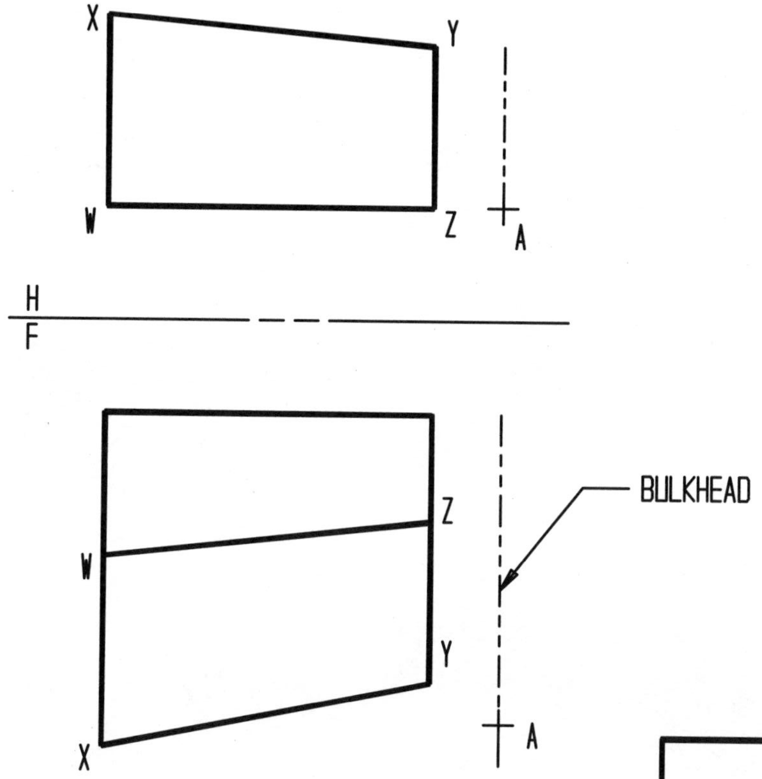

BULKHEAD

TRUE LENGTH =

ADG INC.

METAL TANK

SIZE	DATE	DWG NO.	SCALE	REV
			3/8' = 1'0"	
DRW. BY		SECTION #		SHEET

Chapter 7, Problem 5: ADVERTISING IS TO BE PLACED ON A SIGNBOARD, WXYZ. THE METAL SUPPORT FRAME IS SHOWN IN DASHED LINES. FROM POINT A AND B, TWO HORIZONTAL BRACES TO THE SIGNBOARD ARE TO BE INSTALLED. FIND THE TRUE ANGLE THESE HORIZONTAL SUPPORTS MAKE WITH THE SIGNBOARD. SHOW THE SUPPORTS IN ALL VIEWS.

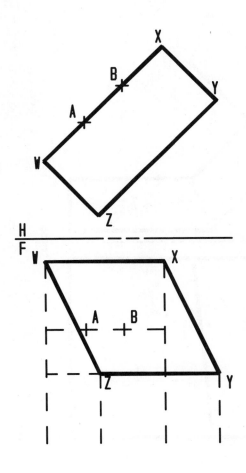

	ADG INC.
ANGLE =	ADVERTISING SIGNBOARD

SIZE	DATE	DWG NO.	SCALE 1" = 20'	REV
DRW. BY		SECTION #	SHEET	

Chapter 7, Problem 6: THE SLOPING SIDES OF THE CONCRETE BRIDGE FOOTING EACH HAVE THE SAME SLOPE. CORNERS SUCH AS THOSE BETWEEN SURFACES X AND Y AND SURFACES Y AND Z ARE USUALLY COVERED WITH STEEL ANGLE IRON EMBEDDED IN THE CONCRETE. DETERMINE THE ANGLE TO WHICH EACH OF THESE SPECIAL IRON ANGLES MUST BE BENT.

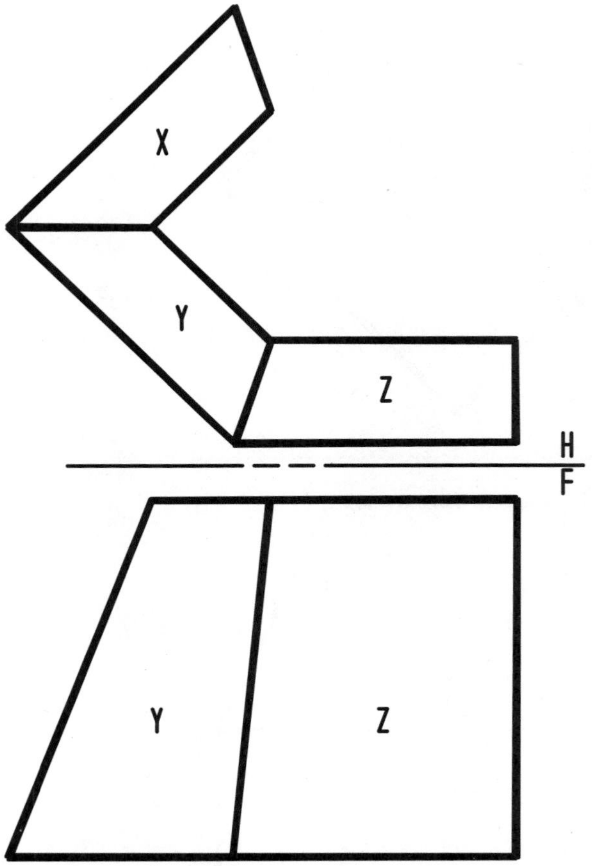

H
F

XY ANGLE =	ADG INC.			
YZ ANGLE =	BRIDGE			
	SIZE	DATE	DWG NO.	SCALE 1/8" = 1'0"	REV
DRW. BY			SECTION #	SHEET

Chapter 7, Problem 7: A SHEET METAL CHUTE IS TO BE
INSTALLED BETWEEN THE SECOND AND THIRD FLOORS OF A
BUILDING. FIND THE TRUE SIZE OF THE ANGLE BETWEEN PLANES
A AND B FROM THE TOP VIEW, AND A AND D FROM THE FRONT VIEW.

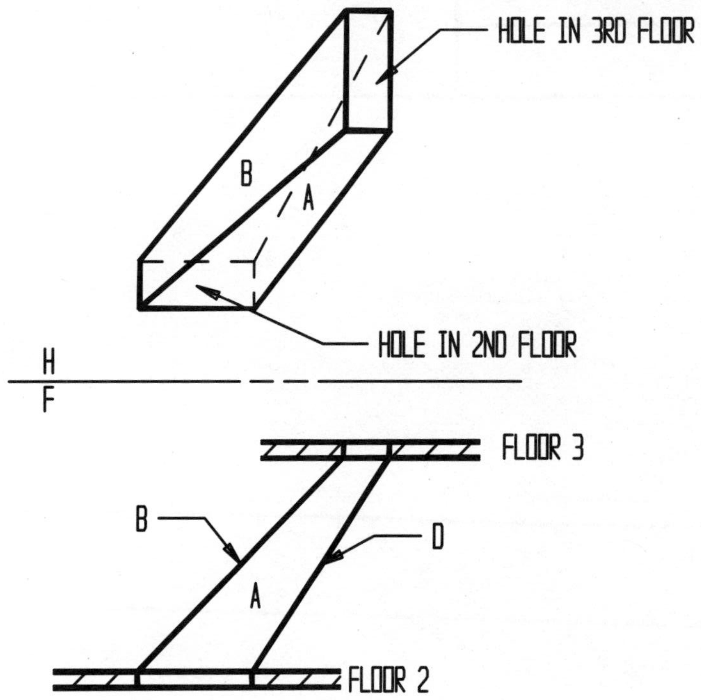

HOLE IN 3RD FLOOR

HOLE IN 2ND FLOOR

B

A

H
F

FLOOR 3

D

B

A

FLOOR 2

ADG INC.

ANGLE AB =
ANGLE AD =

SIZE	DATE		DWG NO.		SCALE 1/4" = 1'0"	REV

CHUTE

	SECTION #		SHEET

DRW. BY

Chapter 7, Problem 8: LINE MN IS THE CENTER OF A PIPE THAT INTERSECTS THE
SIDE OF A HOPPER AT M AND THE FLOOR AT N. FIND THE ANGLE THE PIPE MAKES
WITH THE HOPPER SIDE AND WITH THE FLOOR.

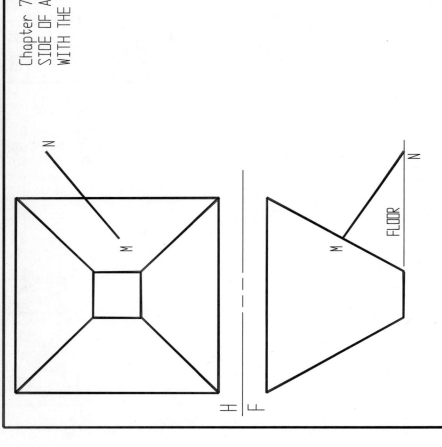

Chapter 7, Problem 9: A HAM RADIO ANTENNA IS SUPPORTED BY TWO SUPPORT RODS, AB AND AC, ATTACHED TO THE ROOF OF A HOUSE. FIND THE ANGLES MADE BY THE SUPPORT RODS AND THE ROOF.

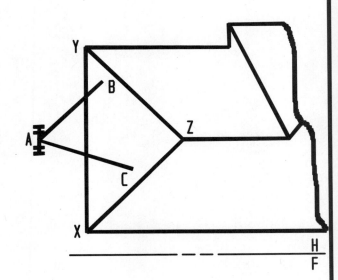

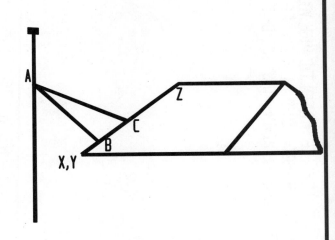

	ADG INC.			
ANGLE B = ANGLE C =	SUPPORT RODS			
	SIZE	DATE	DWG NO.	SCALE 1/8" = 1'0"	REV
DRW. BY		SECTION #		SHEET

Chapter 7 Problem 10: THE CLEARANCE BETWEEN TWO HIGH VOLTAGE
POWER LINES MUST BE AT LEAST 5 FEET. DOES THE CLEARANCE
BETWEEN THE POWER LINES EXCEED THE MINIMUM DISTANCE OF 5 FEET?

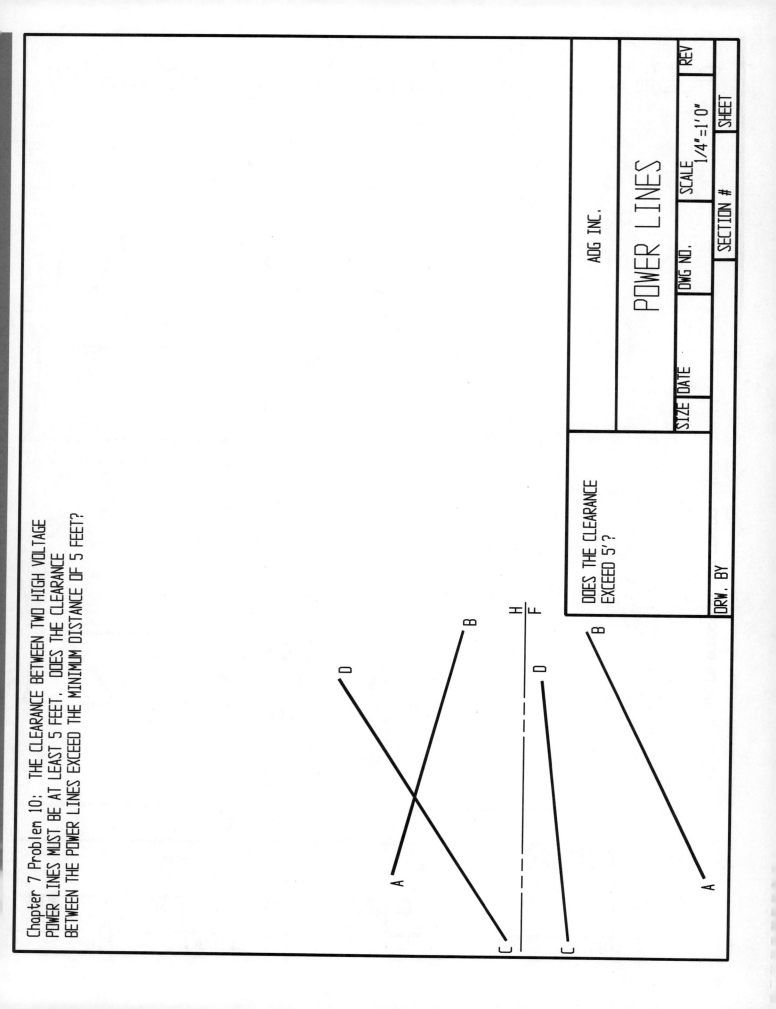

DOES THE CLEARANCE
EXCEED 5'?

ADG INC.

POWER LINES

SIZE	DATE	DWG NO.	SCALE 1/4"=1'0"	REV
DRW. BY			SECTION #	SHEET

Chapter 7, Problem 11: LINES AB AND CD REPRESENT TWO CONTROL CABLES IN AN AIRPLANE. THE MINIMUM CLEARANCE BETWEEN THE CABLES MUST BE 4 INCHES. DOES THE CLEARANCE BETWEEN CABLES AB AND CD MEET THE MINIMUM REQUIRED CLEARANCE OF 4 INCHES?

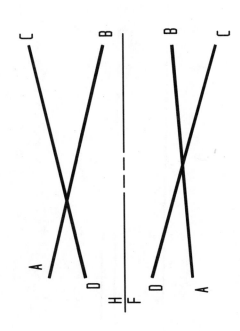

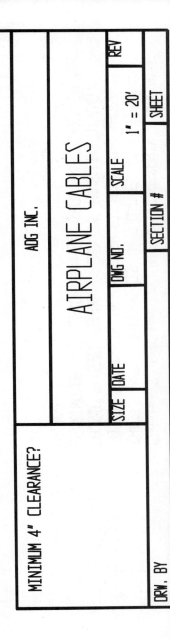

MINIMUM 4" CLEARANCE?

ADG INC.

AIRPLANE CABLES

SIZE	DATE	DWG NO.	SCALE 1" = 20'	REV
DRW. BY		SECTION #	SHEET

Chapter 7, Problem 12: LINE AB REPRESENTS A NEW MINE SHAFT, AND CD REPRESENTS AN OLD ONE. IT IS DESIRED TO CONNECT THESE TWO MINE SHAFTS WITH THE SHORTEST POSSIBLE LEVEL TUNNEL. FIND THE TRUE LENGTH AND BEARING OF THIS LEVEL TUNNEL. SHOW THE TUNNEL IN ALL VIEWS.

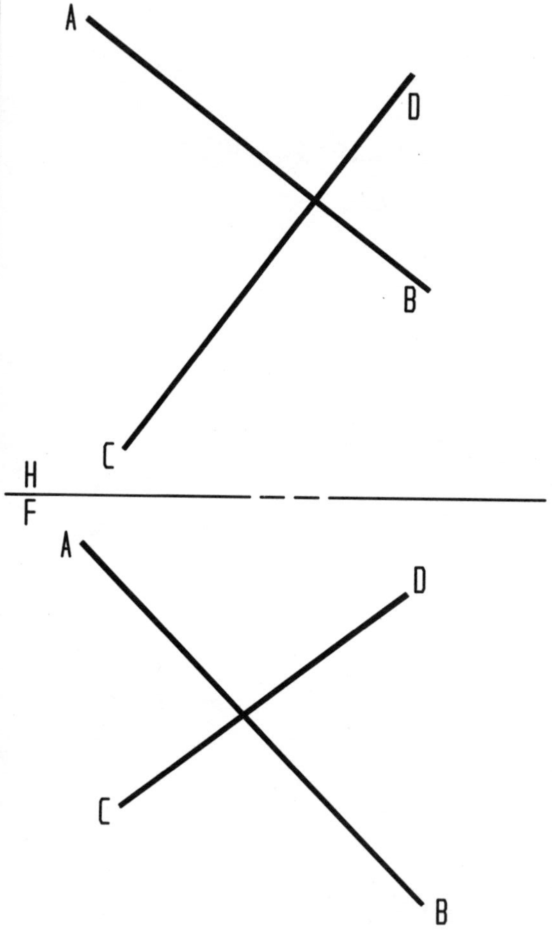

A

D

B

C

H
F

A

D

B

C

ADG INC.

MINE SHAFT

Chapter 7, Problem 13: THE CENTER LINES OF TWO PIPES ARE
SHOWN IN THE TWO VIEWS BELOW. FIND THE TRUE LENGTH OF THE
SHORTEST LEVEL CONNECTING PIPE (CENTER LINE) BETWEEN
THEM IN ORDER THAT A BYPASS VALVE MAY BE INSTALLED.
BE SURE TO SHOW THE CENTER LINE OF THE LEVEL PIPE
IN ALL VIEWS.

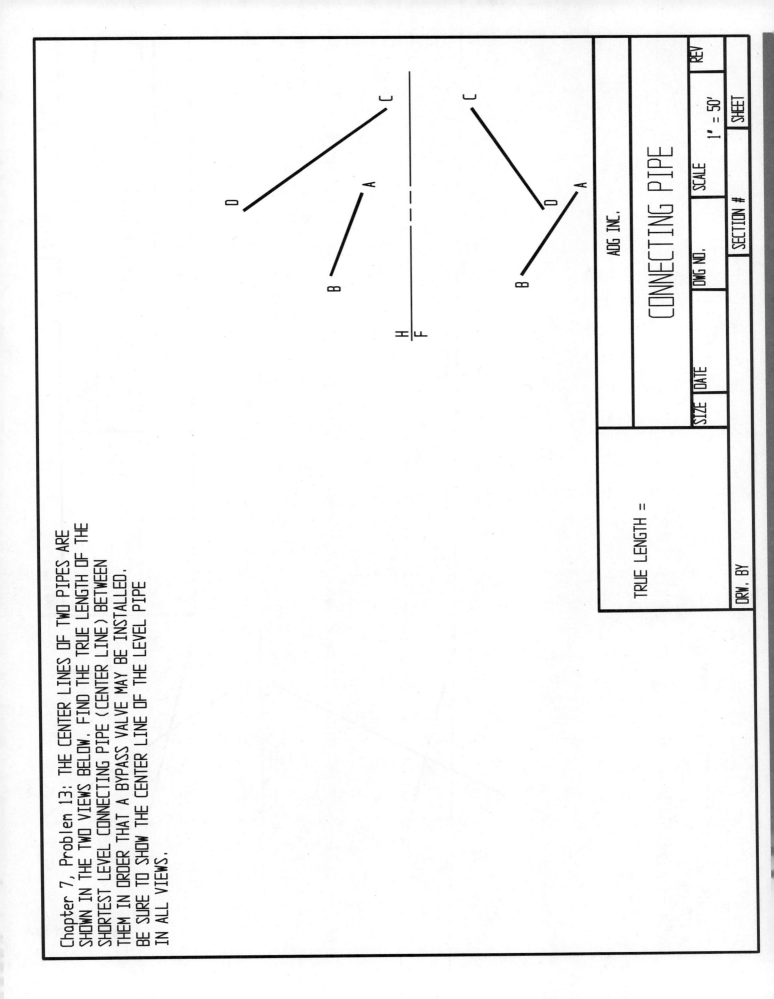

TRUE LENGTH =

ADG INC.

CONNECTING PIPE

SIZE	DATE	DWG NO.	SCALE 1" = 50'	REV
		SECTION #	SHEET	

DRW. BY

Chapter 7, Problem 14: LINES WX AND YZ ARE THE
CENTER LINES OF TWO NATURAL GAS LINES.
FIND THE SHORTEST CONNECTING PIPE HAVING
A 20° DOWNHILL SLOPE FROM WX TO YZ.
SHOW THE CONNECTOR IN ALL VIEWS.

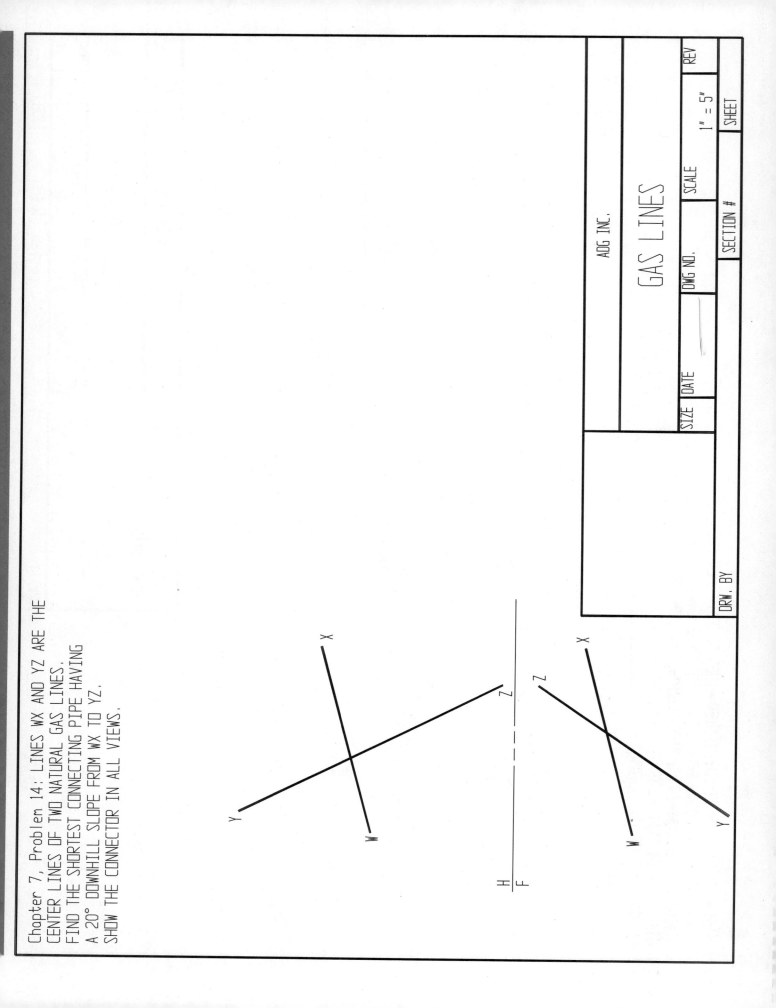

Chapter 7, Problem 15: TWO PIPES ARE DESIGNATED BY
THEIR CENTER LINES, AB AND CD. IT IS NECESSARY TO
CONNECT THEM WITH THE SHORTEST PIPE AT A 29 PERCENT
GRADE DOWN FROM CD. FIND THE TRUE LENGTH OF THIS
CONNECTOR, AND SHOW IT IN ALL VIEWS.

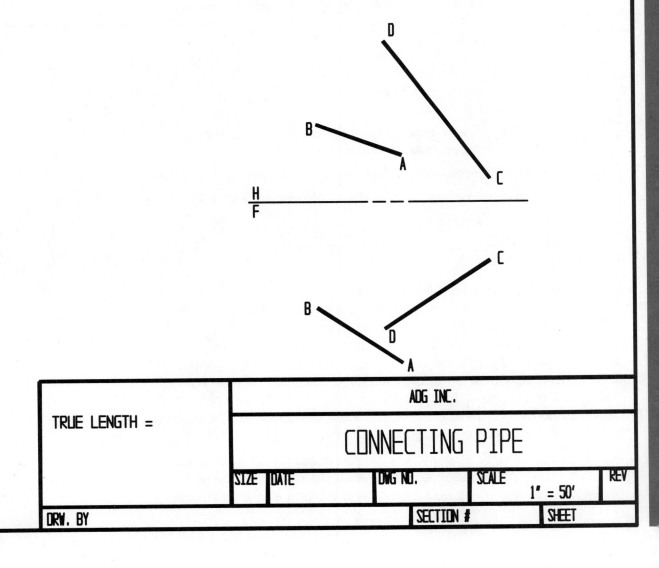

TRUE LENGTH =	ADG INC.			
	CONNECTING PIPE			
	SIZE	DATE	DWG NO.	SCALE 1" = 50'	REV
DRW. BY			SECTION #	SHEET

Piercing Points and the Intersections of Planes

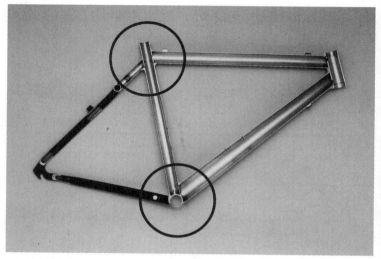

Locate the bike frame tubes that are cut to the intersecting shape of the tube they touch.

After completing this chapter, you will be able to:

▸ *Determine the piercing point of a line with a plane using the edge view method and the cutting plane method.*

▸ *Determine the line of intersection between two nonparallel planes using the edge view method and the cutting plane method.*

▸ *Determine the intersection of a three-dimensional object and a line.*

▸ *Determine the intersection of a three-dimensional object and an oblique plane.*

▸ *Determine the intersection of two solids.*

▸ *Determine the correct visibility.*

We live in a world of three-dimensional objects. These objects are made up of lines and planes. The relationship between these elements is very important to the manufacturer. For example, in order to cut the tubing for a bicycle frame, the shape, or intersection, where two pieces of the tubing meet must be determined. In order to machine an injection mold to produce the outer case or housing of a hair dryer, the shape of the housing needs to be determined. The intersection of the round hair dryer nozzle (a cylinder) and the area that incases the motor (a sphere) needs to be determined. Sometimes the information will be in a given view, sometimes additional work is needed to determine the intersection of two elements.

Rather than jumping straight into the intersection of two solids, this chapter will break it down into smaller steps by starting with the intersection of a line and a plane, two planes, a line and a solid, and a plane and a solid.

▼ PIERCING POINTS OF A LINE AND A PLANE

If a line neither lies in a particular plane nor lies parallel to it, the line will intersect the plane at a single point. This point of intersection, called a **piercing point,** must lie in the plane and on the line. The problem of finding the piercing point of a line in a plane is the basic element in so many descriptive geometry problems that it could well be considered as important as the four fundamental views.

The two methods described in this section will locate the piercing point. Either method may be used.

The Piercing Point of a Line and a Plane— Auxiliary View Method

This method, also called the edge view method, requires that the plane appear as an edge. *The piercing point is a point common to both the line and the plane.* Because the edge view of the plane contains all points on the plane, the view that shows the edge view of the plane will also show the point where the line pierces the plane.

The simplest case occurs when the given plane appears as an edge in one of the primary views. Figures 8–1 and 8–2 show the pictorial and the orthographic views, respectively, of pipe M-N, which intersects wall A-B-C-D. In the orthographic drawing (Figure 8–2) the top view shows the edge view of the wall and the center line of pipe M-N. The intersection of the pipe with the wall is determined at point P in the top view, which is common to both. The piercing point is marked with a ⅛" perpendicular line at the point of intersection. This point is projected to the front view at P. The visibility is determined by simple observation. You can see that line P-M is behind the wall, as you view the top; therefore, line P-M is drawn hidden in the front view. (Review visibility in Chapter 3.)

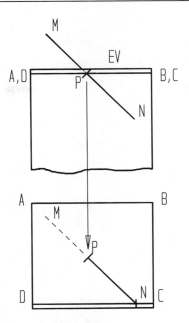

Figure 8–2   Piercing point P, found in a principal view (top), is projected to the adjacent view (front view).

A portion of an aircraft bulkhead plane (E-F-G-H) is intersected by cable X-Y, in Figure 8–3. By locating the bulkhead as an edge in view 1, the piercing point, P, is identified and can be projected to line X-Y in the top view and front view (Figure 8–4).

Visibility may be determined by the method explained in Chapters 3 and 5. Figure 8–5 demonstrates locating the visibility of line X-Y in the top view. Locate the crossing point of the line X-Y, and a line in the plane, E-F, in the top view. Which is higher? The front view shows line X-Y is higher than E-F; therefore, line X-Y

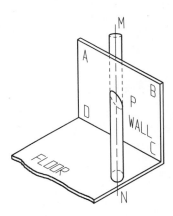

Figure 8–1   Pipe M-N, piercing wall, A-B-C-D at point P.

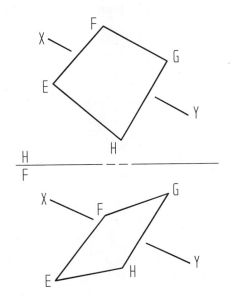

Figure 8–3   Locate where the line pierces the plane, using the edge view method.

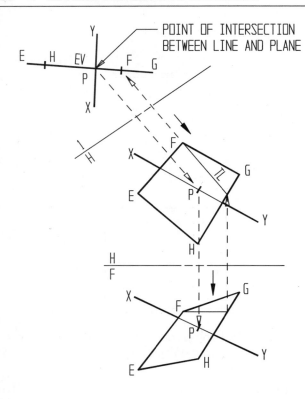

Figure 8–4 Piercing point P, located using the edge view method.

will be drawn as an object line in the top view on the left side of the piercing point. If you chose the crossing point of X-Y and G-H in the top view, you found that G-H is higher than X-Y; therefore, line X-Y is drawn as a hidden line in the top view on the right side of the piercing point.

Figure 8–6 demonstrates locating the visibility of line X-Y in the front view. Locate the crossing point of the line X-Y, and a line in the plane, E-F, in the front view.

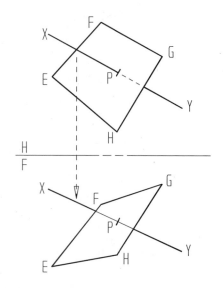

Figure 8–5 Visibility of line X-Y in top view.

Which line is in front of the other? The top view shows line X-Y is in front of line E-F on the left side of the piercing point; therefore, line X-Y is drawn as a visible line in the front view on the left side of the piercing point. If you chose the crossing point of X-Y and G-H in the front view, you found G-H is in front of X-Y; therefore, line X-Y is drawn as a hidden line in the front view on the right side of the piercing point.

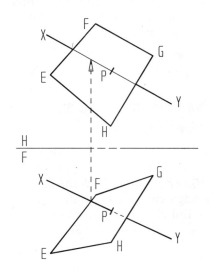

Figure 8–6 Visibility of line X-Y in the front view.

Figure 8–7 demonstrates what the finished drawing should look like.

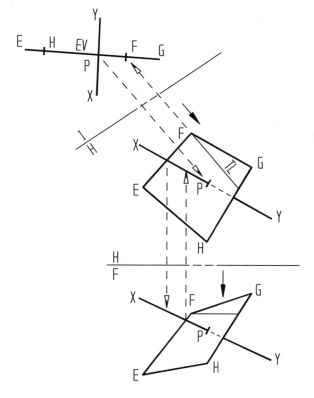

Figure 8–7 Visibility of line X-Y.

The Piercing Point of a Line and a Plane— Cutting Plane Method

It is possible to find the piercing point of a line and a plane by using only two views and a cutting plane. This is an important and frequently used technique based on the following principle: *The intersection of a line with a plane will lie on the intersection of a given plane and a cutting plane that contains the line.*

Figure 8–8 shows the pictorial and Figure 8–9 shows the orthographic views of an oblique plane, A-B-C and a line X-Y. First, examine the pictorial drawing. A vertical cutting plane has been placed so the line X-Y lies entirely within it. The vertical cutting plane and plane A-B-C must intersect on a straight line. This line is

labeled D-E. The given line X-Y and the intersection line D-E both lie in the vertical cutting plane, and must be either parallel or intersecting. In this case, they intersect at P, which is the desired intersection between the line X-Y and plane A-B-C.

Now, apply this to the orthographic drawing. The vertical cutting plane, CP-1, is shown in the top view, where it appears as an edge view and contains line X-Y (Figure 8–10). This cutting plane could be extended indefinitely, but is drawn lightly and only as long as necessary.

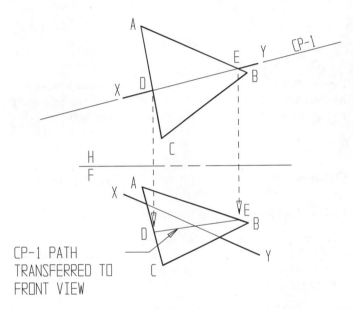

Figure 8–10 Cutting plane 1 is shown as an edge view in the top view. Line D-E shows the line of intersection between plane A-B-C and cutting plane 1. Project line D-E, CP-1 path, to adjacent view (front view).

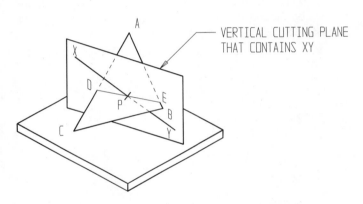

Figure 8–8 A vertical cutting plane, including line X-Y, is used to locate the line of intersection, D-E. The piercing point is located where line X-Y intersects line D-E.

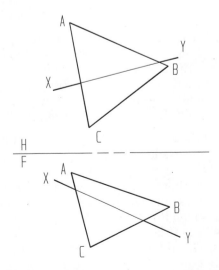

Figure 8–9 Locate where the line pierces the plane using the cutting plane method.

The next step is to find the line of intersection, D-E. Note that line A-C intersects the cutting plane at point D and line A-B intersects the cutting plane at point E. These points are projected to the front view to create line D-E. The piercing point P is located where the CP-1 cutting path, line D-E, intersects with line X-Y (Figure 8–11). This is the required piercing point line X-Y makes with plane A-B-C. (Imagine cutting the three-dimensional plane A-B-C, with a pair of scissors, along line X-Y in the top view. Line D-E shows the path in the front view where the scissors cut the plane into two pieces. The piercing point is located where the scissors cut the plane and the line X-Y at the same time.)

Point P is located on line X-Y in the adjacent view by simple projection (Figure 8–11).

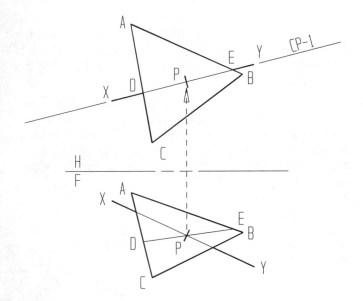

Figure 8–11 The piercing point is located where the intersection line D-E crosses line X-Y. Project the piercing point to the adjacent view using the projection method.

Complete views—remember visibility. Figure 8–12 shows visibility for the top view.

Figure 8–13 shows visibility for the front view.

The cutting plane may also be selected so it appears as an edge in the front view. In the pictorial representation, Figure 8–14, a cutting plane is placed so as to contain the line X-Y and intersect plane A-B-C.

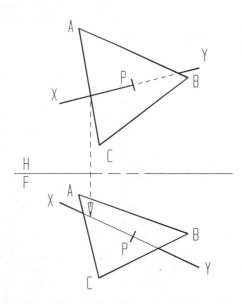

Figure 8–12 Line X-Y is higher than line A-C; therefore, line X-Y is visible in the top view on the left side of the piercing point.

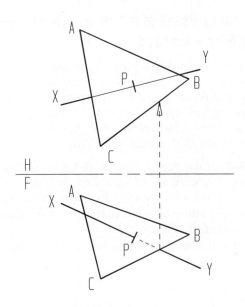

Figure 8–13 Line B-C is in front of line X-Y; therefore, line X-Y will be hidden in the front view on the right side of the piercing point.

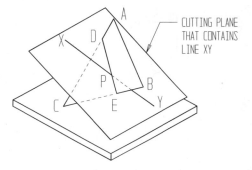

Figure 8–14 Piercing point of line X-Y and plane A-B-C. The piercing point is located where intersecting line E-D crosses line X-Y.

Figure 8–15 demonstrates the cutting plane line orthographically. The line of intersection D-E, is projected to the top view. The piercing point P, is located where the line of intersection D-E, crosses line X-Y. The front view of the piercing point P, is found by projection (Figure 8–16). Notice the answer is the same as Figure 8–13.

While either the edge view method or the cutting plane method may be used, the cutting plane method has several advantages. It uses less space, takes less time to complete, and there is less chance of error because an auxiliary view is not needed. However, if it should happen that the given line is a profile line, the cutting plane method will not produce a solution in either the top or front views. This is because the cutting plane appears as an edge in both views. Since a new view must be drawn, it is usually preferable to use the edge view method in such a case.

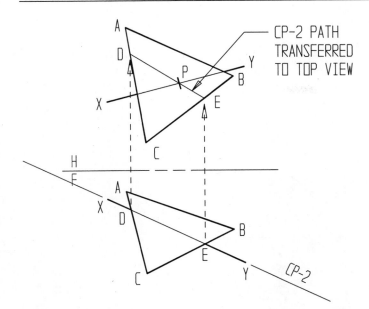

Figure 8–15 Cutting plane 2 intersects plane A-B-C at points D and E. Project D-E to adjacent view. The piercing point is located where line X-Y crosses line D-E.

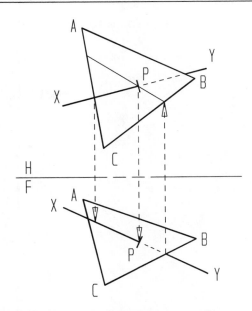

Figure 8–16 Project piercing point to adjacent view and determine visibility.

EXERCISE

Problem: Locate the piercing point using the edge view method. Go ahead and sketch right on Figure 8–17. Locate the intersection of point P in view 1, using the edge view method. Sketch the solution on the figure. The correct answer will be shown on the next figure.

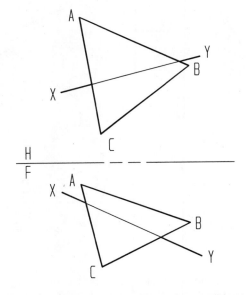

Figure 8–17 Using the edge view method, locate the piercing point of line X-Y and plane A-B-C.

Project point P back to the front view (Figure 8-18).

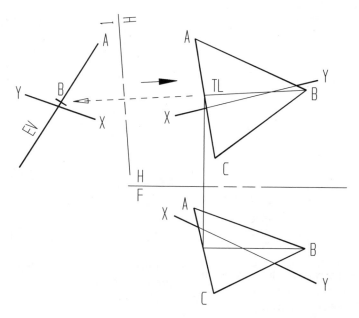

Figure 8-18 Edge view method of solving for the piercing point. Now project point P back to the front view.

Determine visibility.

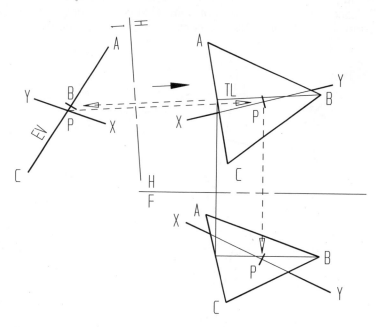

Figure 8-19 Point P projected back to the front view. Now determine visibility.

Answer

Notice the problem is the same one as in Figure 8–9; therefore, the location of point P and visibility are the same. This method is different, but the answer is the same (Figure 8–20).

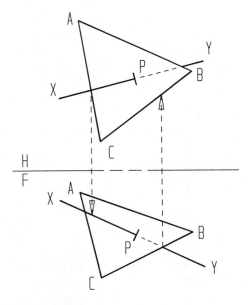

Figure 8–20   Visibility of line X-Y.

Problem: Locate the piercing point of line X-Y and plane A-B-C, using the cutting plane method. Place the cutting plane in the top view. Go ahead and sketch right on Figure 8–21. While the top view is the same as the problem in Figure 8–17, the front view is different.

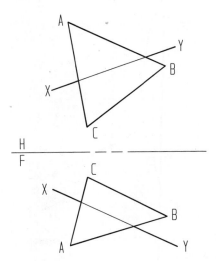

Figure 8–21   Using the cutting plane method, locate the line of intersection and piercing point between line X-Y and plane A-B-C.

Figure 8–22 demonstrates locating the line of intersection D-E, and point P in the front view. Project point P to the adjacent view (top view).

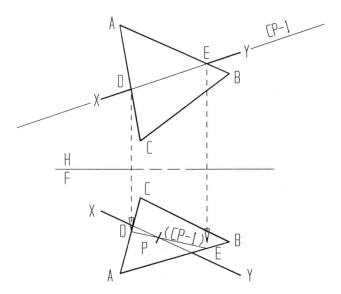

Figure 8–22 Piercing point located using the cutting plane method. Project point P to the top view.

Figure 8–23 demonstrates point P being projected to the adjacent view (top view).

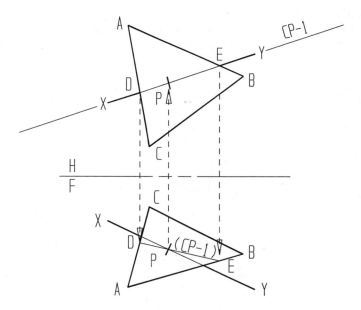

Figure 8–23 Point P projected to the adjacent view. Determine visibility.

Answer

Figure 8–24 demonstrates determining visibility one view at a time.

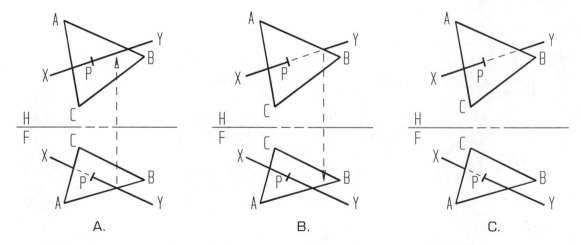

Figure 8–24   Visibility. (A) Visibility for front view—X-Y is in front of A-B. (B) Visibility for top view—A-B is higher than X-Y. (C) Completed problem.

Piercing Points of a Line Passing Through a Three-Dimensional Object

A line, such as a cable or a pipe, cutting through a three-dimensional object uses the piercing point of a line and a plane method demonstrated in the previous sections (Figure 8–25). The key to success is to work with only one plane at a time.

where CP-1, line 9-10, crosses plane 2-2'-3'-3 at points A and B to the front view (Figure 8–27). Piercing point P-1 is where line A-B (CP-1 path) crosses line 9-10 in the front view. Project P-1 to the cutting plane line in the top view. Project where CP-1, line 9-10, crosses plane 6-6'-7-7' at points C and D to the front view (Figure 8–28). Piercing point P-2 is where line C-D (CP-1 path) crosses line 9-10 in the front view. Project P-2 to the cutting plane line in the top view. Complete the visibility for line 9-10 in all views (Figure 8–26).

Figure 8–25   Line passing through a solid.

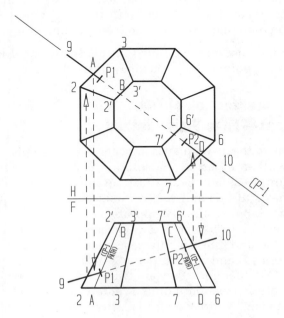

Figure 8–26   Intersection of three-dimensional solid object and a line.

Figure 8–26 shows the piercing points and visibility of the line. Remember to take one plane at a time. Project

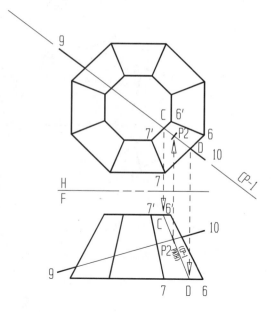

Figure 8–27 Intersection of line 9-10 and plane 2-3-3'-2'.

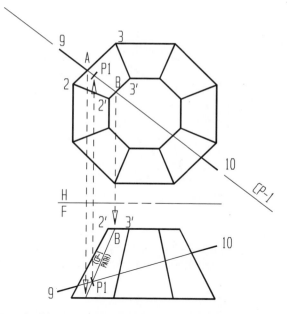

Figure 8–28 Intersection of line 9-10 and plane 6-7'-7-'7.

▼ THE INTERSECTION OF TWO PLANES

If two planes are not parallel, they will intersect along a straight line that is common to both planes. Because the position of this shared line may be fixed by any two points on the line, you need only to find two points that lie on both planes to find the intersection.

The two methods described in this section will locate piercing points. Either one may be used.

The Intersection of Two Planes—Edge View Method

If one of the given planes appears as an edge in a given view, then a part of the solution is already available. Figures 8–29 and 8–30 illustrate this type of problem. Planes A-B-C-D and X-Y-Z intersect. In the top view, plane A-B-C-D already appears as an edge view. It is evident that the line X-Z intersects plane A-B-C-D at point R, and line X-Y intersects plane A-B-C-D at point S. Points R and S lie in plane X-Y-Z and also in plane A-B-C-D; therefore, they are two points on the required line of intersection. Points R and S are located in the front view by projection. Visibility is then determined.

When both of the planes are oblique, as seen in Figure 8–31, the solution may be found by showing one of the planes as an edge.

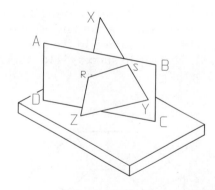

Figure 8–29 Intersection of two planes.

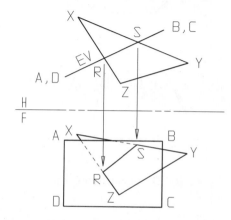

Figure 8–30 Plane A-B-C-D is an edge view in a principal view; therefore, the intersection is found in the adjacent view.

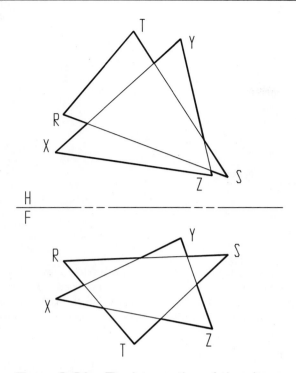

Either plane may be selected for projection to an edge view. The auxiliary view may be projected from any principal view. In Figure 8–32, plane X-Y-Z is found as an edge in view 1. Plane R-S-T is also shown in view 1.

The intersection points of lines R-T and S-T, with the edge view of plane X-Y-Z, are shown at points A and B, respectively (Figure 8–33). These two points can be projected to the top view to locate A and B. Remember that A lies on line R-T and B lies on line S-T. The line of intersection A-B, is similarly located in the front view.

Determine visibility (Figure 8–34 through Figure 8–36). Shading is used in this text only to help you envision the visibility. Figure 8–36 shows the finished drawing. Select one line from each plane that crosses to test for visibility. Ignore all other lines, except for the two you are working with.

Figure 8–31   The intersection of the planes X-Y-Z and R-S-T is required.

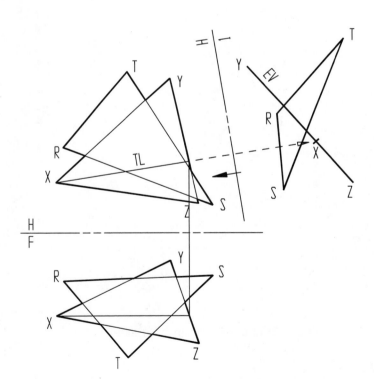

Figure 8–32   Locate edge view of one of the planes.

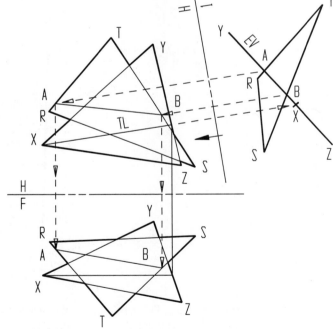

Figure 8–33   Project the intersecting points A and B back to the front view.

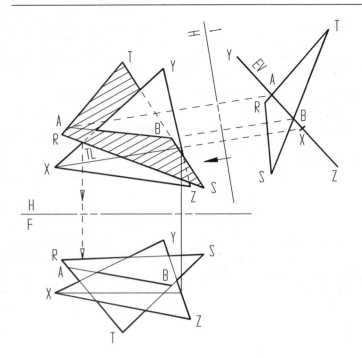

Figure 8–34 Determine visibility for the top view. In the top view, select one line from each plane that cross, for example R-S and X-Y. Look at the front view to determine height. R-S is higher than X-Y. Test the other side of the piercing point—line X-Y is higher than T-S.

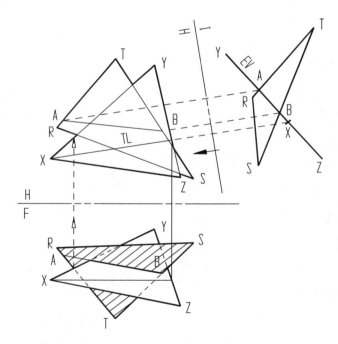

Figure 8–35 Determine visibility for the front view. X-Y is in front of R-T. On the other side of the piercing point, X-Y is behind R-S.

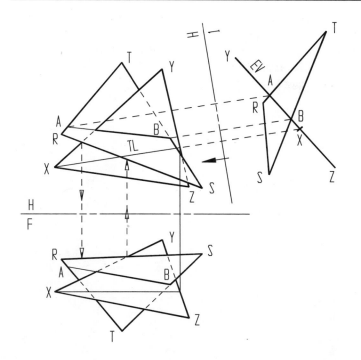

Figure 8–36 Intersection of planes X-Y-Z and R-S-T.

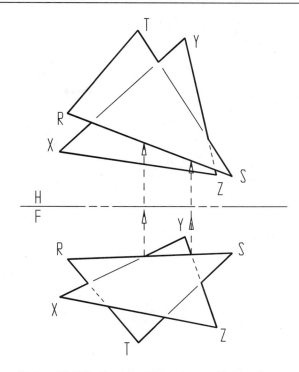

Figure 8–37 Intersecting planes. Determine as much visibility as possible.

The Intersection of Two Planes—Cutting Plane Method

It is important to understand that when you are finding the line of intersection between two planes, it is simply a matter of finding the piercing point of a line and a plane twice, thus locating two piercing points which form the line of intersection.

In Figure 8–37 planes X-Y-Z and R-S-T are intersecting. Their line of intersection will be found by using two cutting planes to locate two piercing points.

Cutting plane lines may be arbitrarily chosen; however, some lines make better cutting planes than others. Some lines may not cross the other plane and have to be extended, while other lines will only cross a small portion of the other plane. One excellent suggestion is to first determine as much visibility between the two planes as possible. This process identifies on which lines the piercing points will be located, thereby eliminating the frustration of the time consuming trial and error method.

Figure 8–37 demonstrates that line S-R is in front of both lines X-Y and Y-Z; therefore, proving line R-S does not intersect plane X-Y-Z. Further testing will show that while line X-Y is behind line R-S, it is in front of line R-T. This proves that line X-Y pierces plane R-S-T, resulting in one of the piercing points being located on line X-Y. The other piercing point will be located on line T-S.

Line T-S and its intersection with plane X-Y-Z is found by using the vertical cutting plane, CP-1 (Figure 8–38). This cutting plane intersects plane X-Y-Z at points A and B. Point A is on line X-Y and is projected to X-Y in the front view. Point B is on line Y-Z, and is projected to Y-Z in the front view. Line A-B, CP-1 path, in the front view intersects S-T (CP-1) at P-1. This point, P-1, is where S-T intersects plane X-Y-Z and is one point on the required line of intersection.

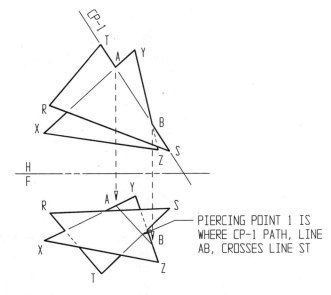

Figure 8–38 Vertical cutting plane CP-1 crosses plane X-Y-Z at points A and B, locating P-1 in the front view.

Figure 8–39 demonstrates projecting P-1, located on line S-T (CP-1) to line S-T (CP-1) in the top view.

A second vertical cutting plane, CP-2, containing line X-Y, is now selected (Figure 8–40). CP-2, line X-Y, intersects plane R-S-T creating points C and D. Line C-D, CP-2 path, in the front view intersects line X-Y (CP-2) at point P-2. This point P-2, where line X-Y (CP-2) intersects plane R-S-T, is the second point on the required line of intersection (Figure 8–41).

Visibility will be the same as in Figure 8–36.

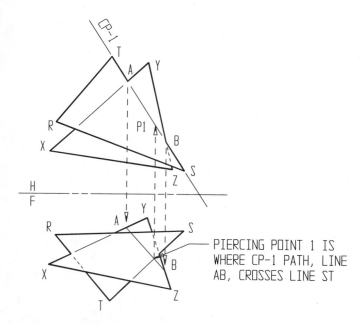

PIERCING POINT 1 IS WHERE CP-1 PATH, LINE AB, CROSSES LINE ST

Figure 8–39 Project piercing point to adjacent view.

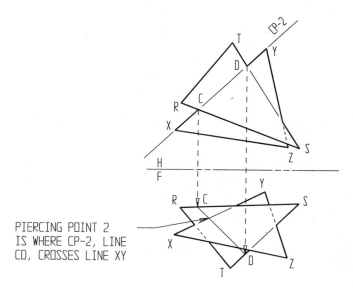

PIERCING POINT 2 IS WHERE CP-2, LINE CD, CROSSES LINE XY

Figure 8–40 Vertical cutting plane CP-2 crosses plane R-S-T at points C and D, locating P-2 in the front view.

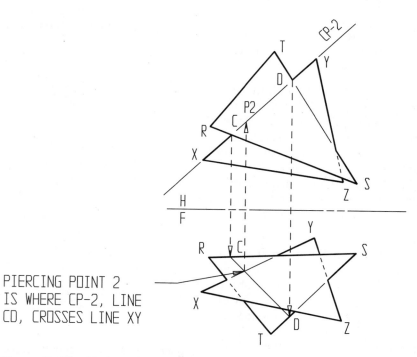

PIERCING POINT 2 IS WHERE CP-2, LINE CD, CROSSES LINE XY

Figure 8–41 Project piercing point to adjacent view.

EXERCISE

Problem: Locate the intersection between planes A-B-C and R-S-T, using the edge view method. Sketch each step on the following figures. The answer for each step will be shown in the following figure.

The first step is to locate the edge view of one of the planes. While either plane may be taken to an edge view, the answer key finds the edge view of plane R-S-T (Figure 8–42).

Locate intersection points M and N in all views (Figure 8–43).

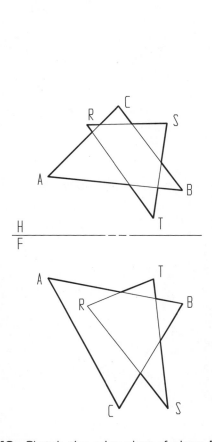

Figure 8–42   Sketch the edge view of plane R-S-T, carry plane A-B-C along.

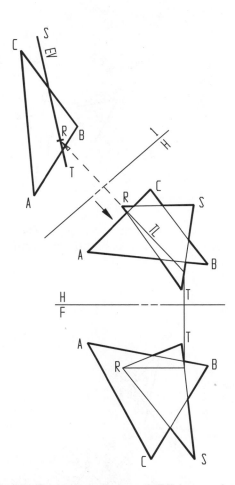

Figure 8–43   Edge view of plane R-S-T. Locate intersecting M-N.

Determine visibility of intersecting planes.

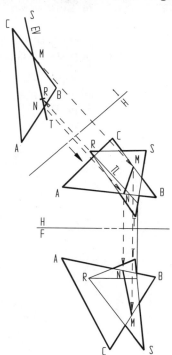

Figure 8-44 The intersection (line M-N) between planes A-B-C and R-S-T located. Determine visibility.

The shading shown in Figures 8-45 and 8-46 is used in this text only to help you visualize the two planes. The finished drawing is shown in Figure 8-47.

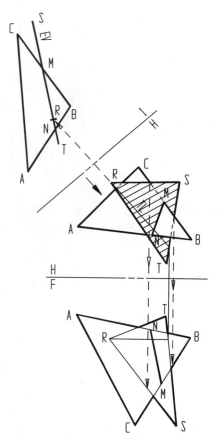

Figure 8-45 Visibility for top view—C-B is higher than T-S, but lower than R-S.

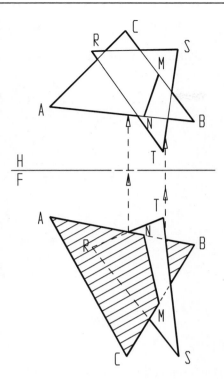

Figure 8–46 Visibility for front view—A-B is in front of R-T, but behind T-S.

Answer

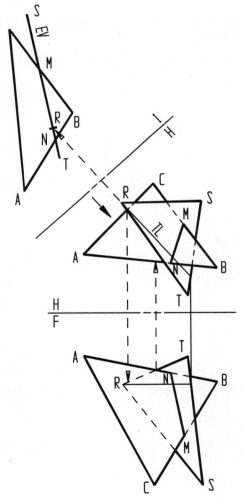

Figure 8–47 Visibility of planes A-B-C and R-S-T.

Problem: Using the same problem as in Figure 8–42, solve for intersection of planes A-B-C and R-S-T using the cutting plane method. Start by determining as much visibility between the two planes as possible. Go ahead and sketch right on Figure 8–48.

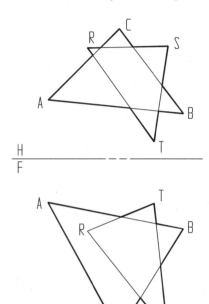

Figure 8–48 Using the cutting plane method, solve for the intersection of planes R-S-T and A-B-C. Determine as much visibility between the two views as possible.

Line R-T is higher than A-C and A-B. This tells us that line R-T does not pierce plane A-B-C and is drawn as a solid line. However, lines A-B and C-B pierce plane R-S-T.

Locate one of the piercing points, between plane A-B-C and R-S-T, using line A-B as the first vertical cutting plane line (CP-1).

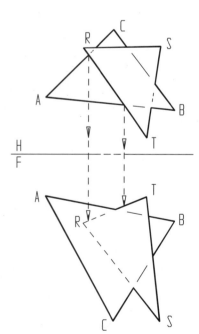

Figure 8–49 Intersection points will be located on lines A-B and C-B. Use line A-B as CP-1 to locate one of the piercing points between the two planes.

Line A-B (CP-1) crosses plane R-S-T at points W and X in the top view. Points W and X are projected to the front view and connected. Piercing point 1 is located where line W-X, cutting plane path, crosses line A-B (CP-1) (Figure 8–50).

Locate the other piercing point, using line C-B as the second vertical cutting plane line, CP-2.

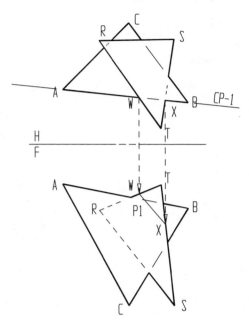

Figure 8–50 Line A-B (CP-1) locates one of the piercing points in the front view. Use line C-B as CP-2 to locate the other piercing point.

Line C-B (CP-2) crosses plane R-S-T at points Y and Z in the top view. Points Y and Z are projected to the front view and connected. Piercing point 2 is located where line Y-Z, cutting path line, crosses the line C-B (CP-2) (Figure 8–51).

Project P-1 and P-2 to the top view.

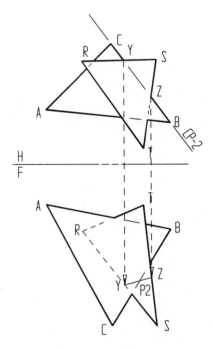

Figure 8–51 Line C-B (CP-2) locates the other piercing point in the front view. Project P-1 and P-2 to the top view.

Answer

Point P-1 is located on A-B (CP-1); therefore, it is projected to A-B (CP-1) in the top view. Point P-2 is located on C-B (CP-2); therefore, it is projected to C-B (CP-2).

Visibility will be the same as in Figure 8–47.

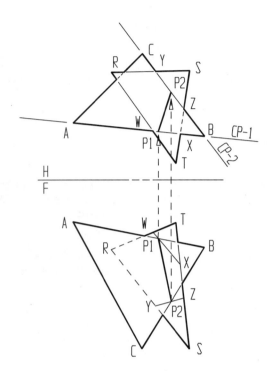

Figure 8–52 Intersection of planes A-B-C and R-S-T using CP-1 and CP-2.

▼ INTERSECTION OF A THREE-DIMENSIONAL OBJECT

The only difference between cutting a plane and cutting a three-dimensional object, is there are more planes to deal with. It is easiest to locate the cut through one plane at a time.

Cutting Path of a Line Passing Through a Three-Dimensional Object

There are two ways to look at the problem shown in Figure 8–53. One way is where the line intersects with the solid object. (See Piercing Points of a Line Passing Through a Three-Dimensional Object.) The second is using the line as a cutting plane.

In this section, line 9-10 will be used as a vertical cutting plane in the top view (Figure 8–53). Imagine you are going to cut the object in two pieces using a hand saw. Line the hand saw up with line 9-10 (CP-1) in the top view and cut from the top of the object to the bottom, cutting the object into two pieces. The object is cut along plane A-B-C-D. Mentally remove the front piece.

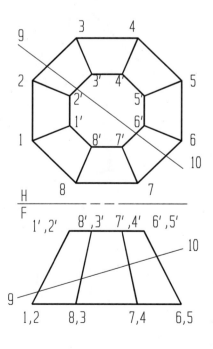

Figure 8–53 Intersection problem.

What would the front view look like (Figure 8–54)? If the two pieces were put back together, what would the visibility of the cutting path (plane A-B-C-D) look like in the adjacent view (front view)? You are correct, line A-B would be hidden (Figure 8–55).

what would the visibility of the cutting path look like in the adjacent view (top view)? You are correct, there would be no change. Figures 8–57 through 8–59 show each cutting path line plane step by step. Figures 8–60 and 8–61 demonstrate working with multiple planes. Figure 8–62 shows there would be no change in visibility.

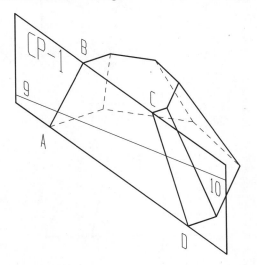

Figure 8–54 Line 9-10 used as a cutting plane in the top view.

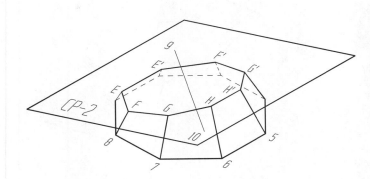

Figure 8–56 Line 9-10 used as a cutting plane in the front view.

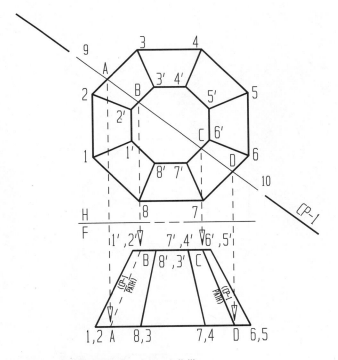

Figure 8–55 Front view visibility.

Use line 9-10 in the front view as your cutting plane (Figure 8–53). Line the hand saw up with line 9-10 (CP-2) in the front view and cut from the front of the object to the back, cutting the object into two pieces. This creates points E,F,G,H,H',G',F', and E'. Mentally remove the top piece. What would the top view look like (Figure 8–56)? If the two pieces were put back together,

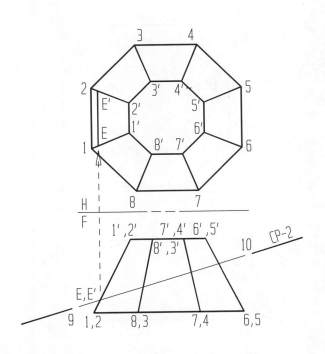

Figure 8–57 Intersection of line 9-10 with plane 1-2-2'-1'.

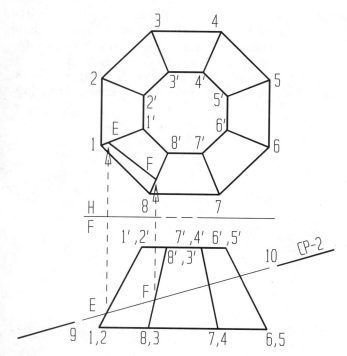

Figure 8–58 Intersection of line 9-10 with plane 1-1'-8'-8.

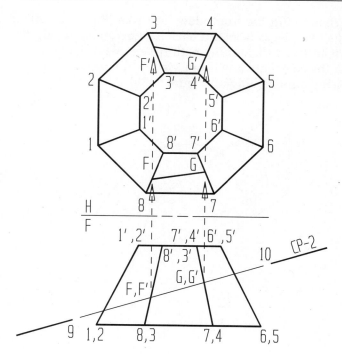

Figure 8–60 Intersection of line 9-10 with planes 8-8'-7'-7 and 3-4-4'-3'.

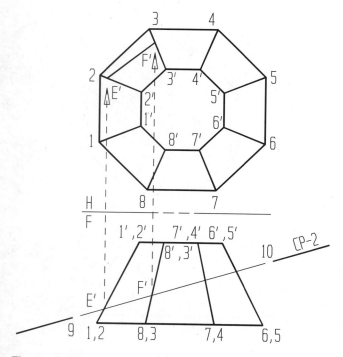

Figure 8–59 Intersection of line 9-10 with plane 2-3-3'-2'.

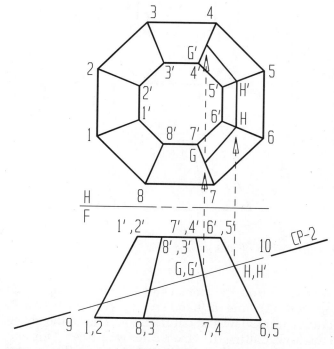

Figure 8–61 Intersection of line 9-10 with planes 7-7'-6'-6, 4'-4-5-5', and 6'-5'-5-6.

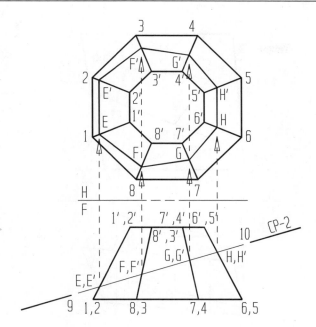

Figure 8–62   Top view visibility of cutting path E-F-G-H-H'-G'-F'-E'.

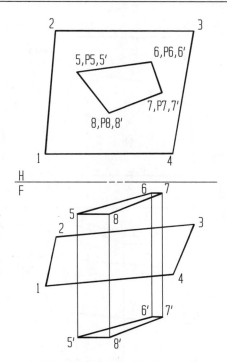

Figure 8–63   Intersection of solid and plane problem.

Intersection of a Three-Dimensional Object and an Oblique Plane

Figure 8–63 shows a skirt (plane) wrapped around a prism. The number of cutting plane lines is determined by the number of piercing points needed and will be chosen arbitrarily. How many piercing points will there be? Circle the lines in Figure 8–63 that the piercing points will be located on. There are four piercing points with only one piercing point located along each of the following lines: 5-5', 6-6', 7-7', and 8-8'.

Which lines will work as cutting planes lines? Lines 1-2, 2-3, 3-4, and 4-1 will not be used as cutting plane lines because they do not cross the prism. If line 5-6 is extended it will cross line 1-2 and 3-4; therefore, it will work as a cutting plane line (Figure 8–64). This cutting plane line, CP-1, locates points A and B in the top view. Project points A and B to the front view. Points P-5 and P-6 are located where lines 5-5' and 6-6' cross CP-1 path (A-B) respectively.

If line 6-7 (CP-2) is extended, it will cross lines 2-3 and 3-4 locating points C and D in the top view (Figure 8–65). Project points C and D to front view. If CP-1 and CP-2 are done correctly, line C-D will cross P-6. Line 7-7' locates point P-7 where it crosses CP-2 path, line C-D.

Extend line 7-8 (CP-3) to cross lines 3-4 and 1-2 locating points E and F in the top view (Figure 8–66). Project points E and F to the front view. Draw a construction line from E to F. It should cross P-7 and create P-8 where 8-8' crosses CP-3 path, F-E.

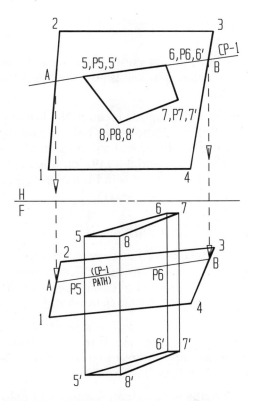

Figure 8–64   CP-1 locates piercing points P-5 and P-6.

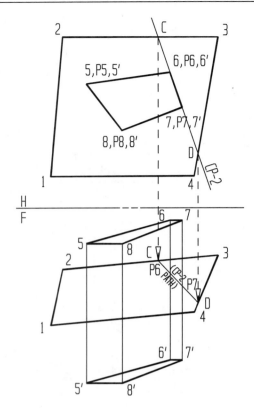

Figure 8–65 CP-2 locates piercing points P-6 and P-7.

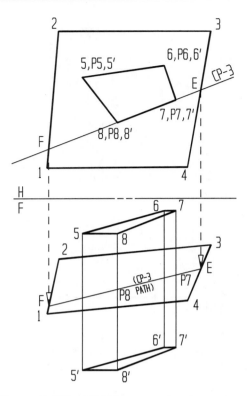

Figure 8–66 CP-3 locates piercing points P-7 and P-8.

All four piercing points could have been located by using only two cutting plane lines, line 5-6 (CP-1) and line 7-8 (CP-3).

With all four piercing points located, the last step is to determine visibility (Figure 8–67). Select crossing lines, one line from each plane, and solve for visibility—for example, test lines 1-4 and 8-8'. Which line is in front? The top view shows line 1-4 is in front of 8-8'; therefore, 8-8' is hidden from P-8 to line 1-4.

Intersection of Two Solids

We notice that each section covered in this chapter shows the same process. The only difference is that the number of piercing points increases; therefore, increasing the number of cutting planes needed.

Imagine Figure 8–68 as being a strangely shaped fake chimney sitting on a roof. The front view needs to be completed. How many intersecting points between the roof and the chimney will there be?

Sometimes it helps to sketch an isometric to help visualize what the objects looks like (Figure 8–69).

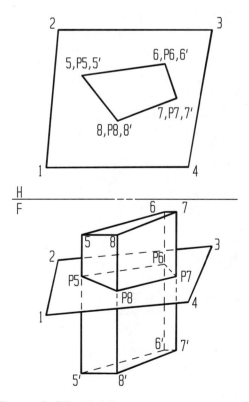

Figure 8–67 Visibility.

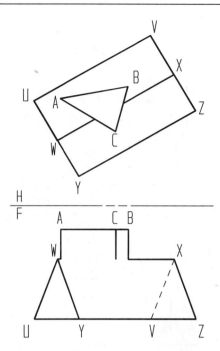

Figure 8–68 Complete the front view.

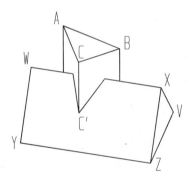

Figure 8–69 Isometric.

The intersecting points are where the chimney rests on the roof. The points are directly below lines A-B, B-C, and C-A. The key to success is to work with one plane at a time. Study the top view, before determining the cutting planes lines. Lines U-V, Y-Z, V-X, X-Z, U-X, and W-Y do not intersect the other solid; therefore, they may not be used as cutting plane lines. Plane A-B-C is higher than line W-X; therefore, line W-X may not be used as a cutting plane line. CP-1 may be arbitrarily chosen from the remaining lines, A-B, B-C, and C-A.

Figure 8–70 constructs CP-1 through points C and B in the top view. This cutting plane line will locate points 2, 3, and B' on plane U-V-W-X. This same cutting plane

line will also locate points 1, 2, and C' on plane W-X-Y-Z (Figure 8–72). The front view in Figure 8–70 shows the location of point 2 on line W-X, point 1 on Y-Z, and point 3 on line U-V, forming CP-1 path—lines 1-2 and 2-3.

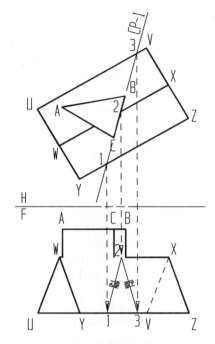

Figure 8–70 CP-1 projected to front view.

Point B' is located in the front view where line B intersects with CP-1 path, line 2-3. Point C' is located in the front view where line C intersects with CP-1 path, line 2-1 (Figure 8–71).

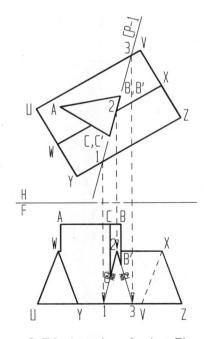

Figure 8–71 Location of points B' and C'.

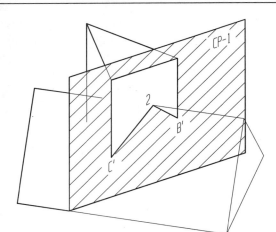

Figure 8–72 Cutting plane 1.

Figure 8–73 constructs CP-2 through points A and C. This cutting plane line will locate points 4, 5, and A' on plane U-V-W-X. This same cutting plane line will also locate points 5, 6, and C' on plane W-X-Y-Z (Figure 8–75). The front view in Figure 8–73 shows the location of point 5 on line W-X, point 4 on U-V, point 6 on Y-Z, forming CP-2 path—lines 4-5 and 5-6.

Point A' is located in the front view, where line A intersects with CP-2, line 4-5. Point C' is located in the front view, where line C intersects with CP-2—line 5-6 (Figure 8–74).

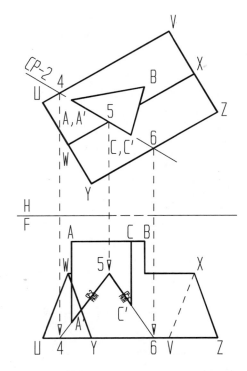

Figure 8–74 Location of points A' and C'.

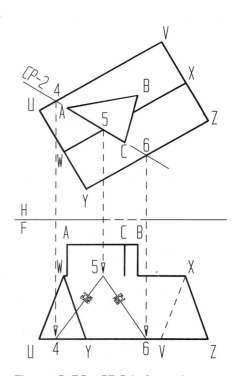

Figure 8–73 CP-2 in front view.

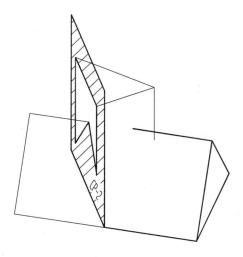

Figure 8–75 CP-2.

Connect each point using the top view to help determine what points will be connected (Figure 8–76). For example, connect A to A', B to B', and C to C'. Study the top view. Come up the back side of the roof, plane U-V-X-W, from A' to point 5. Then follow down the front side of the roof, plane W-X-Z-Y, from 5 to C' and back

up to point 2. Follow from point 2 down the back side of the roof, plane U-V-X-W, to B'. Line B' connects to A'. Now connect these points in the front view.

Determine visibility (Figure 8–77).

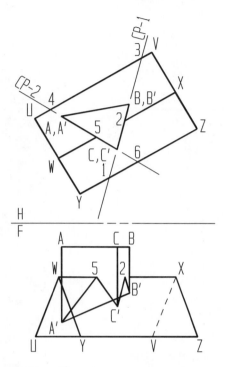

Figure 8–76 Connect endpoints.

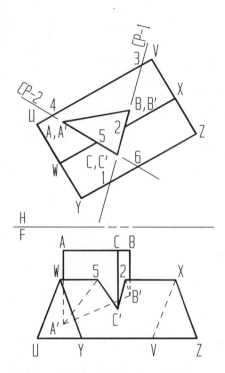

Figure 8–77 Determine visibility.

EXERCISE

Problem: A triangular piece and rectangular piece (Figure 8–78) are cast. One piece does not go through the other. Complete the front view, using the cutting plane method on Figure 8–78.

Answer

There are two ways to approach this problem. One is to use lines T-U, V-W, and R-S in the top view as cutting plane lines. The other is to use lines 1-4, 1-2, and 2-3 in the top view as cutting plane lines. Using T-U, V-W and R-S will be easier.

Figures 8–79 through 8–82 demonstrate locating where the triangular prism intersects with the rectangular plane 2-2'-3'-3.

First the known points are located, then cutting planes are used to locate the rest of the points.

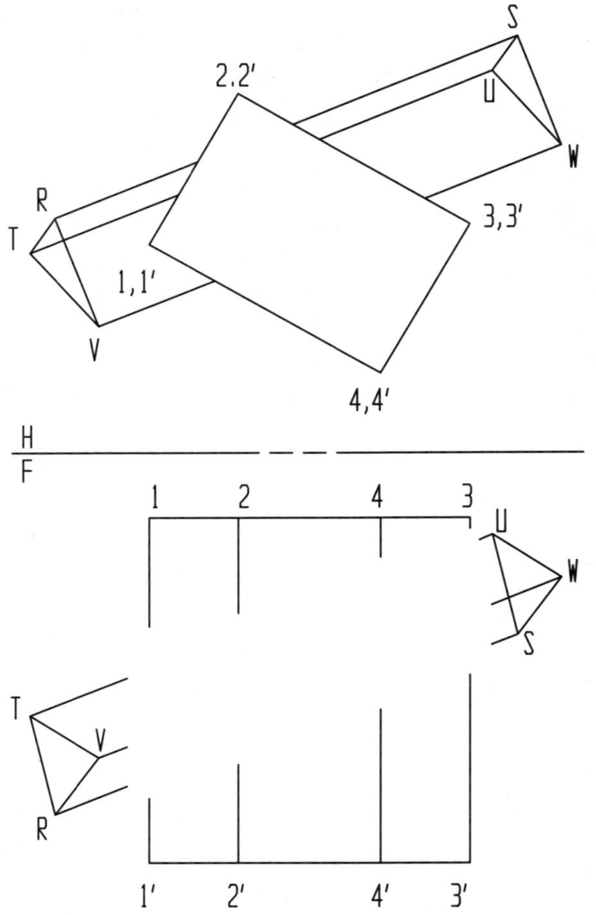

Figure 8–78 Complete the front view, using the cutting plane method.

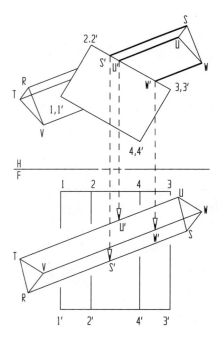

Figure 8–79 Lines R-S, T-U, and V-W intersect with plane 2-3-3'-2' at points S', U', and W'.

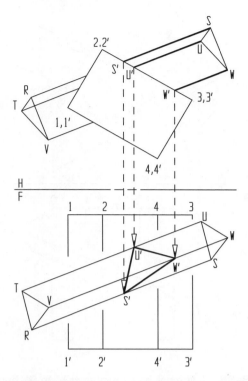

Figure 8–80 Points S', U', and W' form a triangle on plane 2-2'-3'-3.

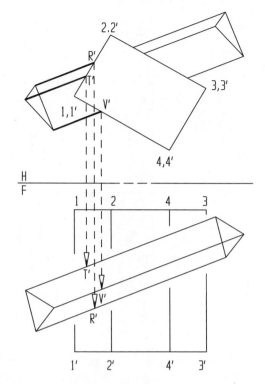

Figure 8–82 Lines R-S, T-U, and V-W intersect with planes 1-2-2'-1' and 1-4-4'-1' at points T', V', and R'.

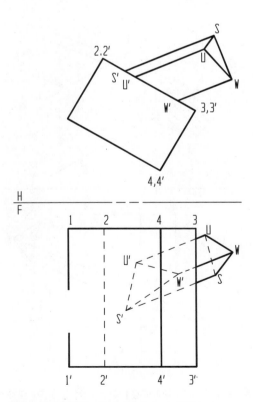

Figure 8–81 Determine visibility.

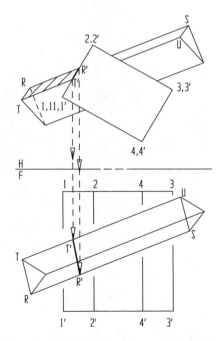

Figure 8–83 After all of the intersecting points have been located, they must be connected. Pick a point, such as R', and imagine walking around the triangular piece in the top view, connecting the points in the front view. In the top view we see that lines R-S and T-U intersect with plane 1-2-2'-1', creating line T'-R' in the front view.

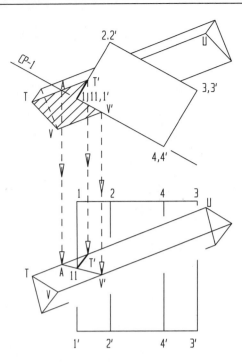

Figure 8–84 In the top view, we see line V-W intersects with plane 1-4-4'-1' creating point V'. However, the top view also shows us line T'-V' must wrap around the corner of the prism somewhere along line 1-1'. This point, 11, may be located by using the cutting plane method. In the front view, connect points T' and 11.

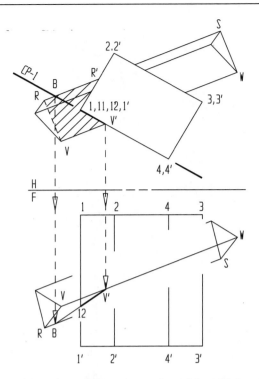

Figure 8–86 A line will be drawn from V' to R', but it must also wrap around the prism where plane R-R'-V'-V intersects with line 1-1'. The same cutting plane line that was used to locate point 11 may be used. In the front view, connect points V' and 12.

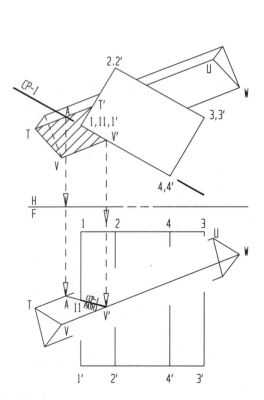

Figure 8–85 In the front view, connect points 11 and V'.

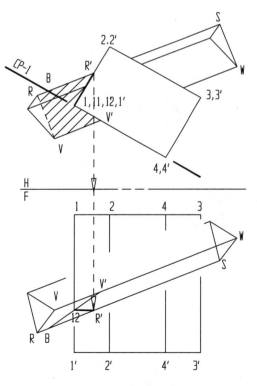

Figure 8–87 Finish closing plane R-R'-V'-V by creating line 12-R'.

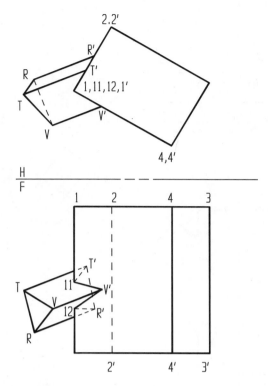

Figure 8-88 Determine visibility.

Figures 8–90 through 8–96 demonstrate using equally spaced frontal cutting plane lines to locate top view ellipse and the intersection between two cylinders. Multiple cutting planes are needed. If the cutting plane lines are spaced too far apart, projected points will appear distorted. Spacing the cutting plane lines too close together will give too many points and result in confusion. The following cutting plane lines were spaced ³⁄₁₆" apart.

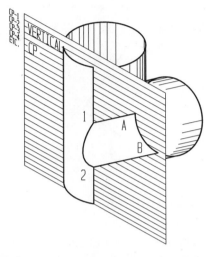

Figure 8-90 Cutting plane 1 locates points 1, 2, A and B.

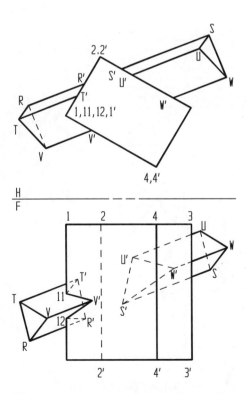

Figure 8-89 Finished orthographic for the intersection of two solids.

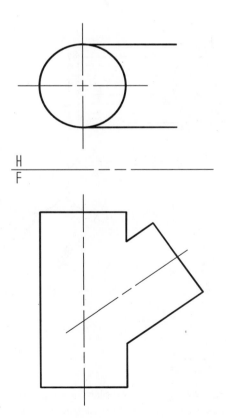

Figure 8-91 Orthographic of two intersecting cylinders.

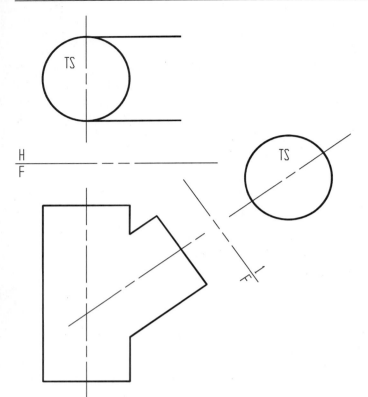

Figure 8–92 Cross sections of both cylinders must be seen true shape.

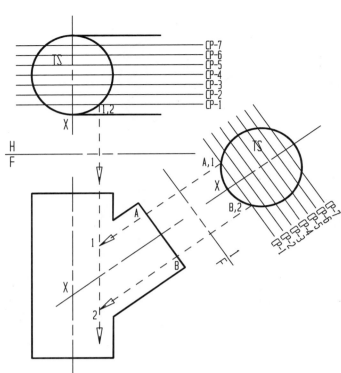

Figure 8–94 Locate the intersection of the two cylinders in the front view by projecting points to front view.

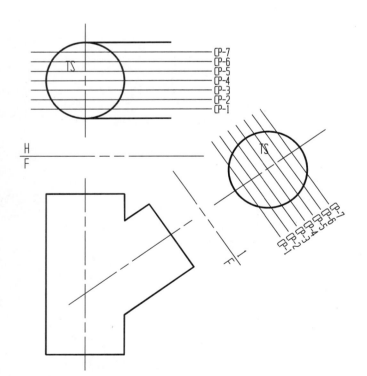

Figure 8–93 Multiple cutting plane lines are necessary in order to locate the intersection of the two cylinders and complete the top view.

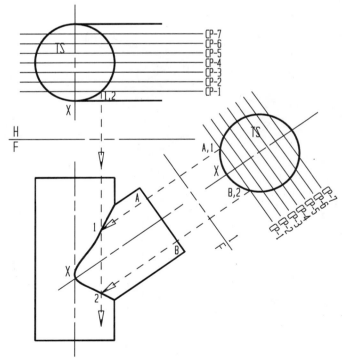

Figure 8–95 Connect points with a spline.

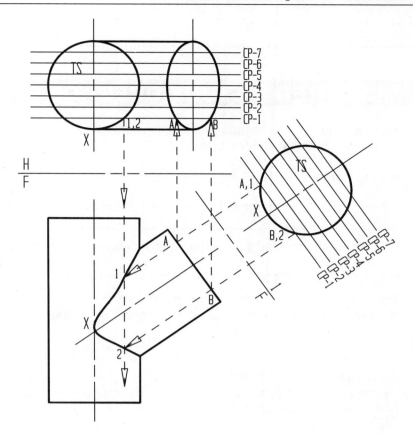

Figure 8–96 Locating intersection of two cylinders in front view and ellipse in top view using multiple cutting planes.

Name _____

Date _____

Course _____

CHAPTER 8 STUDY QUESTIONS

The Study Questions are intended to assess your comprehension of chapter material. Please write your answers to the questions in the space provided.

1. Define a piercing point.

    _____

    _____

2. What are the two methods used to find the point of intersection of a line and a plane?

    _____

    _____

3. Which method of finding the intersection of a point and an oblique plane requires only two views?

    _____

4. To determine the piercing point of a line and an oblique plane, an edge view of a cutting plane can be used. Provided that the top and front views of the given line and plane are shown, in which view can the cutting plane be indicated?

    _____

5. Using the cutting plane method, locate the point of intersection between a line and a plane for the problem below. Label all points and show visibility.

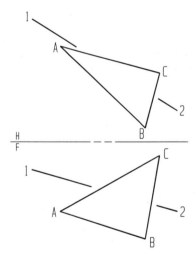

6. The solution to question 5 shows the line hidden behind the plane on the same side of the piercing point in the top and front views. Will this always be the case?

    _____

7. How many points must you find to determine the intersection between two nonparallel planes?

    _____

8. Using the edge view method, locate the line of intersection of the two planes below. Label all points and determine visibility.

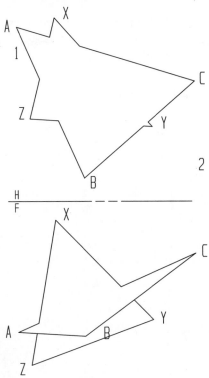

9. Using the cutting plane method, locate the line of intersection of the two planes below. Label all points and determine visibility.

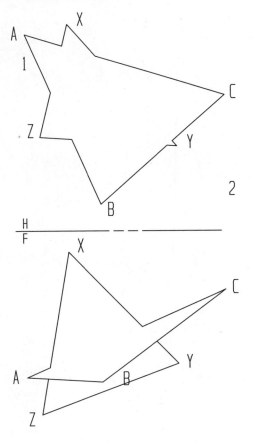

10. Questions 8 and 9 are the same problem; therefore, results should be exactly the same. Are yours the same?

The problems in this chapter are based on a variety of industrial applications. In solving these problems you should try to utilize direct approaches that get to the heart of the problem. At this analysis stage of the design process, many problems use center lines and single lines to represent the problem situation. Yet, when necessary, relevant sizes are given.

The following problems may be drawn using instruments on the page provided, created using a 2D or 3D CAD system, or a combination of drawing board and CAD. Refer to Appendix A for additional dimensions needed to solve problems three-dimensionally.

Chapter 8, Problem 1: LINE XY IS THE CENTER LINE OF A HAWSEPIPE (A PIPE THROUGH THE HULL OF A SHIP THROUGH WHICH THE ANCHOR CHAIN PASSES) INTERSECTING THE HULL AND THE DECK OF THE SHIP. THE HULL AND THE DECK ARE DRAWN AS PLANE SURFACES FOR SIMPLICITY. USING THE EDGE VIEW METHOD, FIND THE INTERSECTION OF THE HAWSEPIPE AND HULL OF THE SHIP. CHECK YOUR ANSWER USING THE CUTTING PLANE METHOD.

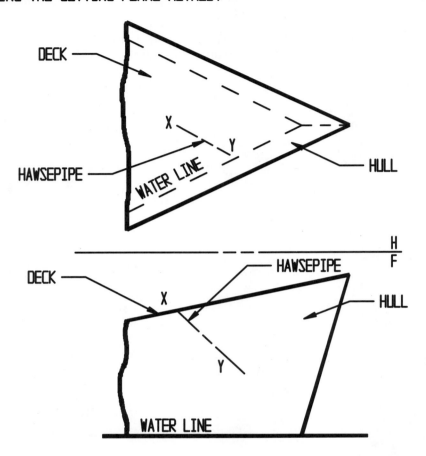

	ADG INC.			
	HAWSEPIPE			
	SIZE	DATE	DWG NO.	SCALE 1/8" = 1'0"	REV
DRW. BY			SECTION #	SHEET

Chapter 8, Problem 2: A VEIN OF ORE HAS BEEN DETERMINED BY PLANE RST. A MINE SHAFT, AB, IS DUG TOWARD THE VEIN. USE THE CUTTING PLANE METHOD TO LOCATE WHERE THE MINE SHAFT WILL INTERSECT THE VEIN OF ORE.

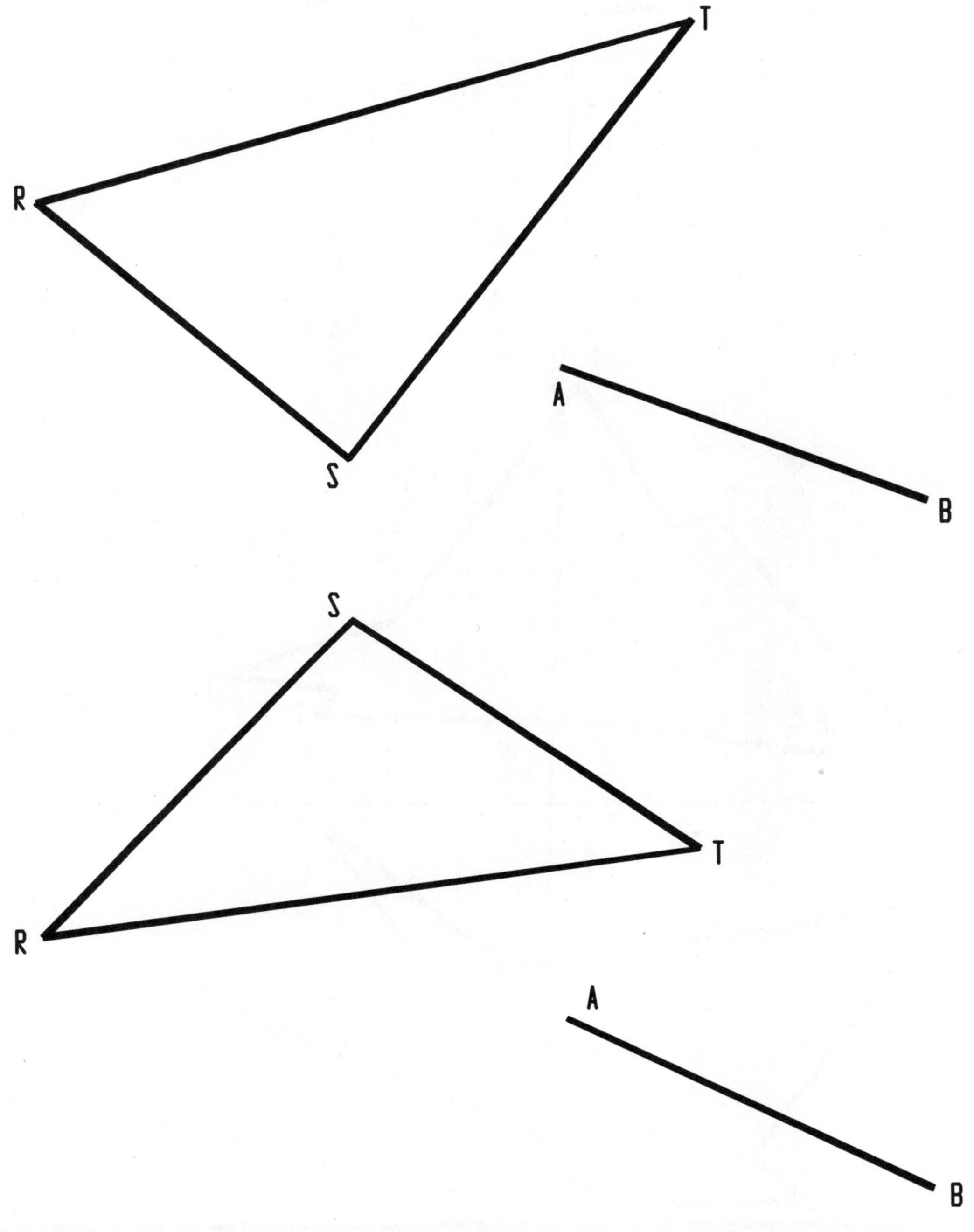

	ADG INC.			
	MINE SHAFT			
SIZE DATE	DWG NO.	SCALE 1" = 200'	REV
DRW. BY		SECTION #	SHEET

Chapter 8, Problem 3: LOCATE, LABEL, AND GIVE THE BEARING OF THE LINE OF INTERSECTION, XY,BETWEEN THE TWO GIVEN PLANES, RST AND ABC. USE THE CUTTING-PLANE METHOD, AND CHECK YOUR SOLUTION WITH THE EDGE VIEW METHOD.
SHOW THE CORRECT VISIBILITY.

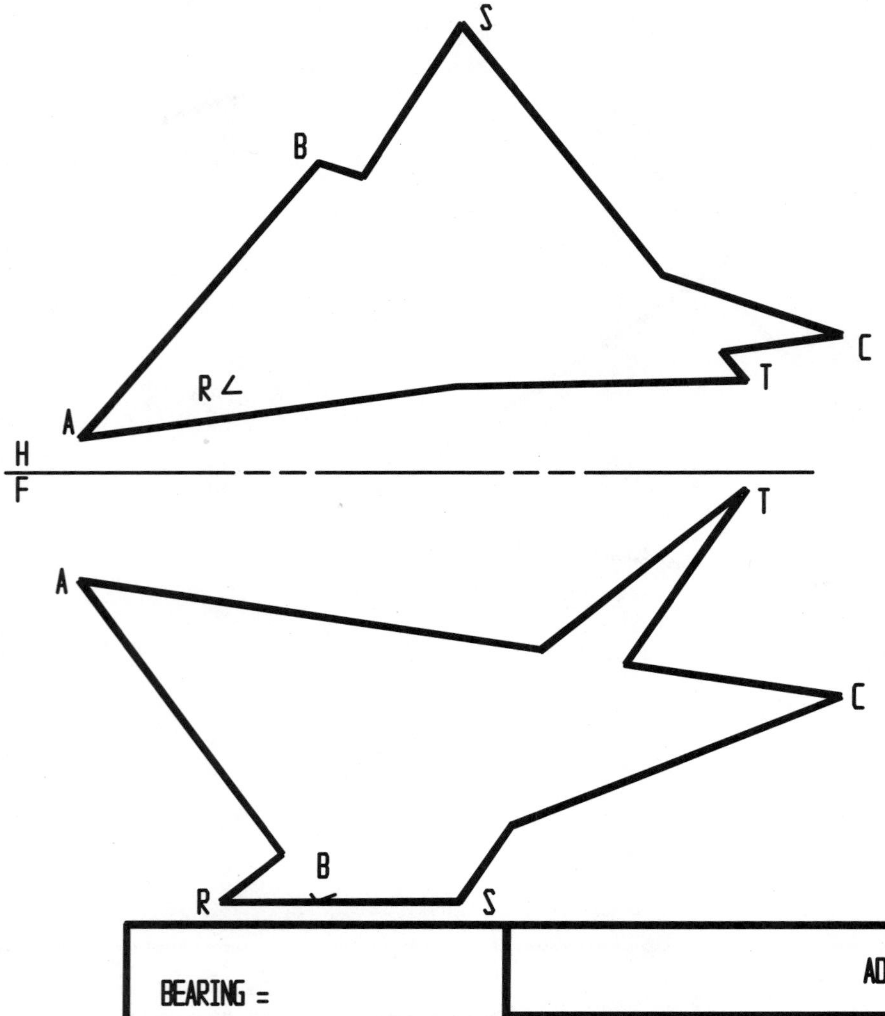

BEARING =	ADG INC.				
	INTERSECTING PLANES				
	SIZE	DATE	DWG NO.	SCALE 1" = 1"	REV
	DRW. BY		SECTION #	SHEET	

Chapter 8, Problem 4: LOCATE, LABEL, AND GIVE THE BEARING OF THE LINE OF INTERSECTION
BETWEEN PLANES RST AND WXYZ, UTILIZING THE CUTTING-PLANE METHOD. CHECK YOUR SOLUTION
WITH THE EDGE-VIEW METHOD. SHOW THE CORRECT VISIBILITY.

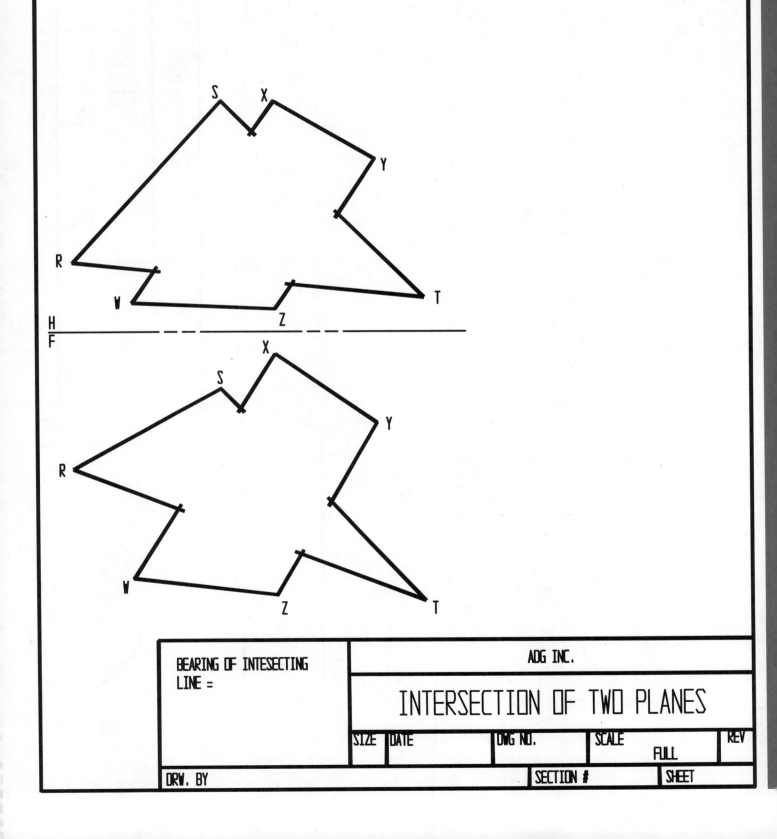

BEARING OF INTESECTING LINE =		ADG INC.			
		INTERSECTION OF TWO PLANES			
	SIZE DATE	DWG NO.	SCALE FULL		REV
DRW. BY			SECTION #	SHEET	

Chapter 8, Problem 5: FIND THE POINTS AT WHICH THE CENTER LINE, XY, OF A CULVERT INTERSECTS THE FILL EMBANKEMENTS HAVING A 1 TO 1 HORIZONTAL TO VERTICAL GRADE. USE THE CUTTING-PLANE METHOD, AND CHECK YOUR SOLUTION WITH THE EDGE-VIEW METHOD. COMPLETE THE PRINCIPAL VIEWS. REMEMBER VISIBILITY.

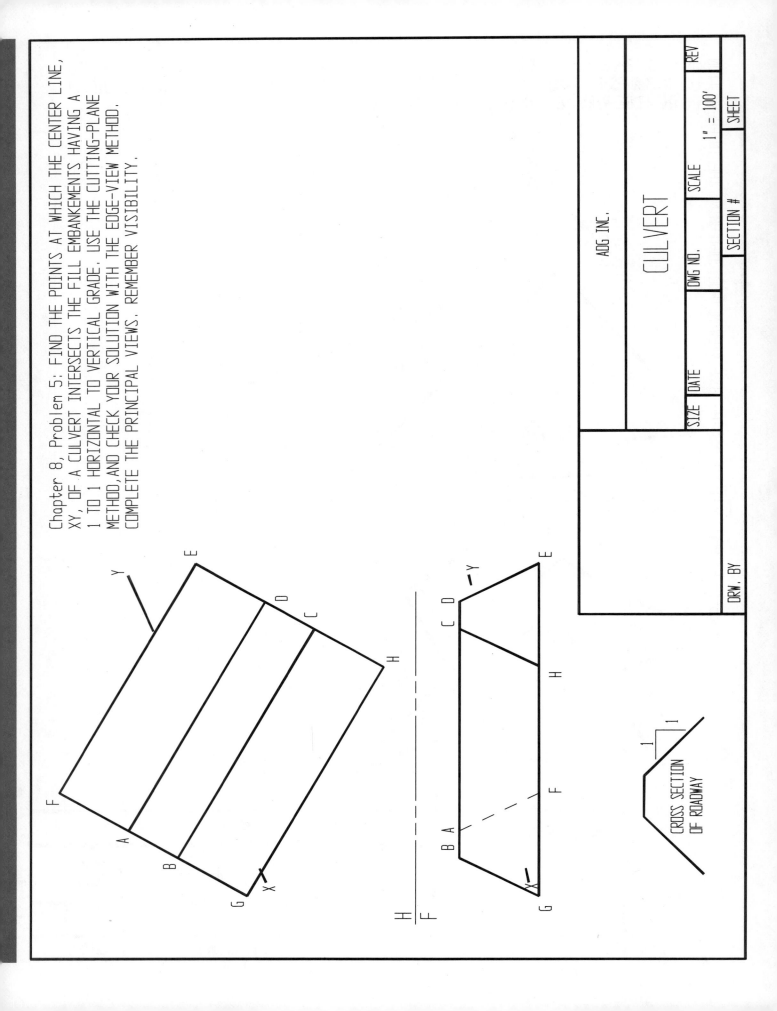

CROSS SECTION
OF ROADWAY

ADG INC.

CULVERT

SIZE	DATE		DWG NO.		SCALE	1" = 100'	REV
DRW. BY				SECTION #		SHEET	

Chapter 8, Problem 6: USING THE EDGE VIEW METHOD, DETERMINE THE INTERESECTION BETWEEN THE 16 X 16 CHIMNEY AND THE GAZEBO ROOF. SHOW THE INTERSECTION IN ALL VIEWS. CHECK YOUR ANSWER USING THE CUTTING PLANE METHOD. ALSO SHOW THE TRUE SHAPE AND SIZE OF THE OPENING TO BE CUT FOR THE CHIMNEY. DIMENSION THE SIZE OF THE OPENING AND ITS LOCATION FROM POINT A.

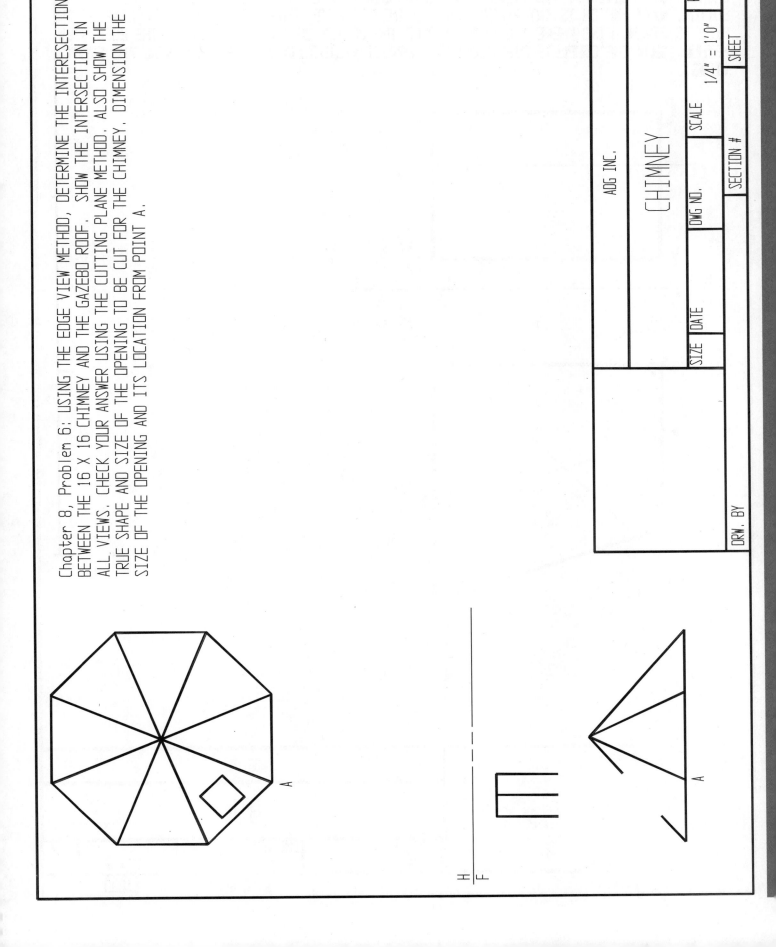

ADG INC.

CHIMNEY

SIZE	DATE	DWG NO.	SCALE 1/4" = 1'0"	REV
		SECTION #		SHEET

DRW. BY

Chapter 8, Problem 7: TWO VIEWS OF A RECTANGULAR-SHAPED BIN ARE SHOWN. IT HAS A SLOPING BOTTOM, WXYZ. WHICH IS INTERSECTED BY A 2-INCH DIAMETER PIPE, CD. THE CENTER LINE OF THE PIPE IS GIVEN. FIND WHERE THE PIPE PIERCES THE BOTTOM OF THE BIN BY USING THE EDGE VIEW METHOD. SHOW THE COMPLETE PIPE, AND ITS CORRECT VISIBILITY IN THE HORIZONTAL AND FRONT VIEWS.

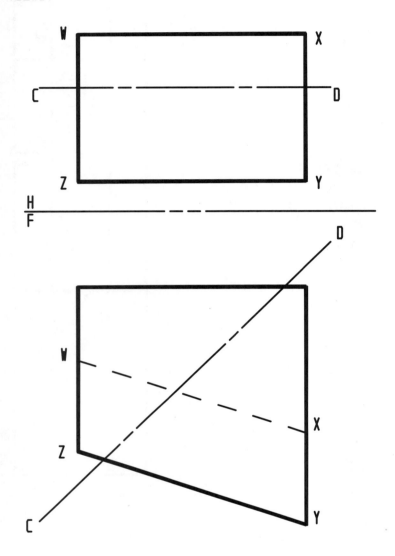

	ADG INC.			
	BIN			
SIZE	DATE	DWG NO.	SCALE 1" = 5'	REV
DRW. BY		SECTION #	SHEET	

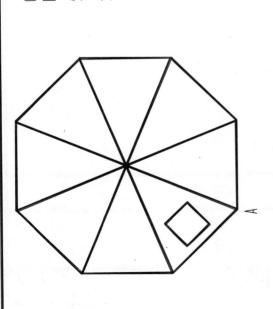

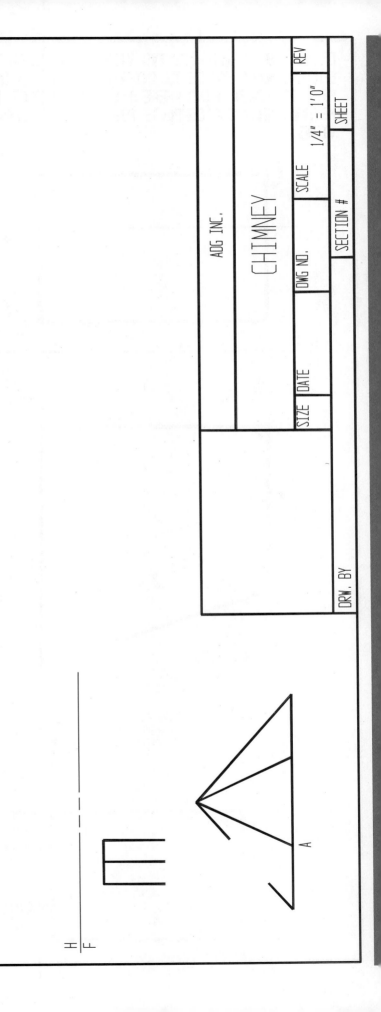

Chapter 8, Problem 8: TWO VIEWS OF A HOUSE AND A DETACHED GARAGE ARE SHOWN. THE HOUSE IS TO BE EXTENDED TO
THE GARAGE, WITHOUT CHANGING THE EXISTING ROOF SLOPES OF BOTH BUIDINGS. FIND THE INTERSECTIONS OF THE EXTENDED
ROOF OF THE HOUSE WITH THE GARAGE. USE THE CUTTING-PLANE METHOD TO FIND THE INTERSECTIONS. INCLUDE THE CORRECT
VISIBLITY IN BOTH VIEWS.

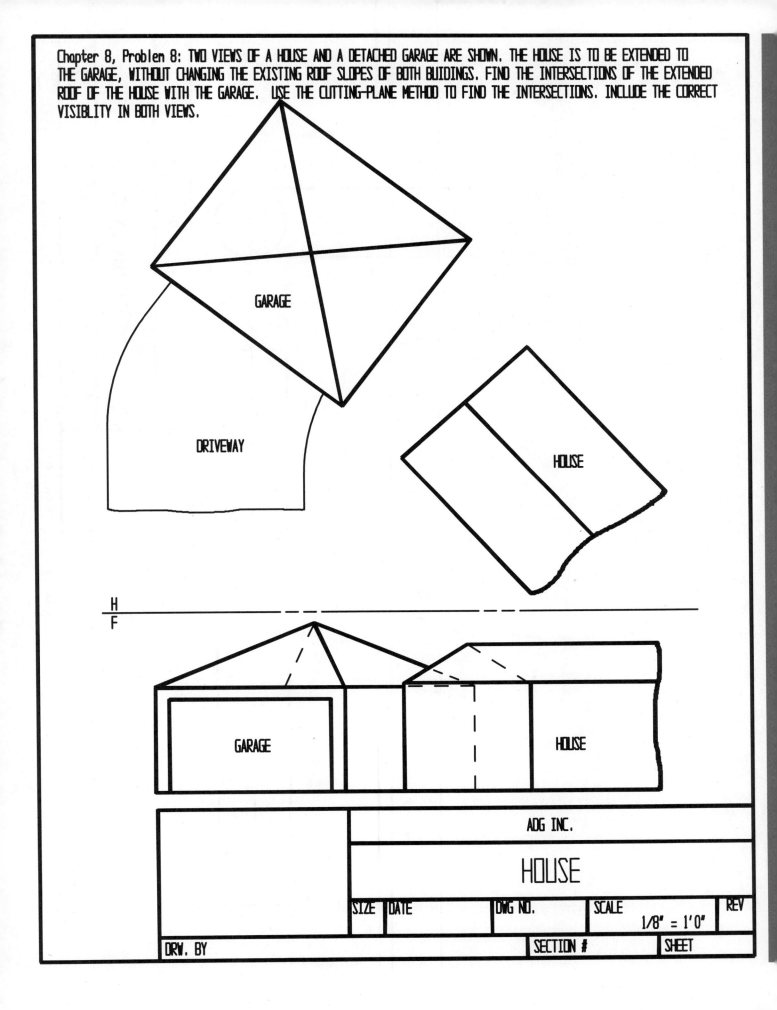

GARAGE

DRIVEWAY

HOUSE

H
F

GARAGE

HOUSE

		ADG INC.		
		HOUSE		
SIZE	DATE	DWG NO.	SCALE 1/8" = 1'0"	REV
DRW. BY			SECTION #	SHEET

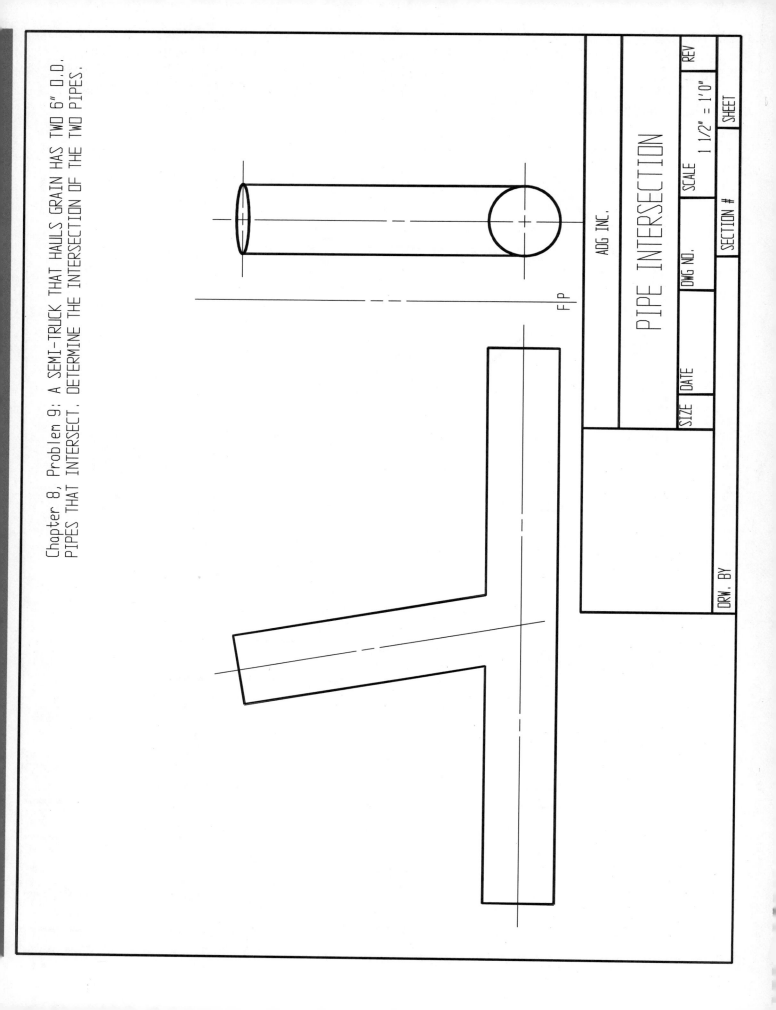

Chapter 8, Problem 9; A SEMI-TRUCK THAT HAULS GRAIN HAS TWO 6" O.D. PIPES THAT INTERSECT. DETERMINE THE INTERSECTION OF THE TWO PIPES.

F|P

ADG INC.

PIPE INTERSECTION

SIZE	DATE		DWG NO.	SCALE 1 1/2" = 1'0"	REV
	DRW. BY		SECTION #	SHEET	

CHAPTER 9

Revolution

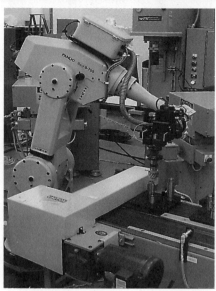

The revolution method may be used to determine the maximum and minimum reach of the robotic arm.

After completing this chapter, you will be able to:

▸ *Explain the primary difference between the change of position methods (auxiliary view methods) and the revolution method of solving descriptive geometry problems.*

▸ *Describe the four fundamental principles of revolution.*

▸ *Determine the true length and true slope of a line using the revolution method.*

▸ *Determine the true shape of a plane using the revolution method.*

▸ *Determine the true angle between a line and a plane and between two intersecting planes using the revolution method.*

There are two methods for solving many descriptive geometry problems. In the preceding chapters, all problems have been explained using the change of position method. When using this method, the drafter imagines that the object is in a fixed position. To obtain a different view, the observer moves to a new position from which to view the object. The object continues to remain stationary, as the observer moves around it. Generally, this method is more direct for many practical drawing problems, and is used almost unconsciously by most drafters.

The alternate method is **revolution**, which requires the observer to remain stationary and the object to be turned to obtain the various views. Some problems are solved more easily by revolution; therefore, students should be familiar with it. The previous chapters have shown you how important it is to view the line, plane, and/or object correctly for information not given in a principal view. With the trend shifting toward computers and the "paperless" world, understanding how to rotate a three-dimensional object is becoming more important from the designer to the machinist on the shop floor.

▼ THE PRINCIPLES OF REVOLUTION

There are four fundamental principles that must be understood before you attempt to solve any problems by revolution. Studying these principles will teach you the concepts of what actually happens in space when revolution is used (Figure 9–1).

1. When a point (C) is revolved in space, it is always revolved around a straight line used as an axis (A-B). It is important to know how the axis actually lies, before you attempt to revolve any point.

2. A point will revolve in a plane that is perpendicular to the axis, and its path is always a circle. The radius of the circle is the shortest distance from the point to the axis.

3. The circular path of the point is seen when the axis appears as a point, as seen in the top view of Figure 9–1.

4. When the axis is shown true length, the circular path of the point will always appear as an edge view at right angles to the axis, as seen in the front view of Figure 9–1.

If the axis is oblique, the view showing the axis true length must be located first (Figures 9–2 and 9–3). Sometimes a point must be rotated to a specific location. For example, drawings of a robot's arm fully extended and compressed may be needed. If a box of parts must be reached at the highest position the robot arm is to reach, knowing the direction of elevation (high or low) in the adjacent views will be important. You

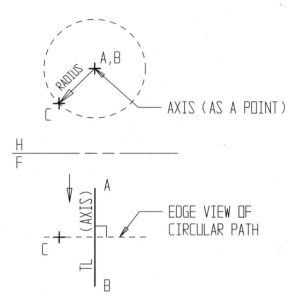

Figure 9–1 Revolution fundamental principles.

know that in an elevation view, the closer an entity is to the top view the higher it is.

To help keep track of the elevation, create an "up" arrow. A vertical line, drawn any size in the front view, with an arrowhead on the highest point will serve as an up arrow. The up arrow is for reference only; therefore, it is drawn lightly, and any size and location in the front view, as long as it is out of the way. In the following figures the highest point of the up arrow is labeled U and the lowest point is labeled P (Figures 9-2 and 9-3).

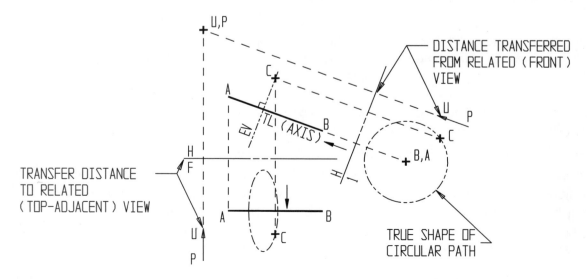

Figure 9–2 View 1 displays the true shape of the circular path that point C will travel about the axis.

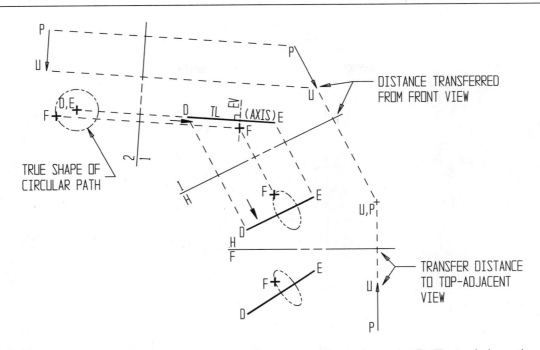

Figure 9-3 View 2 displays the true shape of the circular path that point F will travel about the axis.

The top view of a vertical line is a point. Draw it anywhere in the top view, as long as it does not interfere with the top view and is aligned with the up arrow in the front view. Using the same line of sight used to solve the problem, project the up arrow to the adjacent views.

Figures 9–2 and 9–3 demonstrate that in a top-adjacent view, the highest point of the plane is toward the top view (point U), not the top of the paper. Figure 9–3 demonstrates that the direction of elevation travels parallel to the top-adjacent reference line 1/2. The labels U and P help identify which direction is up and which is down. If you are having trouble visualizing this, photocopy or trace, then cut and fold Figure 9–3, and study the elevation (up arrow) in each view.

The same method for rotating point C in Figure 9–2 to its lowest position, will be used to rotate point F in Figure 9–3 to its highest position. Create a construction line parallel to the up arrow through the center of the true-shape view of the circular path (Figures 9–4 and 9–5). Using the up arrow as a guide to which endpoint of the construction line is higher than the other, rotate the point along the circular path until the desired height is obtained. Project the point to the adjacent view.

Because the point can travel only on the circular path, it will be located somewhere along the edge view of the circular path. (Because the circular path is true shape in one view, the edge view will appear perpendicular to the line of sight in the adjacent view.) A point that is

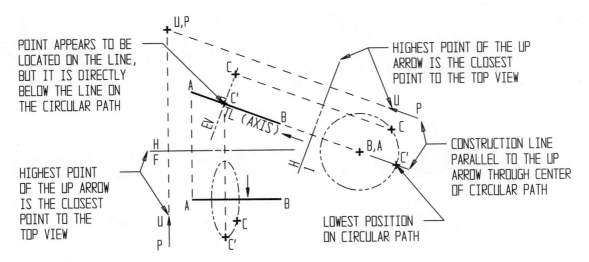

Figure 9-4 Rotated point C-C', projected back to front view.

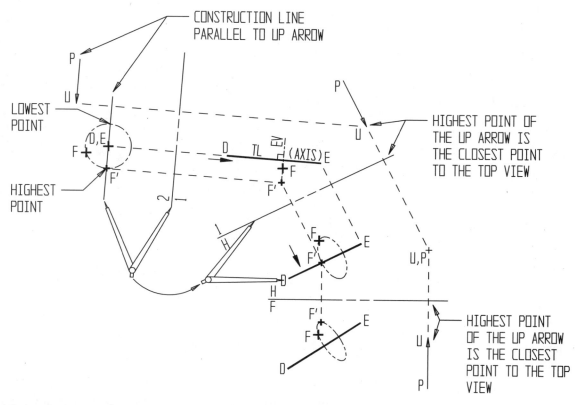

Figure 9–5 Rotated point F-F', projected to front view.

rotated to its highest or lowest position will appear, in the top view, as if it is located on the axis line. Figure 9–4 demonstrates point C revolved to the lowest possible position. Figure 9–5 demonstrates point F revolved to its highest possible position.

HELPFUL HINTS

To help visualize this, hold your pencil in a horizontal position and look directly down on top of it. Using a finger on your other hand, draw an imaginary circle around the pencil. Because you cannot see height in the top view, when your finger is at it's highest or lowest position, it will appear to be located on the pencil.

EXERCISE

Problem: Rotate point I in Figure 9–6 to its lowest position. Sketch each step on the figure provided for that step. The answer will be given in the next figure. Label TL, axis, etc.

The axis must appear true length. Locate it and label it TL (axis). Add an up arrow for future reference.

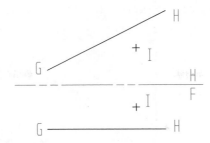

Figure 9-6 If the axis line is true length, label it
TL and sketch an up arrow for future reference.

Locate the point view of the axis and true shape of the circular path the point will follow.

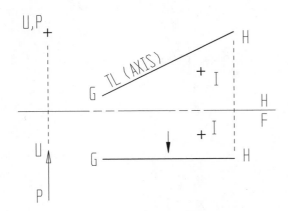

Figure 9-7 TL (axis) labeled and up arrow added.
Locate the point view of the axis and the true shape
of the circular path the point will follow.

Rotate point I to its lowest position and label it I'.

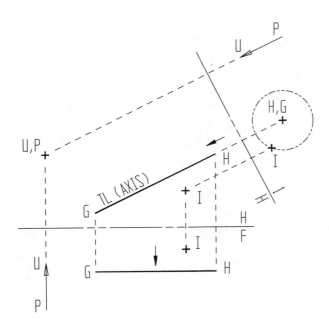

Figure 9-8 True shape of circular path located.
Rotate point I to its lowest position

Create the edge view of the circular path in the adjacent view and project I'(Figure 9–9).

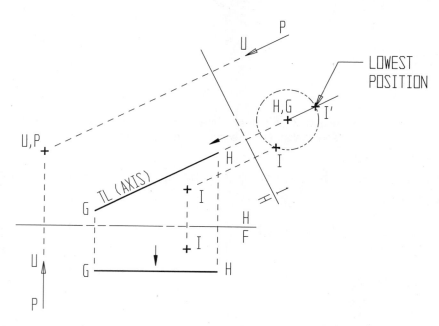

Figure 9–9 Point I rotated to its lowest positions. Project I' to the adjacent (top) view.

Project point I' to the front view (Figure 9–10).

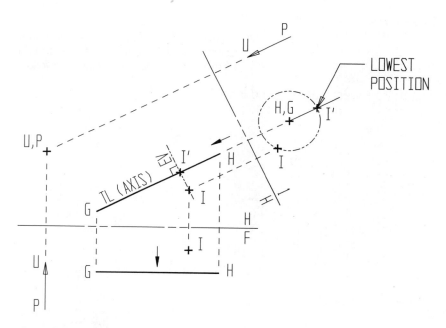

Figure 9–10 Point I rotated to its lowest position. Project point I' to the front view.

Answer

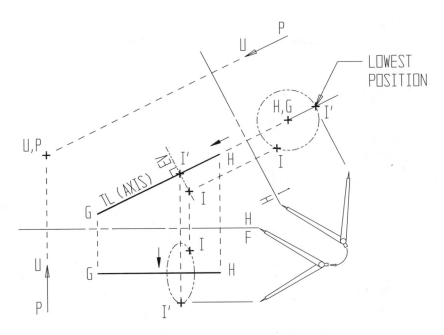

Figure 9–11 Point I revolved to it's lowest position, I'.

▼ THE TRUE LENGTH OF A LINE

In Chapter 4 you learned the rules for locating a true-length line.

I. Determine **line of sight.**

    A. Locate **true-length line.**

        1. Is the true-length line **in** one of the **principal planes**?

            a. If the line of sight is perpendicular to the object line in one view, it will appear true length in the adjacent view and can be measured.

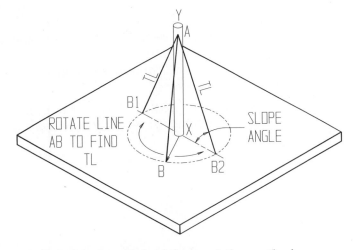

Figure 9–12 Isometric of the revolution method.

When a line revolves about an axis, its position is changed, but its length remains the same. Examine Figure 9–12, which shows guy wire A-B rotated about a vertical pole (axis—X-Y). Point A is fixed, while point B is rotated. The edge view of the rotation path must stay perpendicular to the axis, which means point B must stay on the original plane. In other words, the height of point B does not change. You can see that guy wire A-B is shown true length when it is rotated into a position parallel with the frontal plane, and that its true slope is also shown in this position.

Figure 9–13 shows the orthographic projection of the pole and the guy wire shown in Figure 9–12. Identify

the axis first. If an existing line can double as an axis—use it. If not, lightly construct one. Pole X-Y is already true length in a principal (front) view, so the center line of it may be used as the axis. In the top view, the pole X-Y (axis) appears as a point.

When the guy wire A-B revolves about the pole, point B follows the indicated circular path. Point A, being directly on the axis, stays fixed. When B is revolved to either position B1 or B2, the guy wire is rotated so that it is perpendicular to the line of sight for the adjacent (front) view. When in this position, the adjacent (front) view of A-B will be true length. In the front view, point B moves perpendicular to the axis to either of the

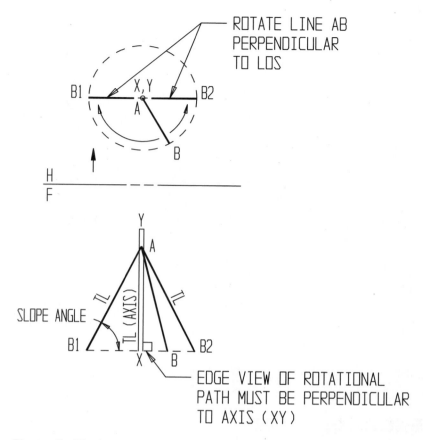

Figure 9–13 Locating true-length of line A-B using the revolution method

revolved positions, B1 or B2. Either position gives the desired true-length solution.

Figure 9–14 shows an oblique line. Solve for the true-length view of line X-Y using the revolution method.

First construct and label an axis. The axis may be true length in either view. Figure 9–15 demonstrates the axis (A-B) constructed through endpoint X. Point X becomes fixed on the axis.

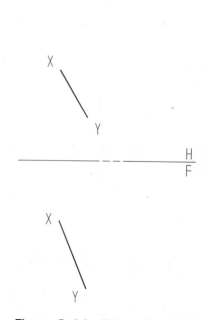

Figure 9–14 Oblique line X-Y.

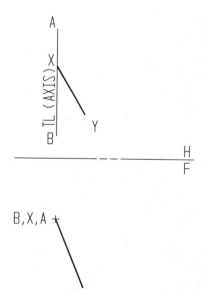

Figure 9–15 Identify axis. It must be viewed true length in one view and a point view in the adjacent view.

Rotate the other end of the line (Y) about the point view of the axis (A-B) until it creates a line perpendicular to the line of sight for the adjacent view (Figure 9–16). Label the rotated endpoint Y'. Note the circular path is true shape in the view where the axis (A-B) appears as a point.

Project Y' to the adjacent view. Remember the projection line must be parallel to the line of sight for the adjacent (top) view (Figure 9–17). Create the edge view of the circular path point Y' followed in the adjacent (top) view. *Remember the edge view of the circular path must be perpendicular to the axis*, because it is true shape in

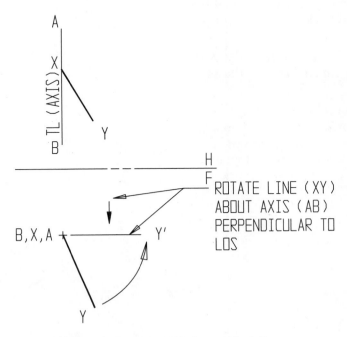

Figure 9–16   Rotate line X-Y about axis A-B.

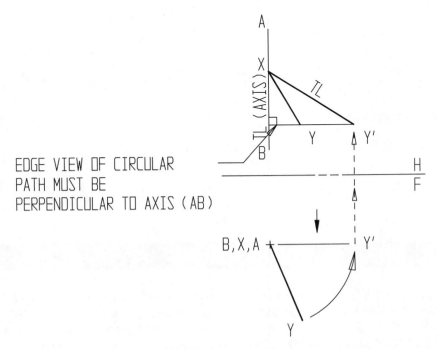

Figure 9–17   Project Y' to adjacent (top) view locating true length of line X-Y', using the revolution method.

the adjacent (top) view, and through point Y. Point Y' is located where the edge view of the circular path and the projection line for Y' intersect. This step is very important. The point cannot be moved off the plane it is resting on. In other words, the depth for Y' must be the same as the depth for Y. The distance in this view (top view) between the fixed point (X) and rotated point (Y') is true-length and can be measured.

Would the answer have been correct if you placed the axis through point Y in the top view? Try it and see (Figures 9–18 through 9–20).

Revolve point X about the point view of the axis (A-B) until it is perpendicular to the line of sight for the adjacent (top) view.

Project rotated line (X'-Y) to the adjacent (top) view. Remember the circular path must be viewed as an edge and perpendicular to the axis in the adjacent (top) view. Figure 9–20 demonstrates the answer is correct whether the axis is placed through endpoint Y or endpoint X.

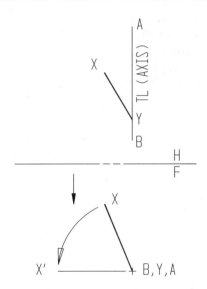

Figure 9–19 Revolve point X about the point view of the axis (BA).

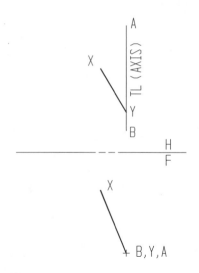

Figure 9–18 Axis through point Y.

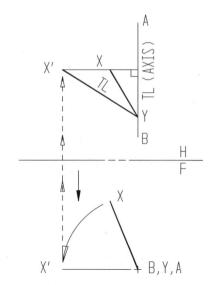

Figure 9–20 Project rotated line (X'-Y) to the adjacent view.

EXERCISE

Problem: Figure 9–21 is the same problem as in Figure 9–14. In Figures 9–17 and 9–20 the true-length of X-Y is found in the top view. Locate the true length of line X-Y in the front view with the axis through point Y. Sketch each step on the figure provided for that step. The answer will be given in the next figure. Label TL, axis, etc.

As per the instructions, locate the axis through point Y so line X-Y will be rotated to true length in the front view.

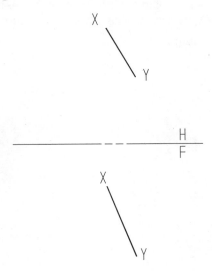

Figure 9–21 Locate axis which line X-Y will rotate about.

If the line X-Y is to be true length in the front view, it must rotate about the point view of the axis; therefore, the point view of the axis must be in the top view (Figure 9–22).

Review

I. Determine **line of sight.**

    A. Locate **true-length line.**

    B. Locate the **point view of the true-length line.**

        1. The line of sight must be drawn parallel to the true-length line.

Rotate line X-Y about the axis.

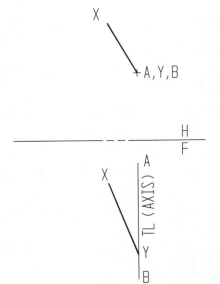

Figure 9–22 Rotate line (X-Y) about the point view of the axis (A-B).

Rotate line X-Y about the point view of the axis (A-B) in the top view until it is perpendicular to the line of sight for the adjacent (front) view (Figure 9–23). It does not matter if the line is rotated clockwise or counterclockwise. In Figure 9–23, the line is revolved in a counter-clockwise direction.

Project rotated line (X-Y) to the adjacent view and label TL

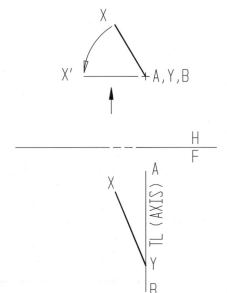

Figure 9–23   Project rotated line X'-Y to the front view.

Draw projection lines to the adjacent (front) view, parallel to the line of sight for the adjacent (front) view. Locate the edge view of the circular path X', which is located perpendicular to the axis and through point X. Point X' is located where these two lines intersect.

Review

I. Determine **line of sight.**

    A. Locate **true-length line.**

    B. Locate the **point view of the true-length line.**

    C. Locate the **edge view of a plane** (all points fall in a line).

        1. The line of sight must be drawn parallel to a true-length line that lies in the plane. (Edge view of circular path is seen in the front view.)

    D. Draw the **true shape of a plane.**

        1. The line of sight must be drawn perpendicular to the edge view of a plane. (True shape of circular path is seen in the top view.)

Answer

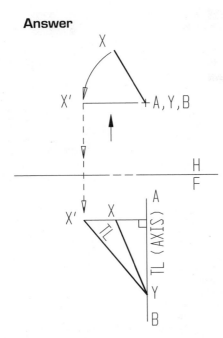

Figure 9–24   Line (X-Y) rotated into true length position (X'-Y).

The Slope of a Line

In Chapter 4 you also learned that the slope angle of a line will be seen in its true size when the line appears true length in an elevation view. The fact that the slope angle will be seen only in an elevation view is true for both the revolution method and the auxiliary-view method. Figure 9–25 shows again the vertical pole X-Y, and guy wire A-B.

When B is rotated to B1, perpendicular to the line of sight to view the front view, wire A-B appears true

length in the front view. This is an elevation view and as such shows the true slope of the wire (Figure 9–25).

If you create the point view of the axis line (C-D) in the front view and rotate point B about it, as shown in Figure 9–26, the true length appears in the top view, as B2. Line A-B2 may be true length, but it is not true length in an elevation view; therefore, the true slope of line A-B is not seen. Be careful when studying this problem, line C-D is the axis—not X-Y.

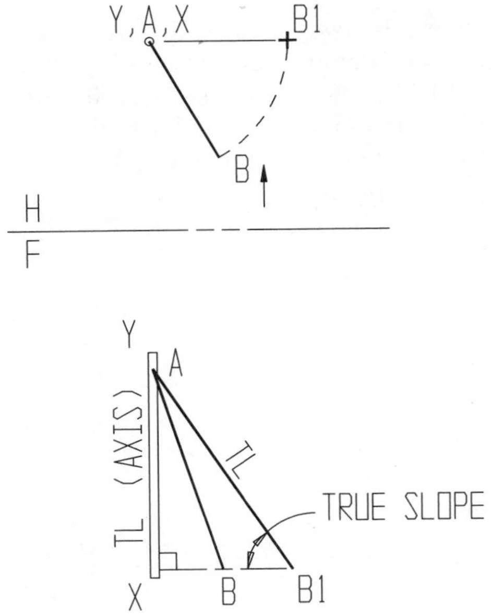

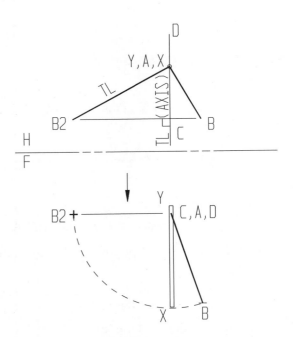

Figure 9–25   True length of line A-B is located in the front view; therefore, slope may be measured.

Figure 9–26   True length of line A-B is rotated about axis C-D. Because true-length line A-B2 is not true length in an elevation view, the slope cannot be measured.

EXERCISE

Problem: Look at Figures 9–17, 9–20 and 9–24. In which figure could you correctly measure slope? Measure the slope angle for line X-Y, in that figure.

Answer

True slope can be measured in Figure 9–24, where line X'-Y is true length in an elevation (front) view (Figure 9–27). The slope angle is 51°.

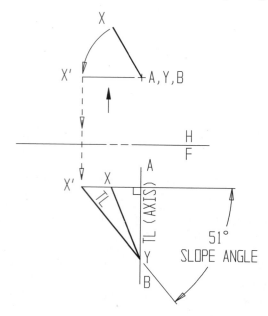

Figure 9–27 Slope is located when the line appears true length in an elevation view.

▼ THE TRUE SHAPE OF A PLANE

In Chapter 6 you learned how to locate the true shape of a plane.

I. Determine **line of sight.**

A. Locate **true-length line.**

B. Locate the **point view of the true-length line.**

C. Locate the **edge view of a plane** (all points fall in a line).

1. The line of sight must be drawn parallel to a true-length line that lies in the plane.

D. Draw the **true shape of a plane.**

1. The line of sight must be drawn perpendicular to the edge view of a plane.

When a plane is seen true shape in a view, it *must* appear as an edge view in all views adjacent to the true-shape. In Figures 9–28 and 9–29, the isometric and orthographic views of a wedge block are shown respectively. The problem is to revolve plane A-B-C until it appears true shape in the top view.

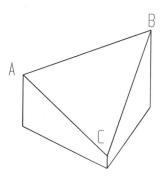

Figure 9–28 Isometric of wedge block.

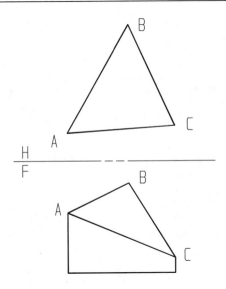

Figure 9–29 Orthographic of wedge block.

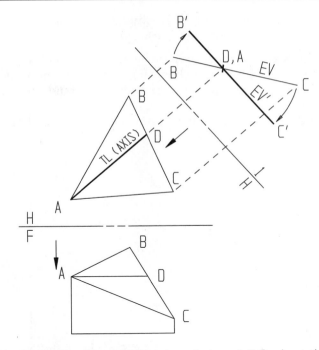

Figure 9–31 Revolve edge view of plane A-B-C, about the axis A-D.

The edge view of the plane will be revolved about an axis that lies in the plane and that appears true length in the view where the true shape is to be shown.

Study Figure 9–30, note the true-length line (A-D) for plane A-B-C, is in the top view and is used as the axis line. View 1 is then drawn showing the axis A-D as a point, and plane A-B-C as an edge.

The edge view of plane A-B-C is revolved about axis A-D until it is perpendicular to the line of sight (Figure 9–31). Points B and C follow the circular paths shown to their new positions, B' and C'.

Create projection lines from points B' and C' parallel to the line of sight used to observe view 1 (Figure 9–32). In the adjacent top view, points B and C must move perpendicular to the axis, along the edge view of circular path, to their new positions B' and C' until they intersect with the correct projection lines. If the points do not move perpendicular to the axis line they will no longer be resting on the edge view of the circular path,

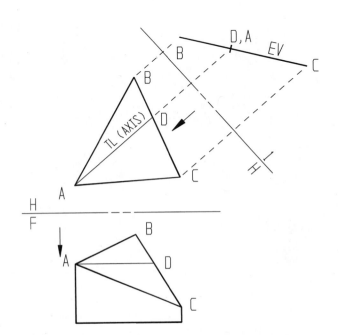

Figure 9–30 Draw the edge view of plane A-B-C.

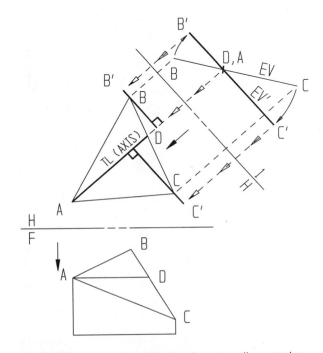

Figure 9–32 Project revolved points to adjacent view.

resulting in an incorrect answer. Remember points A and D lie on the axis and do not move.

Plane A-B'-C' is the true shape required (Figure 9–33). It is not conventional to show the revolved plane in the front view. Notice that the chief accomplishment of revolution is to eliminate drawing one view. As you continue through the problems in this chapter, you should

continue to see that the process of revolution is only a substitute for the final auxiliary view.

The axis may appear as a point anywhere along the edge view. Figure 9–33 demonstrates the axis (A-D) in the middle of the edge view. Figure 9–34 demonstrates the axis (A-B) through an endpoint of the edge view.

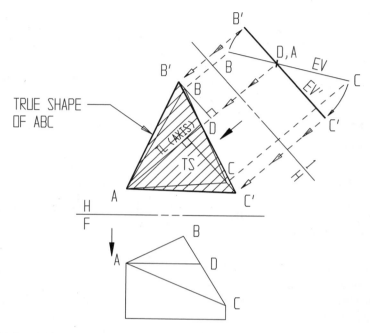

Figure 9–33 True shape of plane located using the revolution method.

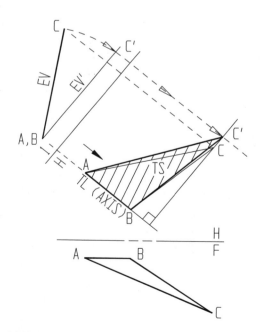

Figure 9–34 Locating true shape of plane A-B-C, using the revolution method with line A-B as the axis.

EXERCISE

Problem: Locate the true shape of plane R-S-T in the top view, using the revolution method. Sketch each step on the figure provided for that step. The answer will be given in the next figure. Label TL, axis, etc.

Locate the edge view of plane R-S-T (Figure 9–35).

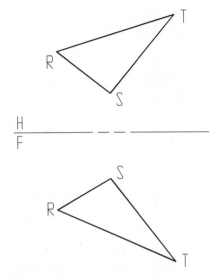

Figure 9–35 Locate the edge view of plane R-S-T.

Revolve the edge view of plane R-S-T about the axis (R-U) in view 1 (Figure 9–36).

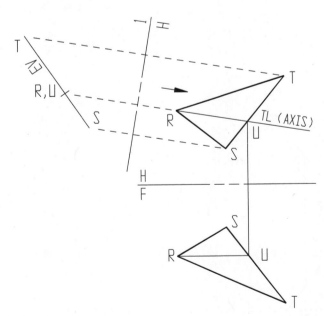

Figure 9–36 Edge view of plane R-S-T located in view 1. Revolve edge view about the axis.

The circular path for point T is the radius from the point view of the axis (R-U) to point T. The circular path for point S is the radius from the point view of the axis (R-U) to point S.

Project points T' and S' to the adjacent view (Figure 9–37).

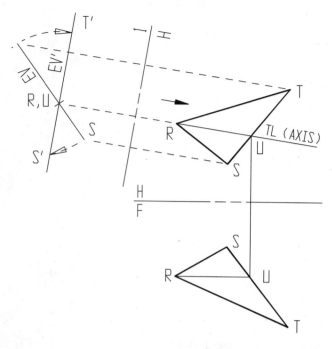

Figure 9–37 Points T and S revolved about the point view of axis (R-U). Project points T' and S' to the adjacent (top) view.

Locate and label true-shape plane (Figure 9–38).

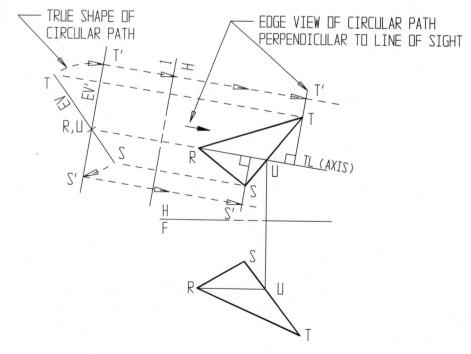

Figure 9–38 Points S' and T' projected to top view. Locate and label true shape.

Review

I. Determine **line of sight.**

   A. Locate **true-length line.**

   B. Locate the **point view of the true-length line.**

   C. Locate the **edge view of a plane** (all points fall in a line).

      1. The line of sight must be drawn parallel to a true-length line that lies in the plane.

   D. Draw the **true shape of a plane.**

      1. The line of sight must be drawn perpendicular to the edge view of a plane. (True shape of circular path is seen in the top view.)

Answer

Sectioning is added only to help you visualize true-shape plane R-S'-T'.

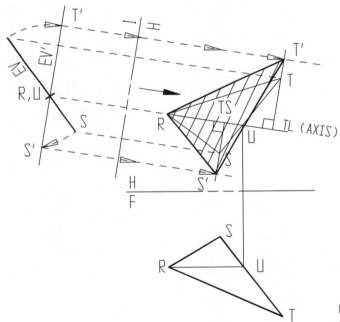

Figure 9–39 True shape of plane R-S-T using the revolution method.

▼ THE ANGLE BETWEEN A LINE AND A PLANE

The angle between a line and a plane may be seen where the plane appears as an edge and the line is revolved to true length in the same view. Figure 9–40 shows an A-frame, X-Y-Z, and two guy wires, A-Y and B-Y, fastened to the frame at Y. It is necessary to find the true angle between line A-Y and the plane of the A-frame, X-Y-Z.

To help visualize the A-frame and guy wires, build a 3D model using pencils, pipecleaners, or other materials. Pick the model up and rotate it until plane X-Y-Z appears as an edge. Now rotate the model, keeping the plane in the edge position, until line A-Y appears true length. This is where the true angle can be measured.

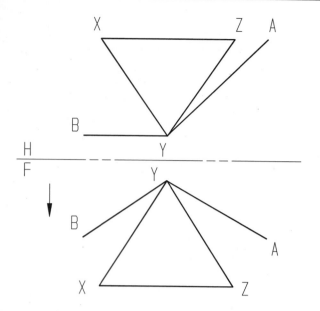

Figure 9–40 Orthographic views of an A-frame and two guy wires.

Whether moving the model or drawing the object, the edge view of plane X-Y-Z must be located first. View 1 in Figure 9–41 shows plane X-Y-Z as an edge. The line (A-Y) must also be true length in the same view that plane (X-Y-Z) appears as an edge view. Revolving the line (A-Y) in the view where the circular path and the plane (X-Y-Z) appear true shape will accomplish this. View 2 shows the plane in true shape.

The axis of revolution for line A-Y must be created first. The axis must be positioned through point Y, because it is part of the line and the plane. Axis M-N is a point in view 2, and true length in view 1 (line M-N) (Figure 9–42).

Figure 9–43, view 2 demonstrates point A revolved about the axis (M-N), until it is perpendicular to the line of sight creating line A'-Y. This creates the circular path true shape in the same view plane X-Y-Z is true shape.

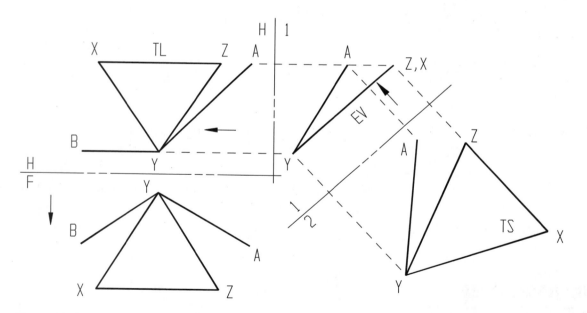

Figure 9–41 True shape of plane X-Y-Z in view 2.

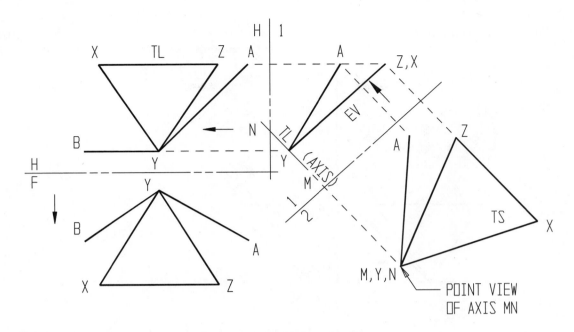

Figure 9–42 Determine axis (M-N) needed to rotate line (A-Y).

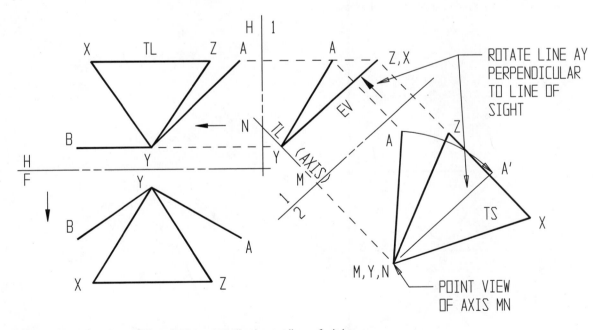

Figure 9–43 Rotate line A-Y perpendicular to line of sight.

Project the line (A'-Y) back to the adjacent view (view 1), where it will appear true length (A'-Y) (Figure 9–44).

Since the line (A'-Y) now appears true length in the view that shows the plane as an edge, the true size of the angle between them can be seen (Figure 9–45).

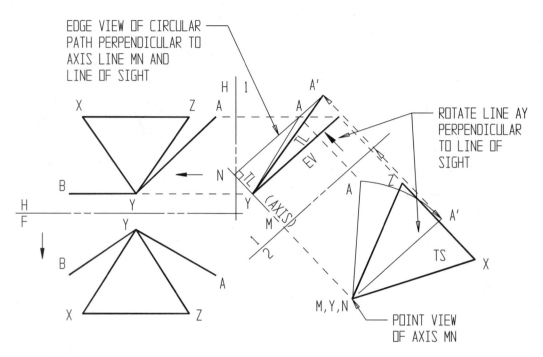

Figure 9–44   Line A'-Y is true length in view 1.

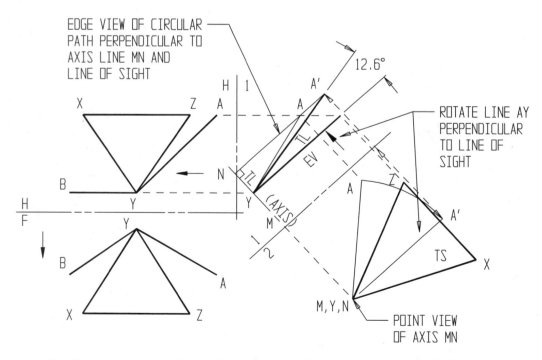

Figure 9–45   The angle between line A-Y and plane X-Y-Z is found where line A'-Y is true length and plane X-Y-Z appears as an edge.

EXERCISE

Problem: Figure 9–45 solves for the angle between the A-frame X-Y-Z, and guy wire A-Y. Solve for the true angle between the A-frame X-Y-Z, and guy wire B-Y, fastened to the frame at Y. Draw right on Figure 9–46. The answers will be given in Figures 9–47 through 9–50. Label TL, axis, etc.

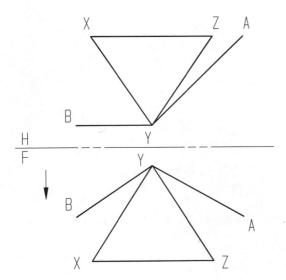

Figure 9–46 Determine the true angle between the A-frame (plane X-Y-Z) and guy wire B-Y.

Answer

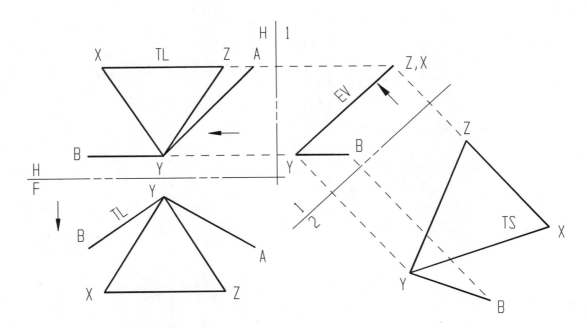

Figure 9–47 Locate the true shape of plane X-Y-Z.

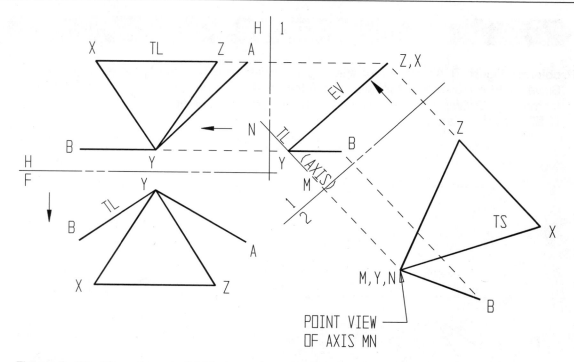

Figure 9–48 Place the axis (M-N) through a common point between the line and plane.

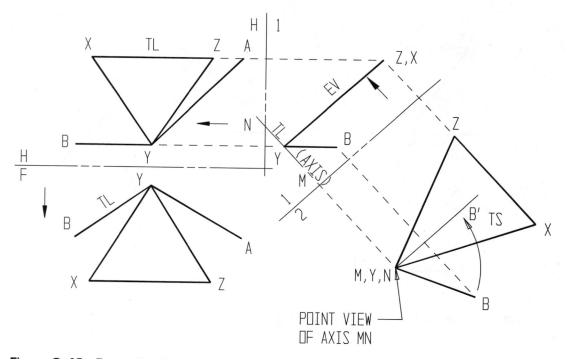

Figure 9–49 Rotate line B-Y, creating true-length line B'-Y.

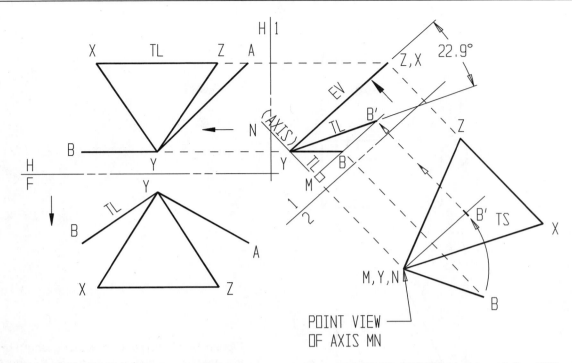

Figure 9–50 Measure the angle between the edge view of plane X-Y-Z and true-length line B'-Y.

▼ THE ANGLE BETWEEN TWO PLANES (DIHEDRAL ANGLE)

The dihedral angle between two planes may be seen on a plane that is perpendicular to the line of intersection of two given planes. If we create a perpendicular cutting plane and find its intersection with each plane, then revolution may be used to show the true dihedral angle. Figure 9–51 illustrates a hopper for which angles will be bent.

It is necessary to find the dihedral angle between planes A-B-C-D and A-B-F-E. View 1, in Figure 9–52, is constructed so the line of intersection, A-B, is shown true length.

Next, draw a cutting plane perpendicular to intersection line A-B, assumed to be an edge in view 1. Although the cutting plane must be perpendicular to the line of intersection, A-B, it may intersect A-B at any desired location along its length. The cutting plane intersects plane A-B-C-D along X-Y and plane A-B-F-E along Y-Z, creating a new plane X-Y-Z. Project plane X-Y-Z to the top view (Figure 9–53).

Revolve the cutting plane, in view 1, so that plane X-Y-Z will be perpendicular to the line of sight for adjacent (top) view. Point X was randomly chosen for the axis line (M-N) to intersect (Figure 9–54).

Project Y' and Z'. Draw projection lines parallel to line of sight from points Y' and Z'. Point Y' is located in the top view where the Y' projection line intersects with a construction line drawn perpendicular to axis M-N through point Y. Point Z' is located in the top view where the Z' projection line intersects with a construction line drawn perpendicular to axis M-N through point Z (Figure 9–55).

Plane X-Y'-Z' will be true shape in the top view. Line X-Y' is the edge view of plane A-B-C-D and X-Z' is the edge view of plane A-B-E-F. These two edge views allow us to measure the dihedral angle between the two planes in the top view. Again, notice in comparison to the auxiliary view method, that the process of revolution simply eliminates the need for the final auxiliary view.

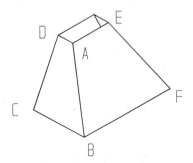

Figure 9–51 Isometric of hopper.

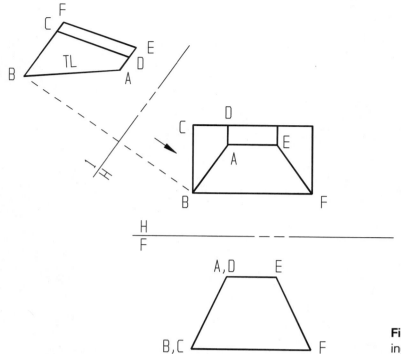

Figure 9–52 View 1 shows true length of connecting line (A-B) between planes A-B-C-D and A-B-E-F.

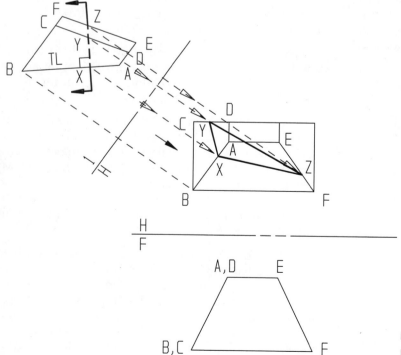

Figure 9–53 Axis M-N is needed to rotate plane X-Y-Z.

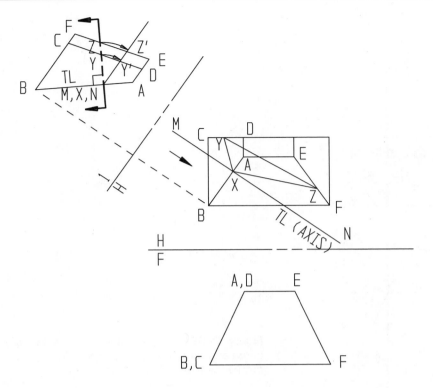

Figure 9–54 Cutting plane locates plane X-Y-Z in view 1 and the horizontal plane.

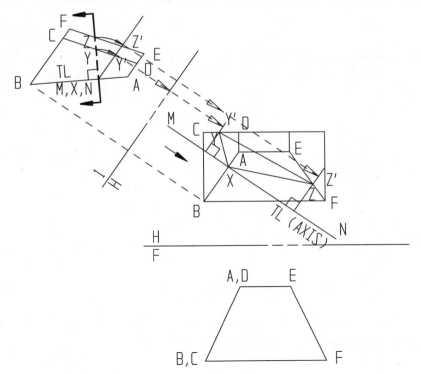

Figure 9–55 Project Y' and Z' to top view.

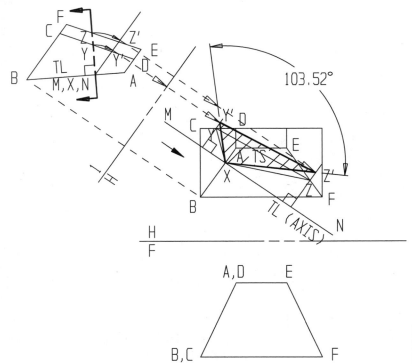

Figure 9–56 The dihedral angle is found between two planes in the view where they both appear as edge views.

EXERCISE

Problem:

In Chapter 7 you found the dihedral angle between the face (plane 1-2-3-4) and the flank (plane 1-4-5-6) of a tool bit (Figure 9–57). Now try the revolution method on the same problem and see if the answer is the same.

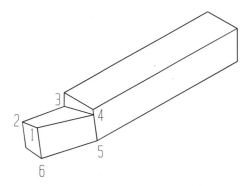

Figure 9–57 Tool bit.

Draw right on the figures. The solution to each step will be in the following figure.

Locate the appropriate true-length line.

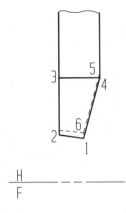

Figure 9–58 Find the dihedral angle between planes 1-2-3-4 and 1-4-5-6 using the revolution method. Locate TL of intersecting line between planes 1-2-3-4 and 1-4-5-6.

The intersecting, connecting, line between plane 1-2-3-4 and 1-4-5-6 is line 1-4. Figure 9–59, view 1, shows the true length of line 1-4.

Create a cutting plane line needed to create plane X-Y-Z and project plane X-Y-Z to the adjacent view.

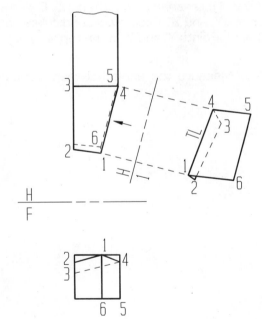

Figure 9–59 True length of the connecting line 1-4, between planes 1-2-3-4 and 1-4-5-6. Create cutting plane line.

The cutting plane line may be placed anywhere along true-length line 1-4, as long as it is perpendicular to the true-length line. Newly created plane X-Y-Z is projected back to the adjacent (top) view.

Revolve plane X-Y-Z into true shape, so the angle between planes 1-2-3-4 and 1-4-5-6 can be measured.

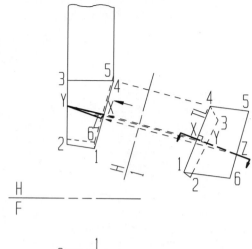

Figure 9–60 Cutting plane line is perpendicular to true-length line, 1-4, creating plane X-Y-Z. Rotate points Y and Z and project them to the top view.

The cutting plane line you created is the edge view of plane X-Y-Z. Rotate plane X-Y-Z perpendicular to the line of sight used to observe this view (view 1), creating points Y' and Z'. Create projection lines parallel to the line of sight through points Y' and Z'. Create perpendiculars to the axis through points Y and Z in the adjacent (top) view. Points Y' and Z' in the top view are where these line intersect.

Connect points, creating true shape of plane X-Y'-Z'. Measure the angle between planes 1-2-3-4 and 1-4-5-6.

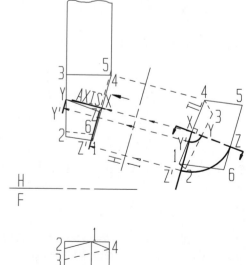

Figure 9–61 Points Y' and Z' are located. Create TS.

Answer

Figure 9–62 locates the dihedral angle between planes 1-2-3-4 and 1-4-5-6. This was accomplished by taking a slice out of the middle of the object, plane X-Y-Z, and rotating it until it is true shape. The answer is the same as in Chapter 7; therefore, either method may be used or to redraw the necessary views at a larger scale.

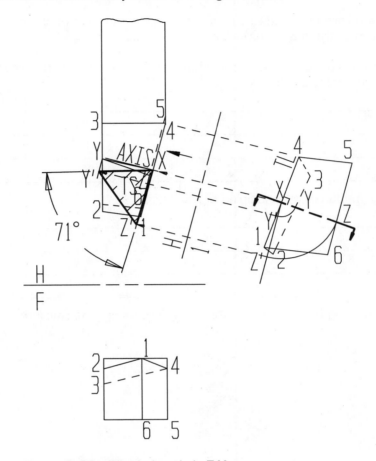

Figure 9–62 Dihedral angle is 71°.

Sometimes a problem becomes too complicated and cluttered using the revolution method, resulting in a plotting error. When this occurs, it is better to use the auxiliary view method.

CHAPTER 9 STUDY QUESTIONS

The Study Questions are intended to assess your comprehension of chapter material. Please write your answers to the questions in the space provided.

1. Explain the difference between the change of position and the revolution method of solving descriptive geometry problems.

    _____

    _____

    _____

    _____

2. List the four fundamental principles of revolution.

    _____

    _____

    _____

    _____

3. Draw the true-length of line 1-2 in the front view using, the revolution method.

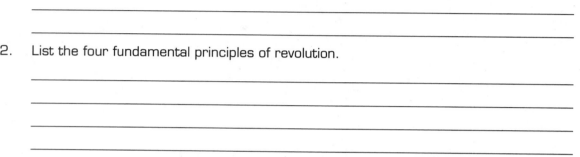

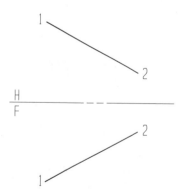

4. List the conditions that must be met when finding the slope or grade of a line using the revolution method.

    _____

    _____

    _____

    _____

5. Locate the true shape of plane A-B-C in the front view, using the revolution method.

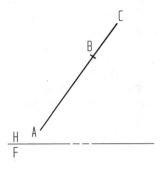

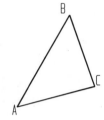

6. How must a plane appear in any view adjacent to its true shape view?

    _____

7. If a plane is to be revolved into true shape in a top view, how will the axis of rotation appear in the top view?

    _____

8. What conditions must be met to show the true angle between a line and a plane using the rotation method?

    _____

    _____

9. Explain in your own words how you find the dihedral angle between two planes using the revolution method.

    _____

    _____

10. In general, what is accomplished by the process of revolution as compared to the auxiliary view method?

    _____

    _____

The problems in this chapter are based on a variety of industrial applications. In solving these problems you should try to utilize direct approaches that get to the heart of the problem. At this analysis stage of the design process, many problems use center lines and single lines to represent the problem situation. Yet, when necessary, relevant sizes are given.

The following problems may be drawn using instruments on the page provided, created using a 2D or 3D CAD system, or a combination of drawing board and CAD. Refer to Appendix A for additional dimensions needed to solve problems three-dimensionally.

Chapter 9, Problem 1: REVOLVE POINT A ABOUT LINE 1-2 TO ITS LOWEST POSITION.

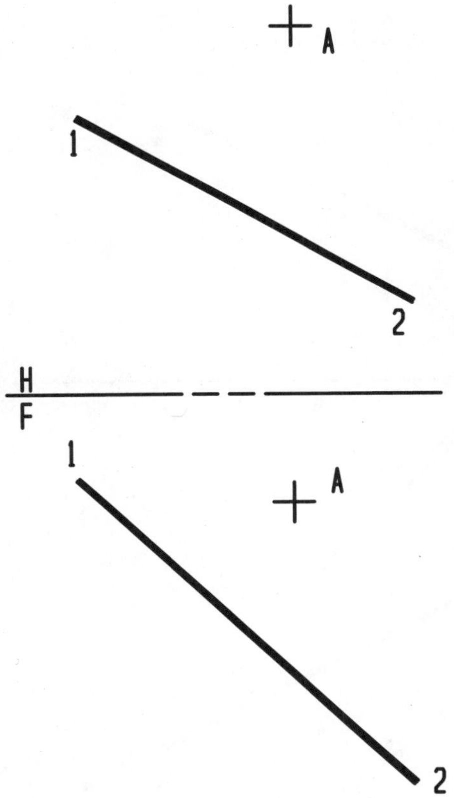

+ A

1

2

H
F

1

+ A

2

ADG INC.

REVOLUTION OF A POINT

	SIZE	DATE	DWG NO.	SCALE 1/4"=1'0"	REV
DRW. BY				SECTION #	SHEET

Chapter 9, Problem 2: REVOLVE POINT C TO
ITS HIGHEST POSITION.

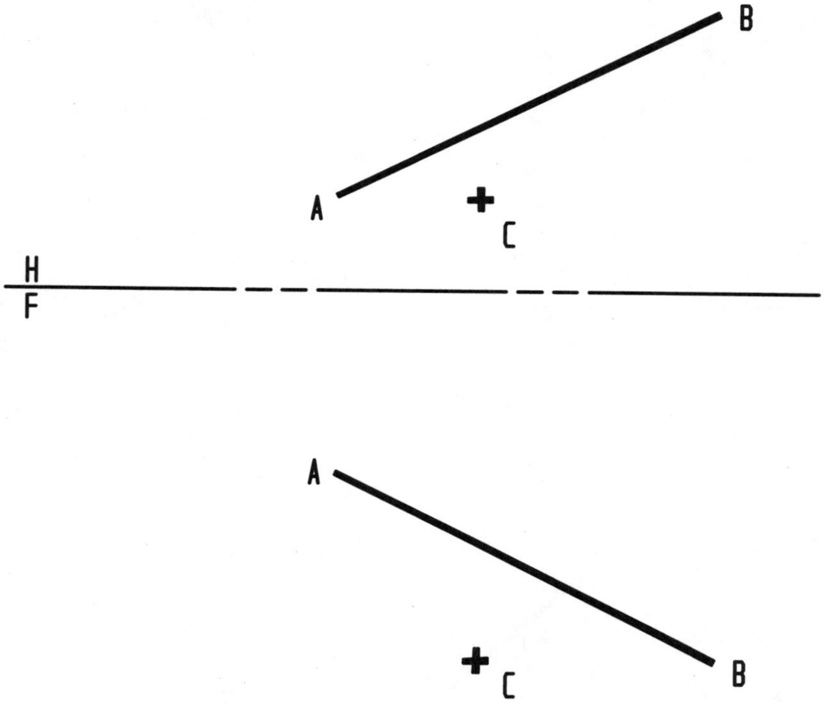

B

A
+
C

H
F

A

+
C
B

ADG INC.

REVOLUTION OF A POINT

SIZE	DATE	DWG NO.	SCALE	REV
A			1' = 1'	
DRW. BY KEY			SECTION #	SHEET 1 OF 1

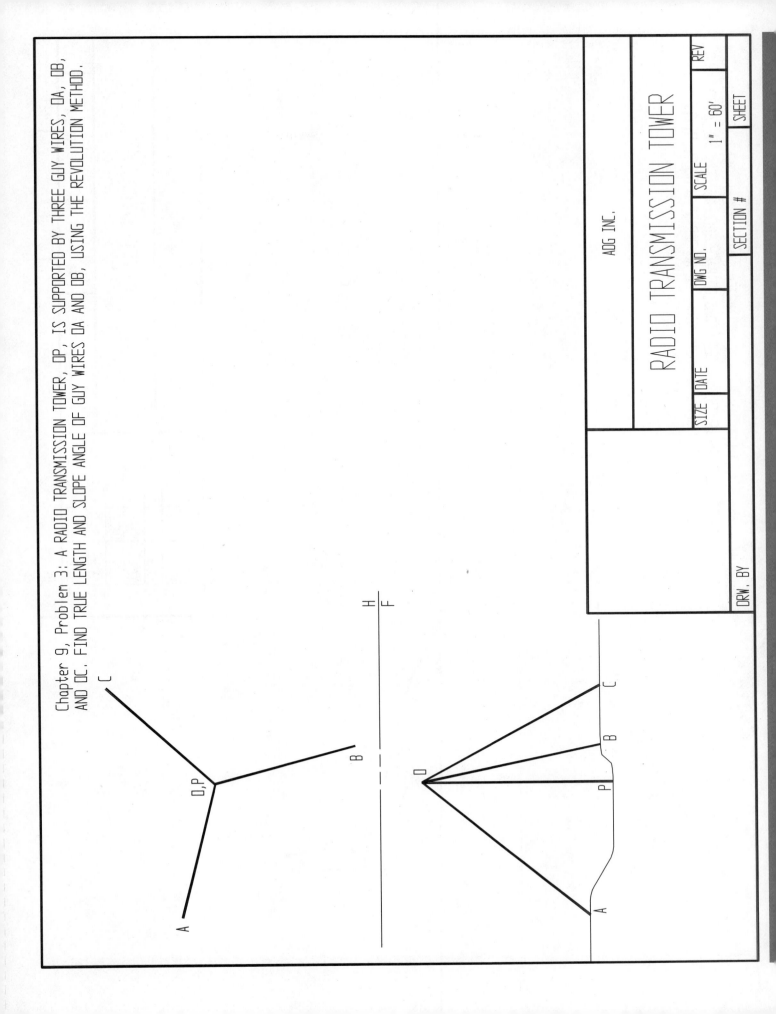

Chapter 9, Problem 3: A RADIO TRANSMISSION TOWER, OP, IS SUPPORTED BY THREE GUY WIRES, OA, OB, AND OC. FIND TRUE LENGTH AND SLOPE ANGLE OF GUY WIRES OA AND OB, USING THE REVOLUTION METHOD.

C

O,P

A

B

H
F

O

P

A

B

C

ADG INC.

RADIO TRANSMISSION TOWER

REV

SCALE   1" = 60'

SHEET

DWG NO.

SECTION #

SIZE   DATE

DRW. BY

Chapter 9, Problem 4: A LENGTH OF 2-INCH DIAMETER STEEL PIPE IS BENT AS SEEN BELOW.
DETERMINE THE TRUE LENGTH OF THE CENTER LINES OF EACH SEGMENT. WX, XY, AND YZ AND
INDICATE THE SLOPE OF EACH SEGMENT. DETERMINE THE NUMBER OF DEGREES IN THE
BEND AT X AND AT Y. USE THE REVOLUTION METHOD.

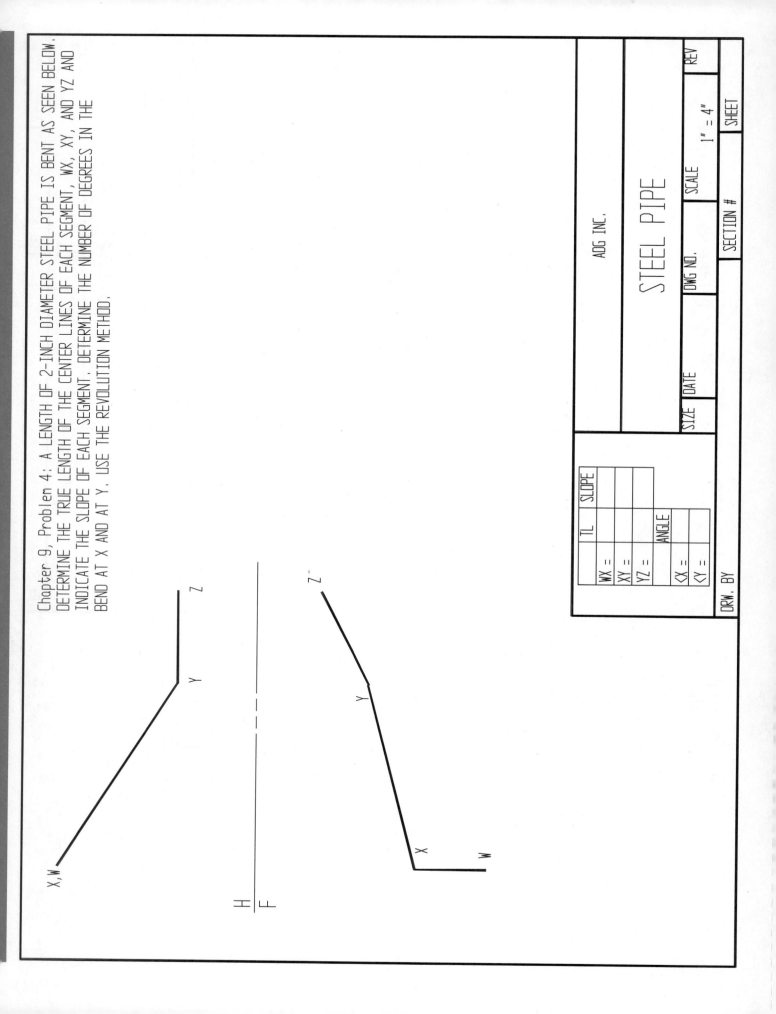

ADG INC.

STEEL PIPE

		SLOPE
TL		
WX =		
XY =		
YZ =		
	ANGLE	
<X =		
<Y =		

SIZE	DATE	DWG NO.	SCALE 1" = 4"	REV

DRW. BY		SECTION #	SHEET

Chapter 9, Problem 5: SURFACE WXYZ MUST BE REVOLVED INTO THE HORIZONTAL POSITION FOR MILLING PURPOSES.
THROUGH WHAT ANGLE MUST THE BLOCK BE ROTATED ABOUT WX IN ORDER TO BRING SURFACE WXYZ INTO THE HORIZONTAL PLANE?
SHOW THE TRUE SHAPE OF THIS SURFACE ON THE TOP VIEW.

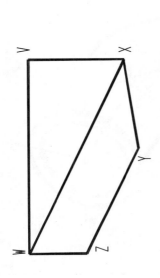

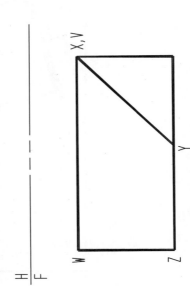

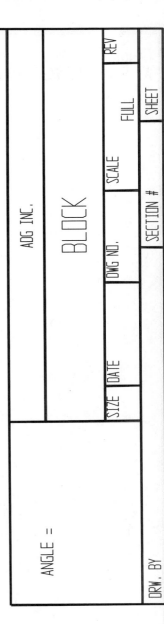

ADG INC.

BLOCK

| SIZE | DATE | | DWG NO. | SCALE | FULL | REV |

SECTION #   SHEET

ANGLE =

DRW. BY

H / F

Chapter 9, Problem 6: TWO VIEWS OF A SPECIAL PIPE FITTING ARE SHOWN. FIND THE BEND ANGLE
AT B BY THE REVOLUTION METHOD.

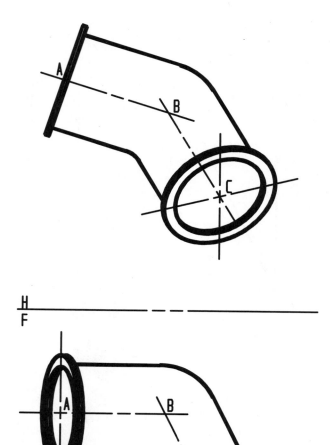

ANGLE =	ADG INC.				
	PIPE FITTING				
	SIZE A	DATE	DWG NO.	SCALE 1" = 2"	REV
DRW. BY KEY			SECTION #	SHEET 1 OF 1	

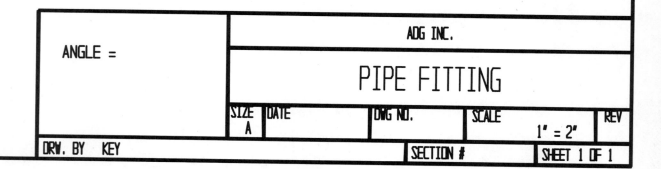

Chapter 9, Problem 7: AN X-FRAME STAMPING IS ILLUSTRATED. FIND THE HOLE CENTER TO CENTER DISTANCES, AB, BC, CD, AND AD, AND THE ANGLE CED. SOLVE THIS PROBLEM BY THE REVOLUTION METHOD.

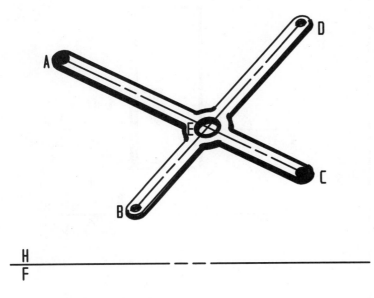

$$\frac{H}{F}$$

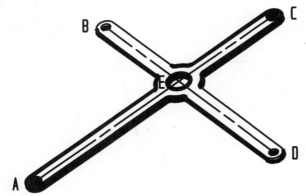

AB =		ADG INC.			
BC =		X-FRAME			
CD =					
DA =					
ANGLE CED =	SIZE DATE	DWG NO. 9-7	SCALE 1" = 4"		REV
DRW. BY		SECTION #		SHEET	

Chapter 9, Problem 8: LINE XY IS THE CENTER LINE OF A 1/4-INCH DIAMETER AIRPLANE CONTROL CABLE THAT PASSES AROUND A PULLEY AT X AND CONTINUES TO Y THROUGH A 1-1/2 INCH DIAMETER HOLE IN THE BULKHEAD. FIND THE ANGLE THAT CABLE XY MAKES WITH THE BULKHEAD SURFACE, USING THE REVOLUTION METHOD.

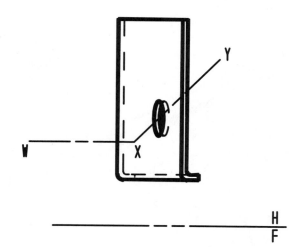

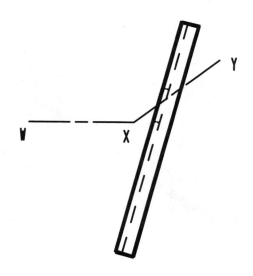

	ADG INC.				
ANGLE =	CONTROL CABLE				
	SIZE	DATE	DWG NO.	SCALE 1/4" = 1'0"	REV
DRW. BY			SECTION #		SHEET

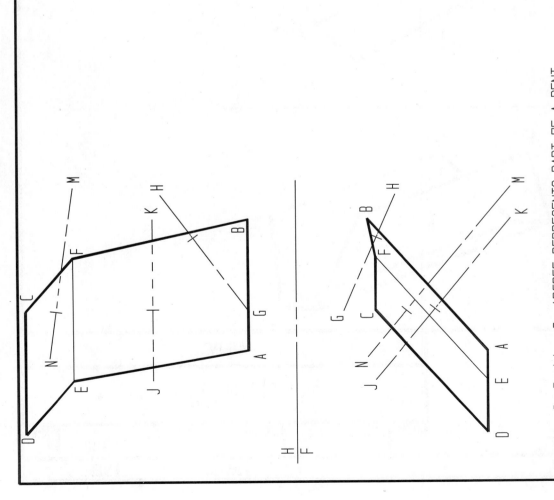

Chapter 9, Problem 9: ABCDEF REPRESENTS PART OF A BENT
FLOOR PLATE. A COLUMN, JK, A CABLE, MN, AND
A BRACE, GH, PASS THROUGH THE FLOOR. USING
THE REVOLUTION METHOD, DETERMINE THE ANGLES
THAT THE COLUMN, THE CABLE, AND THE BRACE
MAKE WITH THE FLOOR PLATE.

ANGLE FLOOR MAKES WITH
  COLUMN =
  CABLE =
  BRACE =

ADG INC.

FLOOR PLATE

SIZE	DATE	DWG NO.	SCALE 1" = 6"	REV
		SECTION #	SHEET	

DRW. BY

A

B

H
F

B

A

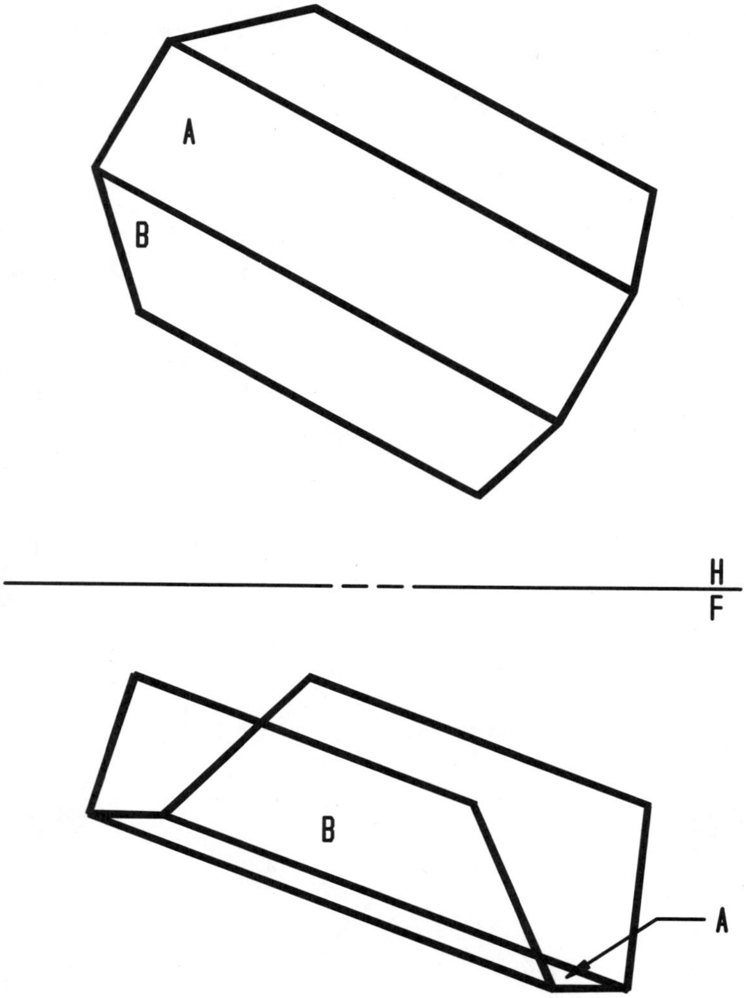

ANGLE =	ADG INC.			
	SLUICEWAY			
	SIZE DATE	DWG NO.	SCALE 1″ = 10′	REV
DRW. BY		SECTION #	SHEET	

Chapter 9, Problem 11: THE HOPPER BELOW IS MADE OF 1/4-INCH STEEL PLATE. THE ADJACENT SIDE PLATES OF THIS HOPPER ARE TO BE RIVETED WITH A SPECIAL CORNER ANGLE PLACED INSIDE THE HOPPER AS SHOWN AT AE. USING THE REVOLUTION METHOD, DETERMINE THE BEND ANGLE FOR THE CORNER ANGLE AE, SO THAT A DETAIL DRAWING CAN BE MADE FOR THE SHOP. DO THE SAME FOR CORNERS, BF, CG, AND DH.

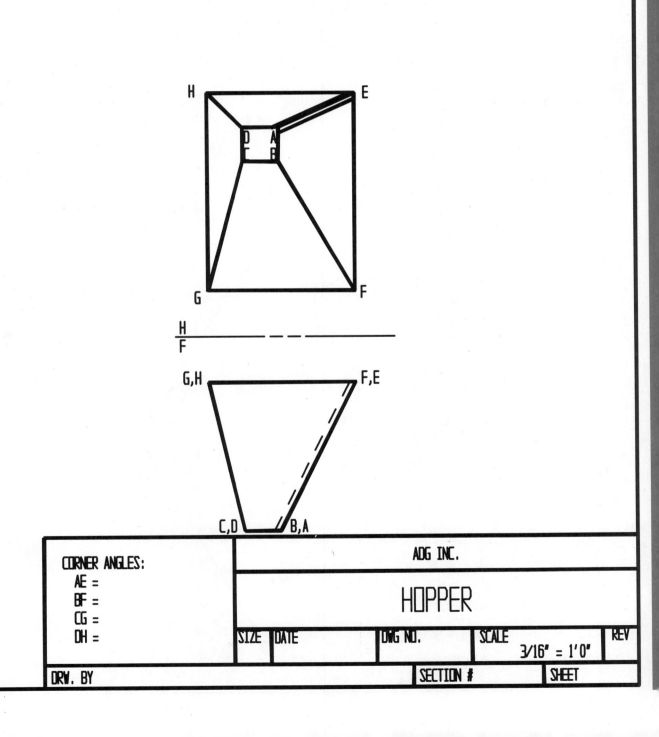

CORNER ANGLES:
AE =
BF =
CG =
DH =

ADG INC.

HOPPER

SIZE	DATE		DWG NO.	SCALE	REV
				3/16" = 1'0"	

DRW. BY | SECTION # | SHEET

Chapter 9, Problem 12: LINES AB-BC-CD, EF, AND GH REPRESENT THE CENTER LINES OF THREE LINKS OF A TOGGLE CLAMP FOR AN INJECTION MOLDING MACHINE. POINT A IS FIXED AND POINTS F AND H ARE FORCED TO FOLLOW A STRAIGHT HORIZONTAL PATH. USING THE ROTATION METHOD, FIND THE POSITION OF POINTS D-E-G, F, AND H WHEN GH IS IN THE VERTICAL POSITION.

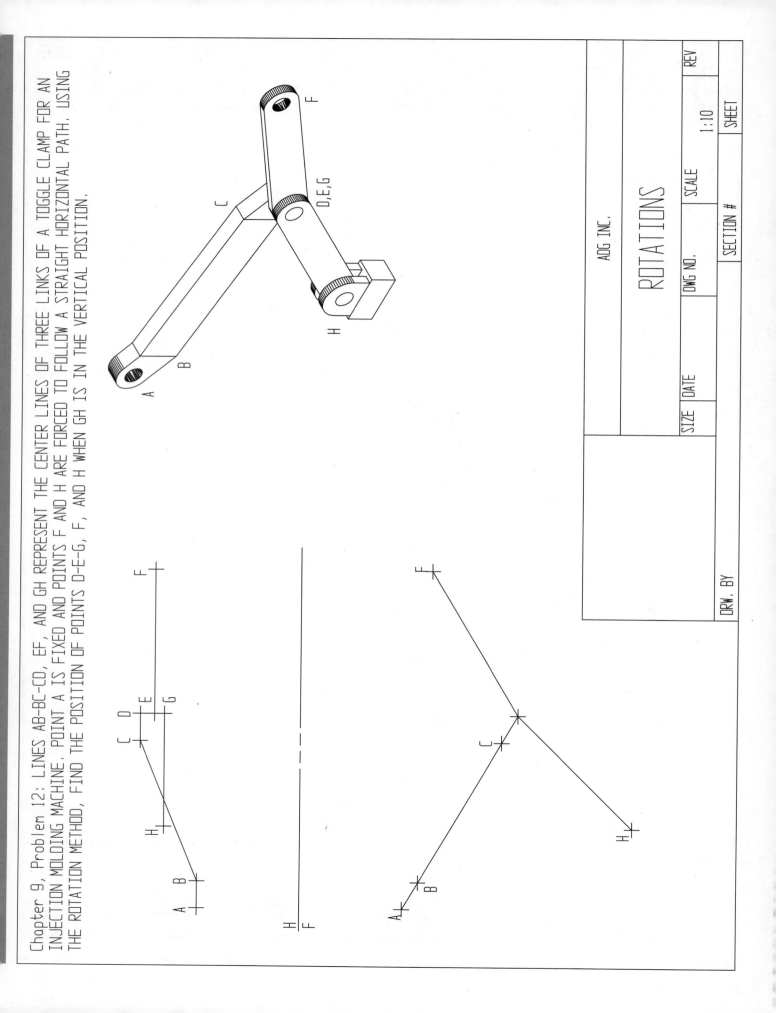

CHAPTER 10

Development

A development (flat pattern) was needed to make the hopper and feeder for the volumatic feeder machine.

After completing this chapter, you will be able to:

▶ *Define the term* development *and explain how developments are used in actual practice.*

▶ *Explain the difference between the three general types of developments.*

▶ *Examine a given object and determine which type of development is appropriate.*

▶ *Accurately construct the necessary development.*

In addition to finding the line of intersection between various surfaces and geometric solids, you must be able to reproduce the shapes of many of these solids as folded planes on a flat surface, such as sheet metal, sheet plastic, or packaging material. The objects that you are probably the most familiar with that use developmental layouts are cereal and pizza boxes, and heating and cooling duct work.

When the geometric shape is laid out in a flat pattern, certain sides are connected so that the item can be rolled or bent into the original shape of the solid. The solid shape that is unfolded or unrolled onto a flat plane is known as a **development**. The important characteristic of these surface developments is that all lines appear true length in the development. Figure 10–1 pictorially illustrates a simple development of a right, rectangular prism that can be assumed to be made of sheet metal. It shows the faces of the prism being unfolded sequentially to form the flat pattern shown in Figure 10–1(D).

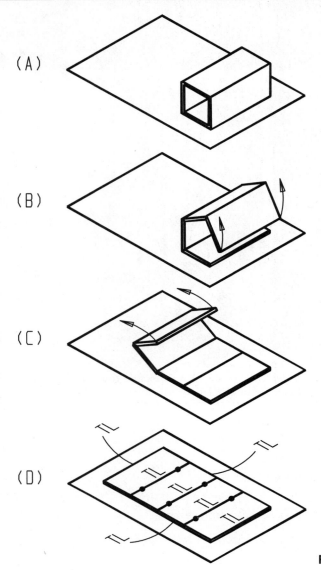

Figure 10-1 Development of a right rectangular prism.

Many manufactured articles are fabricated from sheet metal by cutting and bending the material into the desired shapes. A development of the surfaces of the object is made first, either on paper or directly on the flat surface of the metal. If large quantities of the part are required, a metal pattern, or template, of the development may be made. The outline of the pattern may then be transferred to the flat stock. The **bend lines** are often located by means of small punch marks.

After the metal is cut to the desired shape, it is bent, curved, or pressed into its finished form along the bend lines. The edges are commonly joined by soldering, welding, riveting, or seaming. Additional metal is allowed for these joints. To account for the thickness of the metal, **bend allowances** are made for bending metals thicker than 24 gage (.025 in.). All developments in this chapter will be theoretical, in that we will not allow for these slight additions and modifications necessary for joining and bending.

Developments are divided into three groups, according to the type of surface and the method of development used. Parallel-line developments are those found for prisms and cylinders. Radial-line developments are those obtained for pyramids and cones. Finally, there are triangulation developments, which are found by dividing a given surface into a series of triangular areas.

There are general rules that apply to all groups:

1. In actual practice, it is most economical to join the desired form along its shortest edge, because this would require the least extra material, welding, riveting, and so on. Therefore, start numbering where the seam should be located. The numbering should follow a clockwise direction.

2. With some symmetrical objects the direction the object is unfolded does not matter. For example, the prisms in Figures 10–1 and 10–2 would be the same if they were inside out or upside down. However, asymmetrical objects (as shown in Figure 10–6) must be unfolded correctly and sometimes labeled inside or outside, bend up, and/or bend down to help the craftsperson correctly form the object.

3. Only true-length dimensions may be transferred to the layout.

4. A **stretch-out line** is drawn the length of the perimeter of the prism or cylinder a development is being created for. It may be located at any convenient place on the object as long as it is perpendicular to the lateral (side) lines.

5. Generally, all markings are made on the inside of the development, so the markings will not show on the outside of the object.

6. Small circles are used to denote bend lines. In other words, when cutting out the pattern do not cut along these lines.

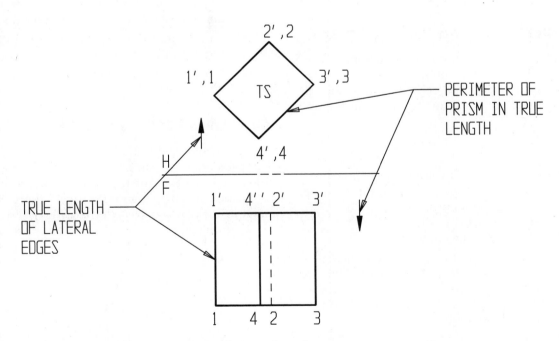

Figure 10–2 Two views of a right rectangular prism.

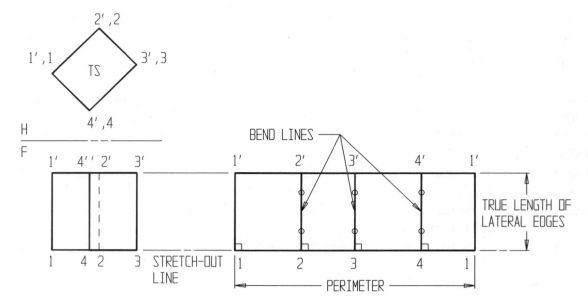

Figure 10–3 Development of a right rectangular prism.

▼ PARALLEL-LINE DEVELOPMENTS

Parallel-line development is used for prisms and cylinders, where the lateral edges, or elements, are parallel to each other.

Development of a Right Rectangular Prism

In Figure 10–2, two views of a right rectangular prism are shown. Notice that the lateral edges appear true length in the front view, and that the perimeter, the actual distance around the prism, appears true length in the top view.

Example

Figure 10–3 shows the development of the prism shown in Figure 10–2. The procedure for obtaining this development is as follows:

1. Number the object in a clockwise direction.
2. Construct a stretch-out line on which the perimeter of the prism will be unfolded. Notice in Figure 10–3 how the stretch-out line is lined up with the front view. This allows the height to be easily transferred using projection lines.
3. Transfer the perimeter distances, 1 to 2, 2 to 3, 3 to 4, and 4 to 1 (found in the top view) to the stretch-out line. (Number 1 must be repeated to complete plane 4-4'-1'-1.) Label points 1, 2, 3, 4, and 1 on the stretch-out line. You are marking the inside of the object, so be sure when the object is folded, the numbering is correct. This is not important on a symmetrical object like this

one, but it helps to practice numbering on easier problems.
4. At each numbered point construct perpendicular lines.
5. Transfer the true length of the lateral edges, 1-1', 2-2', 3-3', and 4-4' (found in the front view) to their respective locations on the stretch-out line.
6. Connect points 1', 2', 3', 4', and 1' with a straight line to complete development.
7. Darken all object lines and mark bend lines with the proper bend line symbol.

Development of a Truncated Right Prism

The isometric and top, front, and auxiliary views of a truncated right prism are shown in Figures 10–4 and 10–5 respectively. The top of the truncated prism, 1'-2'-3', is not true shape; therefore, an auxiliary view is necessary. In this example, the lateral surfaces are developed, with the top and bottom surfaces attached.

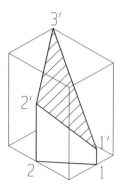

Figure 10–4 Isometric of a truncated prism.

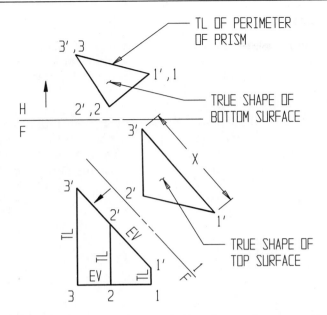

Figure 10-5 Three orthographic views of a truncated prism.

Example

Figure 10–6 shows the development of the truncated right prism shown in Figure 10–5. The procedure for finding this development is as follows:

1. Number in a clockwise direction, starting with the shortest lateral edge (1-1'). Sometimes it is easiest to number an isometric first, then the orthographic.

2. Identify and label all true-length lines.

3. Draw a stretch-out line anywhere on the paper. Transfer the perimeter distance from around the prism to this line. Label points 1, 2, 3, and 1. The marks are made on the inside of the pattern. Be sure to layout the pattern accordingly.

4. Construct lines perpendicular to the stretch-out line at points 1, 2, 3, and 1.

5. Begin with the shortest true-length lateral edge, 1-1' and lay it out on the appropriate perpendicular line. Do the same with the other true-length lateral edges, 2-2', 3-3', and then 1-1' again.

6. Connect 1' to 2', 2' to 3' and 3' to 1'.

7. Attach the true shape of the bottom surface, by transferring the true distances from the top view. Only true-length lines may be measured; therefore, an auxiliary view is needed to create the true shape of plane 1'-2'-3' (Figure 10–5). Attach the true shape of the top surface, by transferring the true-length distances from view 1 using a compass. The true shapes may be attached along any common line. Figure 10–6 uses lines 2'-3' and 2-3.

8. Darken in all object lines and mark bend lines.

Pay particular attention to the fact that this development began at the shortest lateral edge, was numbered in a clockwise direction, and unfolded accordingly.

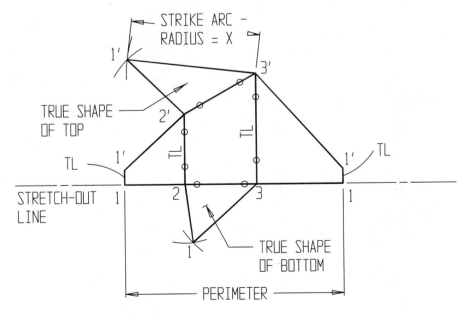

Figure 10-6 Development of a truncated right prism.

Development of an Oblique Prism

Figure 10–7 shows a diagonal offset connector used to connect two rectangular ventilation ducts. Four views—a front, top, primary, and secondary auxiliary, are shown. View 1 is constructed to show the true lengths of the lateral edges, 1'-1", 2'-2", 3'-3", and 4'-4". View 2 is an end view (a right section) which shows the lateral edges as points and shows the true distance between the edges (the perimeter of the prism).

Example

Figure 10–11 shows the development of this connector (an oblique prism). The procedure for constructing the development follows (refer to Figures 10–8 through 10–11):

1. Note that, if not already given, you must construct a view of the prism in which the lateral edges appear true length (Figure 10–7 view 1) and a view in which you can see the perimeter of

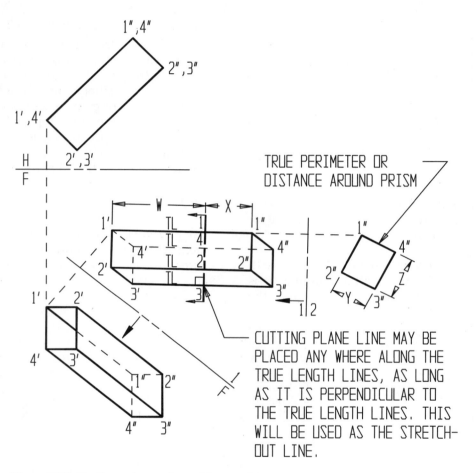

Figure 10–7 Four views of an oblique prism.

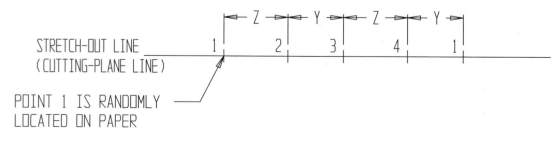

Figure 10–8 Stretch-out line.

the prism (Figure 10–7 view 2). The cutting plane line in view 1 may be placed anywhere along the true-length lateral edges, as long as it is drawn perpendicular to the true-length lateral edges of the prism. This cutting plane cuts the lateral edges at points 1, 2, 3, and 4. This cutting plane line will also serve as the stretch-out line.

2. Draw a stretch-out line anywhere on the paper. Transfer the perimeter distances, points 1, 2, 3 and 4, found in view 2. These are represented as distances Y and Z (Figure 10–8).

3. Construct perpendiculars to the stretch-out line at points 1, 2, 3, 4, and 1 (Figure 10–9).

4. Transfer lateral edge 1'-1, distance W, and 1-1", distance X. Repeat this process to locate edges 2-2', 2-2", 3-3', 3-3", 4-4', 4-4", and 1-1', 1-1" respectively (Figure 10–10).

5. Connect points to complete the development.

6. Darken object lines and mark bend lines (Figure 10–11).

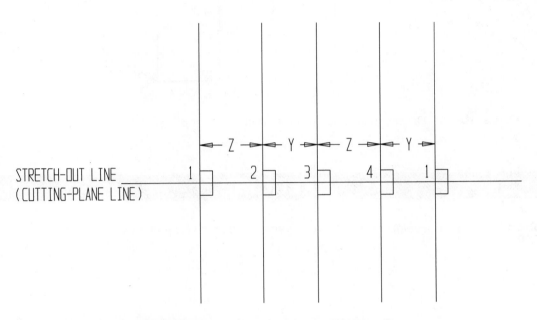

Figure 10–9 Lateral edges drawn perpendicular to stretch-out line.

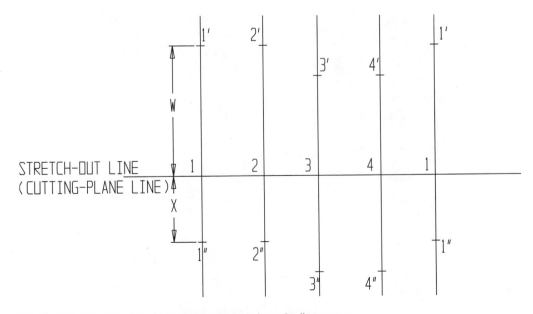

Figure 10–10 Transfer lateral edge true-length distances

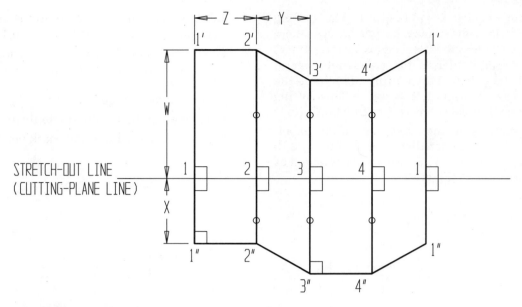

Figure 10–11   Development of an oblique prism.

EXERCISE

Problem: Draw a development for the dryer vent cover shown below in Figure 10–12.

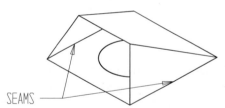

SEAMS

Figure 10–12   Dryer vent isometric.

Create necessary auxiliary views on the Figure 10–13. Draw the development on a separate piece of paper. Figures 10–14 through 10–18 show the step-by-step solutions.

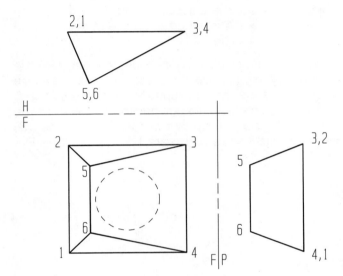

Figure 10–13   Orthographic of dryer vent.

If you are not sure where to start, answer the following questions first.

What am I **solving for**?

What is **given**?

What views are given?

Which lines are viewed true length in the given views?

What other information do I know?

What do I **need to solve** this problem? (True-length lines, stretch-out line, seam(s), what the development will look like, steps necessary to complete the development, etc.)

The answers should be:

What am I **solving for**?

■ Development of dryer vent

What is **given**?

What views are given?

■ Top and front

Which lines are viewed true length in the given views?

■ Lines 1-2, 2-3, 3-4, 4-1, and 5-6

What other information do I know?

■ The object is symmetrical; therefore, lines 1-6 and 2-5 are the same length, as are lines 3-5 and 4-6.

What do I **need to solve** this problem?

■ The location of point 6 and the true-length of the remaining lines, 1-6 and 4-6, are needed. For this problem, to locate the true-length views, the revolution method would be the easiest and fastest procedure. However, for the sake of clarity the answer key will use the auxiliary view method to show plane 3-4-5-6.

■ Identify the seam location. Normally the seam(s) would be located at the shortest line. However, Figure 10–12 specifically identifies lines 2-3 and 1-4 as the seam lines.

■ Identify which line(s) will be laid out on the stretch-out line. It helps to first sketch out what the development will look like (Figure 10–19). The stretch-out line should be drawn the perimeter of the object. Because the object has two seams, any line parallel to line 1-4 or 1-2 may be used. Line 1-4 is a better choice because more of the object can be laid out on the stretch-out line.

■ Complete the following steps: (1) Number in a clockwise direction, starting with the seam. (2) Identify and label all true-length lines. (3) Draw a stretch-out line anywhere on the paper. Transfer the perimeter distance from around the prism to this line. (4) Construct lines perpendicular to the stretch-out line at points identified on the stretch-out line. (5) Lay off distance along the appropriate perpendicular lines. (6) Connect appropriate points, darken all object lines and mark bend lines. (See Figures 10–13 through 10–19.)

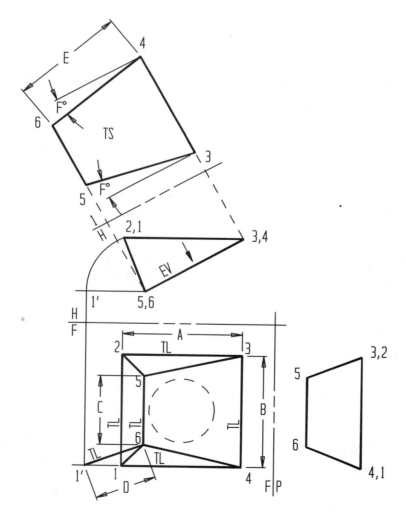

Figure 10–14 Auxiliary views necessary for dryer vent development.

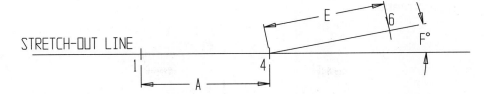

Figure 10–15 Stretch-out line for dryer vent randomly placed on paper.

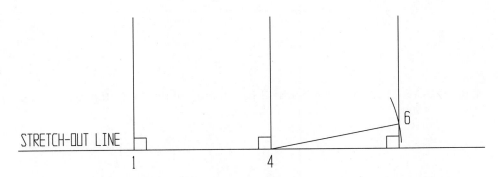

Figure 10–16 Construction lines drawn perpendicular to stretch-out line through points 1, 4, and 6.

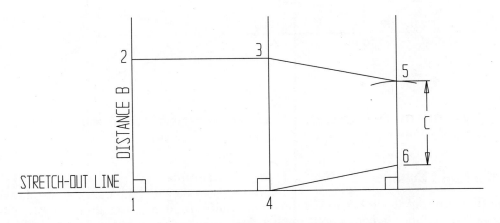

Figure 10–17 Locate points 2, 3, and 5 along lateral edges.

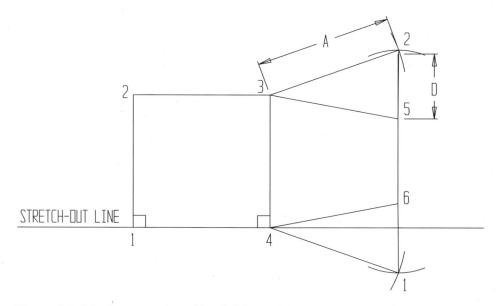

Figure 10–18 Locate points 1 and 2 by striking arcs equal to the appropriate true-length lines.

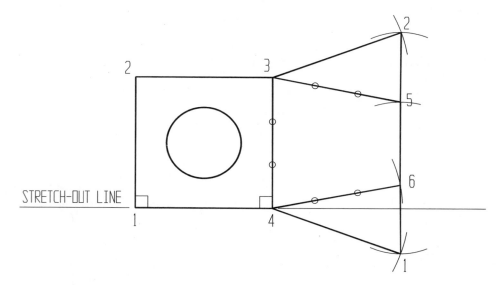

Figure 10–19 Development for dryer vent with holes located and bend lines marked.

Development of a Right Cylinder

The development of a cylinder is similar to the development of a prism. The cylinder is considered to be a prism with an infinite number of faces, or elements, but in practical applications, the number of faces are considered finite. The elements of the cylindrical surface are the same as the lateral edges of the prism. They are parallel, and must be viewed in their true-lengths in order to use them in a development. As with the prism, the perimeter is seen in the end view, and is equal to the circumference (πD) of the circle.

Figure 10–20 illustrates pictorially the development of a cylinder.

Example

Two views of the cylinder are shown along with the development in Figure 10–21. The procedure for drawing the development is as follows:

1. Divide the circumference (seen in the top view of the given cylinder) into a convenient number of radial divisions. For clarity, number these divisions. In this example, twelve divisions at 30° increments are used.

2. Project points 1 through 12 to the base of the cylinder in the front view. At these points construct straight-line elements on the cylinder surface. Since the axis of the cylinder is true length in the

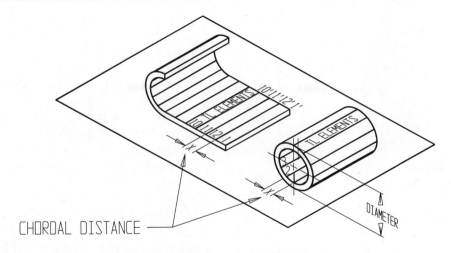

Figure 10–20 Pictorial of the development of a cylinder.

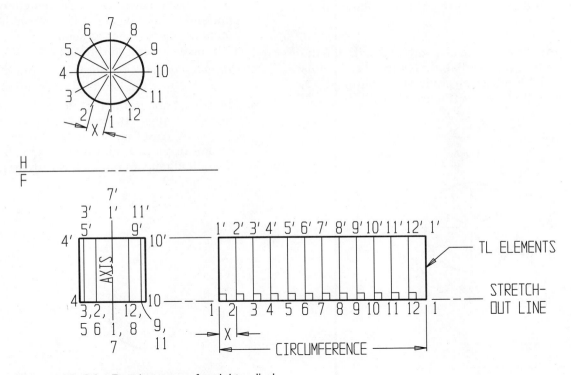

Figure 10–21 Development of a right cylinder.

front view, each element appears true length also.

3. Draw a stretch-out line equal, or a little longer, in length to the circumference of the cylinder. The actual length of the circumference, seen in the top view, can be approximated by measuring the chord distance, X, and transferring it to the stretch-out line. Repeat transferring measurement X for each cord.

4. At each numbered point on the stretch-out line construct a perpendicular. The length of these lines will be equal to the true-length elements of the cylinder.

5. Connect points 1' through 12' and 1' with a straight line to complete the development.

Development of a Truncated Right Cylinder

Like the truncated right prism, the truncated right cylinder has a top surface that is not parallel to its bottom. Because of this, the true lengths of the elements on the cylindrical surface will vary.

Example

Figure 10–22 shows the orthographic drawing and the surface development, including the top and bottom surfaces, of a truncated right cylinder. The pro-

Figure 10–22 Orthographic of truncated right prism.

cedure for finding this surface development is as follows:

1. In the top view, on the circumference, divide the circle into equal radial divisions (12, in this example). Number the divisions 1 through 12, starting with the seam (Figure 10–23).

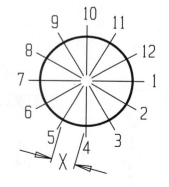

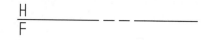

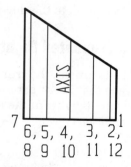

Figure 10–23 Divide cylinder into 12 equal segments.

2. Project these points to the bottom surface in the front view. Because the cylinder axis appears true length in the front view, these numbered elements will also appear true length. Label the elements 1-1', 2-2', etc., as shown, in Figure 10–24. All surfaces must be shown true shape; therefore, a primary auxiliary view will be needed to show the oblique plane true shape.

3. Draw a stretch-out line equal, or a little longer, in length to the circumference of the cylinder (πD). Again you can find the actual length of the circumference in the top view, or you can approximate it by transferring the chord distance, X, to the stretch-out line repeatedly (Figure 10–25).

4. Once you have transferred the circumference to the stretch-out line, number the points. Draw perpendicular lines at each point beginning on the shortest elements at 1. The length of the perpendiculars will be equal to the true-length elements of the cylinder as seen in the front view. This locates points 1' through 12' and 1' again.

5. Connecting these points with a smooth curved line completes the lateral surface development.

6. The true shape of the bottom appears in the top view and may be attached at any convenient tangent point. And finally, the true shape of the top appears in view 1, and may be attached at any convenient tangent point on the upper part of the development.

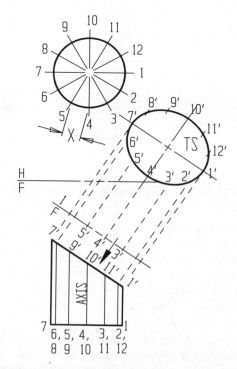

Figure 10–24 True shape of oblique surface.

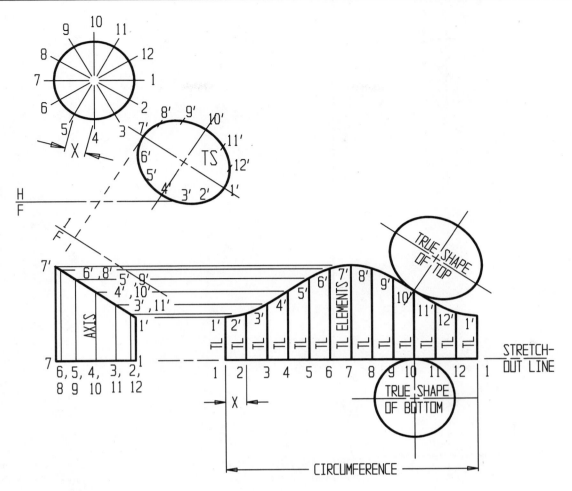

Figure 10–25   Development of a truncated right cylinder.

EXERCISE

Problem: Complete the development for the elbow of a take off transition piece (Figure 10–26).

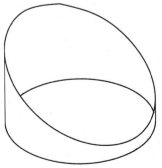

Figure 10–26   Isometric of an elbow.

Create necessary auxiliary views on Figure 10–27. Draw the development on a separate piece of paper. Note the seam, 1-1', location in Figure 10–27, it is not at the narrowest point in the sleeve. The solution is demonstrated in Figures 10–28 through 10–32.

If you are not sure where to start, answer the following questions first.

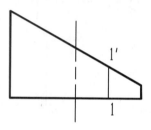

Figure 10–27 Locate information needed to draw development. The location of the seam is located at line 1-1'.

What am I **solving for**?

What is **given**?

    What views are given?

    Which lines are viewed true length in the given views?

    What other information do I know?

What do I **need to solve** this problem? (True-length lines, stretch-out line, seam, what the development will look like, steps necessary to complete the development, etc.)

Answer

    What am I **solving for**?

    ■ Development of an elbow

    What is **given**?

        What views are given?

        ■ Top and front

        Which lines are viewed true length in the given views?

        ■ All radial lines

What other information do I know?

■ The perimeter of the elbow is found in the top view where the diameter is shown true shape.

■ The chordal distance (shortest distance between two elements) is found in the top view.

■ The elbow is hollow; therefore, additional information, such as the bottom shape, is not necessary.

What do I **need to solve** this problem?

■ Identify the seam location. Normally, the seam would be located at the shortest line; however, Figure 10–27 specifically identifies line 1-1' as the seam.

■ Identify which line(s) will be laid out on the stretch-out line. It helps to first sketch out what the development will look like (See Figure 10–32 on page 389). The stretch-out line should be drawn the perimeter of the object.

■ Complete the following steps: (1) Segment circular view into twelve equal segments. Transfer points to the adjacent view, creating the lateral lines. (2) Number in a clockwise direction, starting with the seam. (3) Draw a stretch-out line anywhere on the paper. Transfer the perimeter distance from around the prism to this line. (4) Construct lines perpendicular to the stretch-out line at points identified on the stretch-out line. (5) Lay off distance along the appropriate perpendicular lines. (6) Connect appropriate points, and darken all object lines. (See Figures 10–28 through 10–32.)

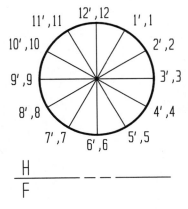

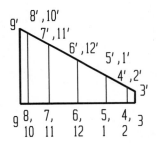

Figure 10–28   Divide circle into equal radial divisions.

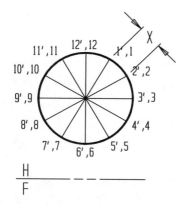

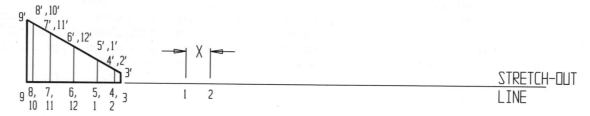

Figure 10–29 Point 1 randomly located on stretch-out line. Point 2 through 12 and 1 located chordal distance X apart.

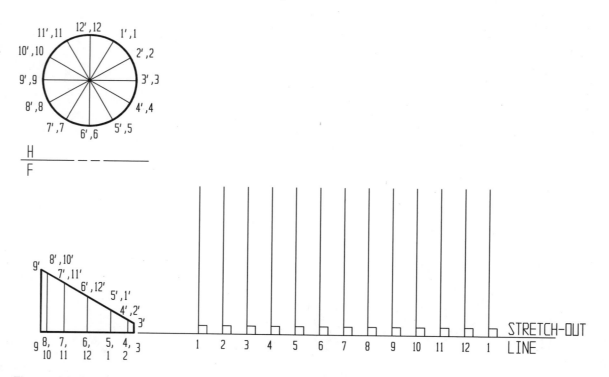

Figure 10–30 Perpendiculars drawn through points 1 through 12 and 1.

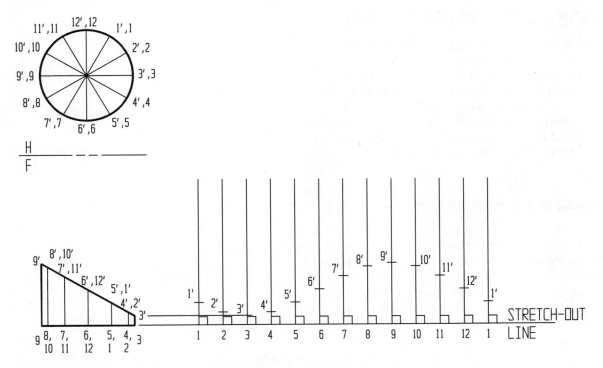

Figure 10–31 Project height to development.

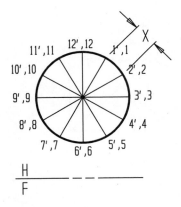

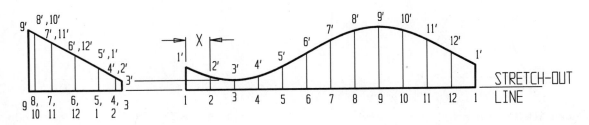

Figure 10–32 Development for elbow.

▼ RADIAL-LINE DEVELOPMENTS

Pyramids and cones are usually made as radial-line developments. The lateral edges of a pyramid and the elements of a cone appear on their respective developments as straight lines that radiate from a point that corresponds to the vertex of the pyramid or cone. When developing truncated pyramids or cones, it is simpler and more accurate to develop the whole pyramid or cone, and then deduct the vertex portion from the whole.

Development of a Truncated Right Pyramid

When developing a pyramid, the true shape of each face of the pyramid must be determined (Figure 10-33). Each face of a right pyramid is a triangle with a common **vertex**, V, which is the highest point of the pyramid. The resulting development will be a series of triangles arranged to give the desired form when folded.

Example

In a right pyramid, all lateral edges, from the vertex to the base, are the same length. In the example shown in Figure 10-34, none of the lateral edges are true length in the given views.

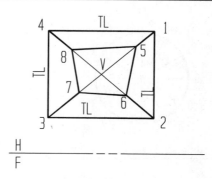

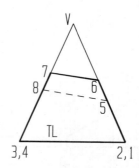

Figure 10-34 Locate vertex and all true-length lines.

Using the axis of the pyramid as an axis of revolution, edge V-1 can be revolved to appear true length in the front view (Figure 10-35). Line V-1 is the true length of all four lateral edges (Figure 10-36).

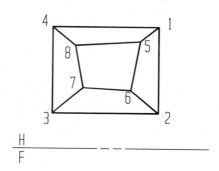

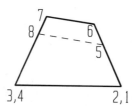

Figure 10-33 Truncated right pyramid.

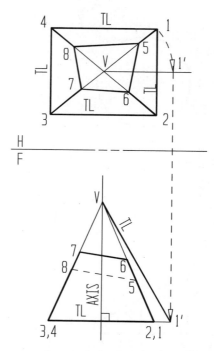

Figure 10-35 True length of lateral edge V-1.

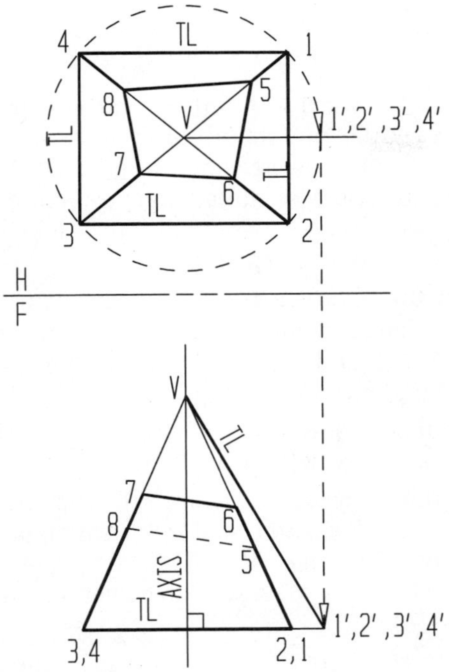

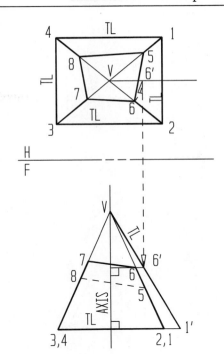

Figure 10–36   All lateral edges are the same length

Figure 10–38   Location of point 6 on line V-1'.

Points 5, 6, 7, and 8 are located on lateral lines; therefore, they may be projected to the true length lateral line (Figures 10–37 through 10–40).

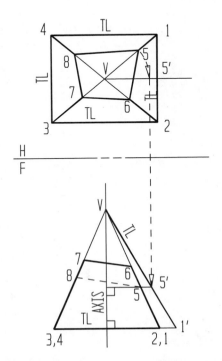

Figure 10–37   Location of point 5 on line V-1'.

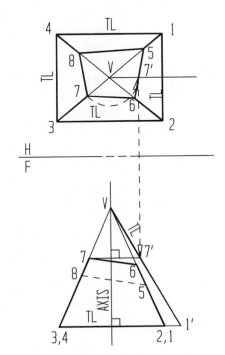

Figure 10–39   Location of point 7 on line V-1'.

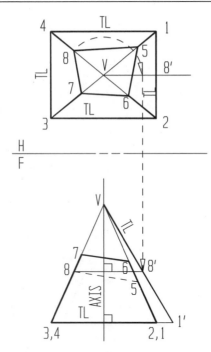

Figure 10–40 Location of point 8 on line V-1'.

The procedure for developing this pyramid is as follows:

1. Select a starting point V, and construct an arc having the true length of lateral edge V-1' as its radius and V as its center (Figure 10–41). Line V-1 is the shortest lateral edge; therefore, it will be the location of the seam. Point V may be located anywhere on the paper.

2. On this arc, starting with point 1, measure the true-length distances around the base, 1 to 2, 2 to 3, 3 to 4, and 4 to 1. (Use your compass to strike the distances.) (See Figure 10–42.)

3. Connect point V to each of points 1, 2, 3, 4, and 1 respectively (Figure 10–43).

4. In the front view, measure the distances from V to 5', 6', 7', 8' and transfer these distances to their respective lines V-1, V-2, V-3, V-4, and V-1. Points 5, 6, 7, 8, and 5 have been located (Figure 10–44).

5. Connect points 1 through 4 and 1, and 5 through 8 and 5 with straight lines to complete the surface development.

6. Darken all object lines and mark bend lines (Figure 10–45).

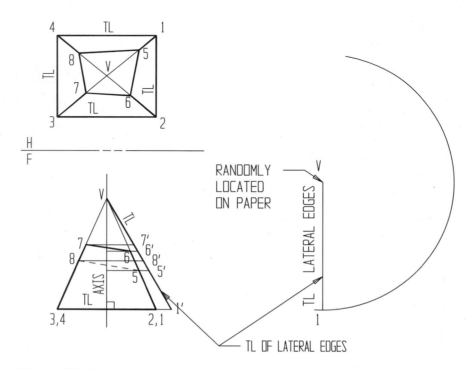

Figure 10–41 Locate line V-1.

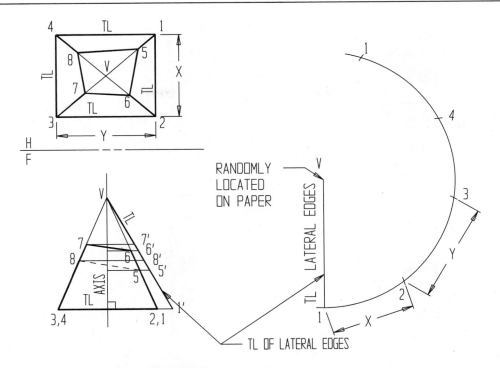

Figure 10-42 Location of points 2, 3, 4, and 1.

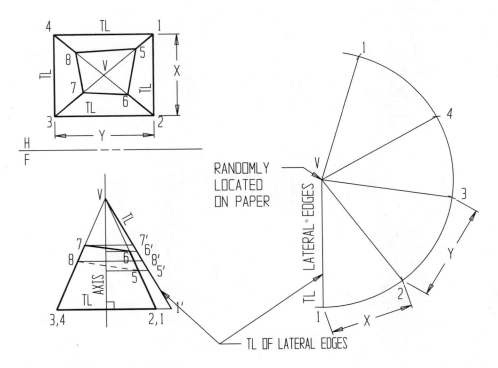

Figure 10-43 Connect points.

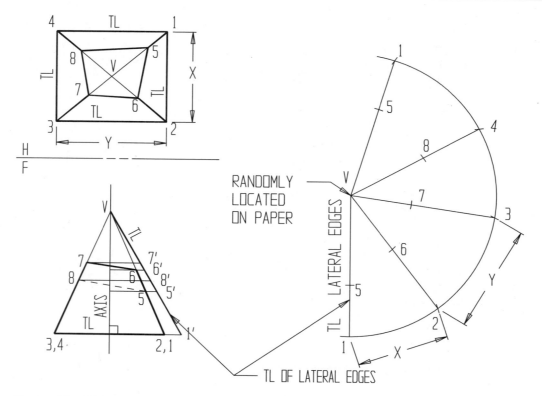

Figure 10–44 Locate points 5, 6, 7, and 8.

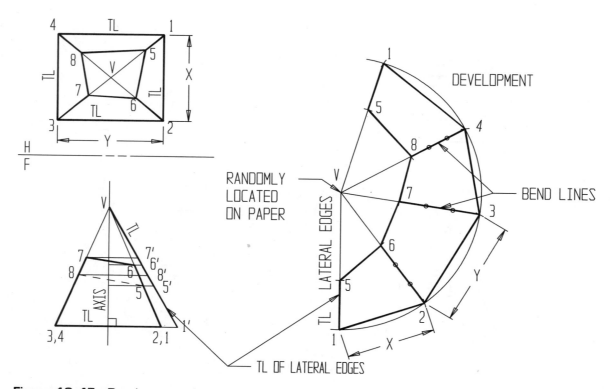

Figure 10–45 Development of truncated right pyramid.

Development of a Truncated Oblique Pyramid

The method of developing an oblique pyramid is essentially the same as that for a right pyramid, except that the lateral edges of the oblique pyramid are different lengths. Because of this difference, the true lengths of each edge must be found and the surfaces are constructed as triangular shapes. Again, revolution is the most efficient method.

Example

In Figure 10–46, two views of a truncated oblique pyramid and its development are shown. The following procedure is used to develop the surface of this pyramid.

1. Using the vertex V as an axis of rotation, find the true length of each lateral edge. The true lengths appear in the front view as V-1', V-2', V-3', and V-4'.

2. Select any point on the paper as the vertex V for the development. Draw an arc equal to V-1', the shortest lateral edge. Point 1 is on the arc and its location on the arc is arbitrary.

3. From point 1 draw an arc equal to side 1-2 of the bottom, which is seen true shape in the top view.

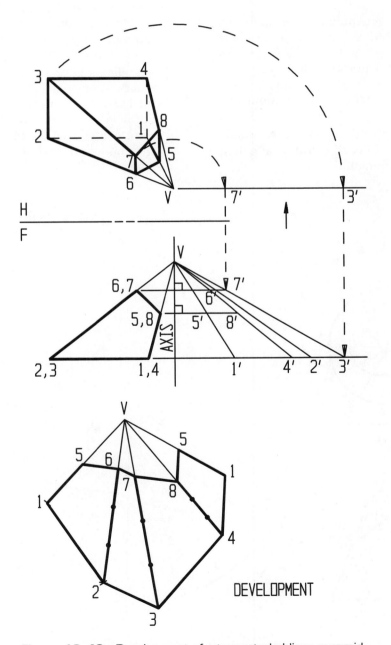

Figure 10–46   Development of a truncated oblique pyramid.

Then from V, draw an arc equal to V-2'. Point 2 is established where these two arcs intersect.

4. From point 2 draw an arc equal to side 2-3 of the bottom. From V draw an arc equal to V-2'. Where this arc intersects arc 2-3, you can locate point 3 on the development.

5. From point 3 draw an arc equal to side 3-4 of the bottom. From V draw an arc equal to V-4' until it intersects the arc 3-4, thus locating point 4.

6. Now, from point 4 draw an arc equal to side 4-1 of the bottom. From V draw an arc equal to edge V-1' until it intersects the arc 4-1. This locates the final point 1.

7. Connect points 1, 2, 3, 4, and 1 with straight lines.

8. From V in the front view project the true-length distances of V-5, V-6, V-7, and V-8 to their respective lines V-1, V-2, V-3, V-4, and V-1 of the development. Connect points 5, 6, 7, 8, and 5 with straight lines. This completes the surface development, excluding the top and bottom surfaces.

9. If you need to develop the top and bottom surfaces, the true shapes of each could be attached to the existing development.

Development of a Right Truncated Cone

Example

In Figure 10–47, the orthographic drawing of a right truncated cone and its development are shown. All elements of the cone, vertex to base, are equal in length. Consequently, the radial lines of the development will each be a constant length equal to the slant height of the cone. The development will form an arc, with V as its center, with the length of the arc equal to the circumference of the base of the cone.

The procedure for drawing the lateral surface development of this cone is as follows:

1. The circumference of the base, as seen in the top view, is divided into 12 equal parts, establishing 12 elements on the cone surface. Number the points 1 through 12. Project the elements to the front view and label them.

2. All elements V-1 through V-12 are the same length. Elements V-7 and V-1 are true length in the front view, and are equal to the slant height S of the cone. Now, using V as the center and the slant height S as the radius, select a convenient starting point and draw an arc of indefinite length.

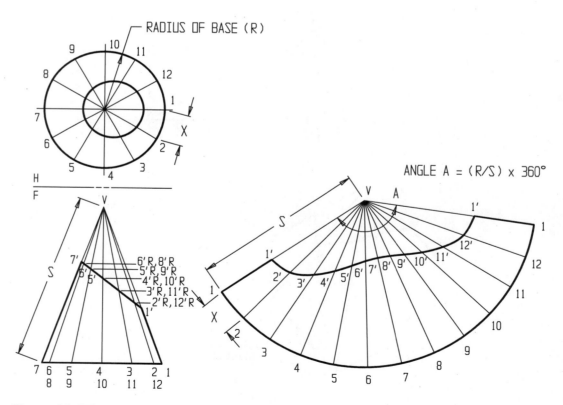

Figure 10–47 Development of a right truncated cone.

3. You may determine the length and divisions of this development in two ways: (A) by stepping off the chord distance X along the arc for the 12 divisions, or (B) by calculating the angle A according to the formula given in Figure 10–47. Radial lines should be drawn from V to each division point on the arc to represent the elements of the cone.

4. To develop the truncated portion of the cone, you must mark off on each element the true length of the segment cut from the element. Transfer true length distance V-1' along element V-1. Element V-2 is cut at 2', element V-3 is cut as 3', etc. The remaining true-length segments are found in the same way, and all are transferred consecutively to their respective elements on the development.

5. Points 1', 2', 3', etc., of the development are connected with a smooth curve to complete the development.

6. Darken all object lines.

Development of an Oblique Cone

Unlike the right cone, the elements of the surface of the oblique cone are not all of equal length; therefore, you must find the true length of each element before you can begin the development.

Example

Figure 10–48 illustrates two views of a truncated oblique cone and its development. The development of such a cone assumes that the surface between any two consecutive elements forms a narrow plane triangle that is a very close approximation to the true shape of this surface. It is very much like an oblique pyramid with 12 sides.

The procedure for constructing the development is as follows:

1. Divide the circumference of the base (as seen in the top view) into 12 equal parts, and number them beginning with the shortest element. Extend all elements to the vertex in the top view. Project the points on the base down to the front view, and extend all elements to V.

2. Using the revolution method, determine the true lengths of all elements V-1 through V- 12. It is usually more convenient to rotate the elements away from the views, and project the true lengths to the front view, forming a **true-length diagram**, as shown.

3. In addition, by revolution, find the location of points 21 through 32 on their respective elements in the true-length diagram. For example, 21 is located on line V-1, 2-2 is located on line V-2, and so on.

4. Begin with the shortest element, V-1', and a convenient location for V. Draw an arc from V with a radius equal to V-1' and mark point 1 on the arc.

5. Using point 1 as center, draw an arc with a radius equal to chordal distance X. Then using true-length element V-2', draw an arc from V. Where this arc intersects the chordal-distance arc from point 1, locate point 2. Repeat this process until all points are located on the development. Connect the numbered points with a smooth curve.

6. From the true-length diagram take the true-length distance from V to the points on the upper portion of the cone (21, 22, etc.,) and transfer these distances to the respective elements on the development. The development will be completed when these numbered points are connected with a smooth curve.

7. Darken all object lines and mark bend lines.

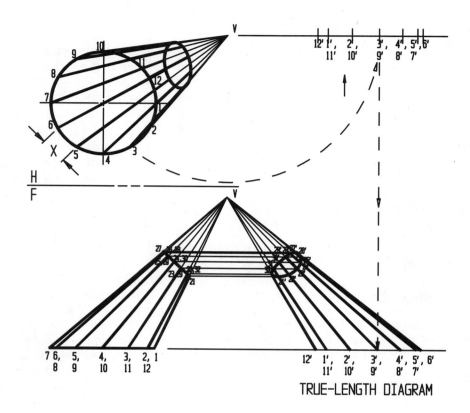

TRUE-LENGTH DIAGRAM

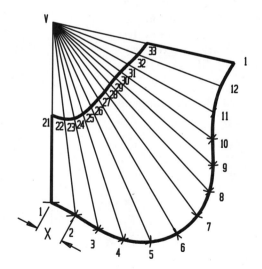

Figure 10–48 Development of an oblique cone.

EXERCISE

Problem: Draw a development for the funnel shown below in Figure 10–49.

Figure 10–49   Isometric of funnel.

Segment the funnel and locate all true-length lines directly on the Figure 10–50. Draw the development on a separate piece of paper. Figures 10–51 through 10–54 demonstrate the solution step by step.

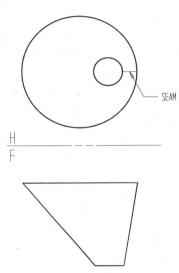

Figure 10–50   Orthographic of funnel.

If you are not sure where to start, answer the following questions first.

What am I **solving for**?

What is **given**?

What views are given?

Which lines are viewed true length in the given views?

What other information do I know?

What do I **need to solve** this problem? (True-length lines, stretch-out line, seam, what the development will look like, steps necessary to complete the development, etc.)

Answer

What am I **solving for**?

■ Development of funnel

What is **given**?

What views are given?

■ Top and front

Which lines are viewed true length in the given views?

■ Lines 1-21 and 7-28

What other information do I know?

■ Chordal distance is found in the top view.

What do I **need to solve** this problem?

■ The true length of the remaining lines are needed.

■ Identify the seam location. The seam should be located at the shortest line, 1-21.

■ Complete the following steps: (1) Segment circular view into twelve equal segments. Transfer points to the adjacent view, creating elements from V to the point. (2) Number in a clockwise direction, starting with the seam. (3) Create a true-length diagram, using the revolution method. Project the rest of points, 21 through 32, to true-length diagram. (4) Locate point V anywhere on the paper. Strike arc from point V the true-length distance line V-1'. (4) Strike arc from point 1 the chordal distance along the first arc. (5) Repeat step 4 for the rest of the elements. (6) Connect points to V. (7) Transfer points 21 to 33 to the appropriate elements. (8) Connect appropriate points, darken all object lines and mark bend lines. (See Figures 10–51 through 10–54.)

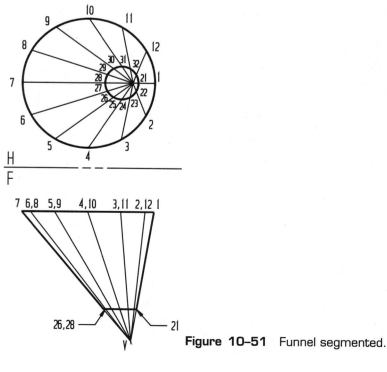

Figure 10–51 Funnel segmented.

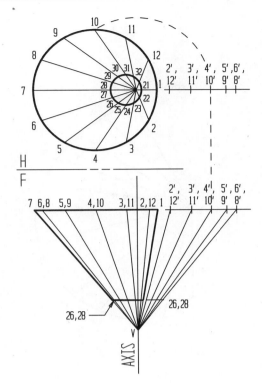

Figure 10–52 Location of true-length lines.

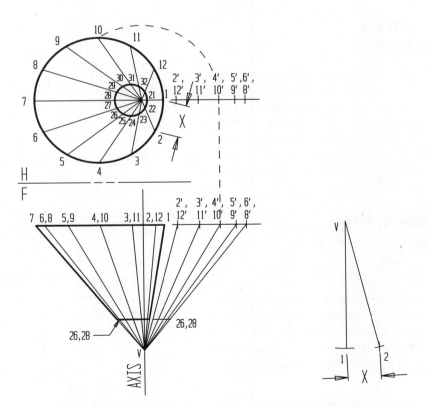

Figure 10–53 Location of seam line V-1, placed any convenient location on paper. Point 2 is the intersection of two arcs. One arc is the chord distance X and one arc the true length of line V-2.

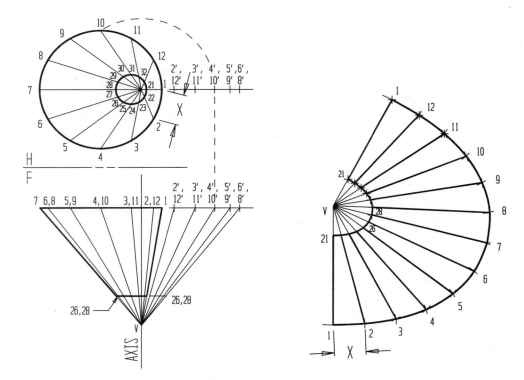

Figure 10–54 Repeat the process for each line. Strike an arc from point 1, equal to true length of line 1-21, on line V-1. This locates point 21. Repeat the process for points 22 through 32.

▼ TRIANGULATION DEVELOPMENT

Often there are surfaces that cannot be developed by the parallel- and radial-line methods. Developments for many of these surfaces may be constructed using the triangulation method. When using triangulation, the plane surfaces are divided into a series of connected triangular areas. Each triangle in the development can be laid out once the true length of each side has been determined.

Development of a Nonpyramidal Sheet Metal Connector

Example

In Figure 10–55, a sheet metal connector joining two offset rectangular ducts is shown isometrically, along with its development. Note that it is not a pyramid, because its edges have no common vertex. The process for constructing the development of this connector is as follows:

1. In the top view of Figure 10–56, the diagonal 1-8 is drawn to divide surface 1-4-5-8 into two trian-

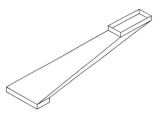

Figure 10–55 Sheet metal connector.

gles. Similarly, diagonals 2-5, 3-6, and 4-7 are drawn to divide the other three sides of the connector into triangles. The four diagonals are also drawn in the front view. The connector surfaces now consist of eight triangles.

2. The true length of each edge and its diagonal must be found. Although each could be revolved individually about a different axis to find its true length, it is more efficient and less cluttered to assume a common axis. A true-length diagram has been constructed in which the true length of each edge has been revolved about a vertical axis. The horizontal span for edge 1-5 is marked off as shown on the true-length diagram, and the vertical rise is projected from the front view to

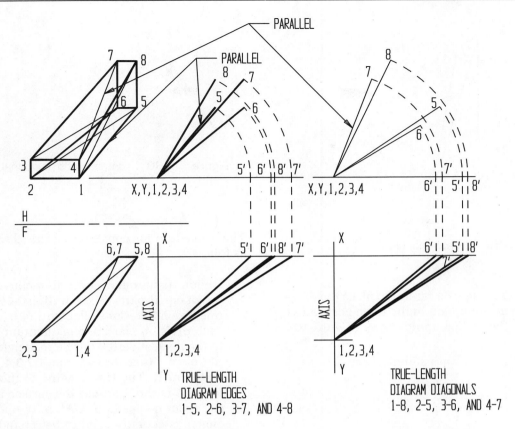

Figure 10–56 Locate all true-length lines.

provide the true length of 1-5. This process is repeated for edges 2-6, 3-7, and 4-8. Study this figure carefully. (Figure 10–56)

3. In a similar manner, the true lengths of the diagonals are revolved about vertical axis Y-Z. The top view lengths (horizontal span) are marked off as shown, and the vertical rise is projected from the front view to provide the true lengths of the diagonals (Figure 10–56).

4. Beginning the development with the shortest edge, the true length of edge 1-5 is drawn in any convenient location on the paper. From point 1, an arc equal to the true length of 1-2 (seen in top view) is drawn. An arc equal to diagonal 2-5 is drawn from point 5. The intersection of these two arcs establishes point 2, creating lines 1-2 and 2-5 (Figure 10–57). From point 2 an arc equal to the true length of edge 2-6 is drawn, and from point 5 an arc equal to the true length of 5-6 is drawn. The intersection of these arcs locates point 6, creating lines 5-6 and 2-6. The first face of the connector is complete (Figure 10–58).

Figure 10–57 Create triangle 5-1-2.

5. The remaining triangles of the development are constructed in the same way and in consecutive order, with the distances needed being taken from the top view and the true length diagrams. The development ends with edge 1-5, since it

Figure 10–58 Create triangle 2-5-6.

began with this edge. Be careful to place the diagonals in the development in the same position as they are in the orthographic views (Figure 10–59).

6. Darken in all object lines and mark bend lines.

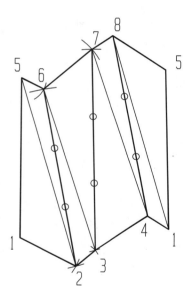

Figure 10–59 Complete remaining triangles.

Development of a Transition Piece

Example

A sheet metal connector that joins two other pipes, or openings, that are different sizes and shapes is called a **transition piece**. Figure 10–60 illustrates a transition piece that connects a round to a rectangular duct. The connecting surface is composed of four triangular planes, W, X, Y, and Z, and four partial oblique cones, R, S, T, and U (Figure 10–61).

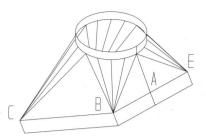

Figure 10–60 Isometric of transitional piece.

The procedure for constructing this development is as follows:

1. Divide the circular end into a convenient number of equal parts. Sixteen divisions are used in this example. Cone elements are drawn from each division mark in one quadrant of the circle to the nearest vertex on the rectangle. For example, elements are drawn from 1,2,3,4, and 5 to the endpoint B. The true lengths of these elements must be found. Because the surface is symmetrical about center line A-A', only one-half of the elements actually need to be rotated to develop the true length diagrams.

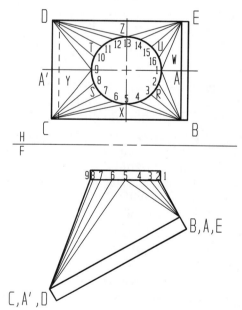

Figure 10–61 Orthographic of transition piece.

2. In order to find the true lengths of the elements from B, they are rotated about B. The true lengths of elements B-1, B-2, B-3, B-4, and B-5 are seen in the true-length diagram next to the front view. Because of symmetry, these lengths will be the same for cone elements E-1, E-16, E-15, E14, and E-13, respectively (Figure 10–62).

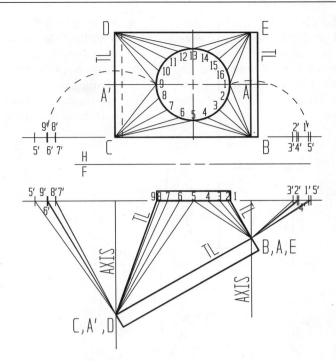

Figure 10–62 True length of all lines located.

3. The elements radiating from C are rotated about C in the top view. The true lengths of C-5, C-6, C-7, C-8, and C-9 are seen in the true-length diagram. These true lengths are equal to the cone elements D-13, D-12, D-11, D-10, and D-9 respectively (Figure 10–62).

4. The development begins along the seam line, which is the shortest line A-1. The true length of

A-1 appears in the front view. Draw a line equal to the length of A-1 any convenient location on paper. From point A, draw an arc equal to the true length of A-B, and from point 1, draw an arc equal to the true length of element B-1. The intersection of these arcs locates corner B. If done correctly lines 1-A and A-B will be perpendicular (Figure 10–63).

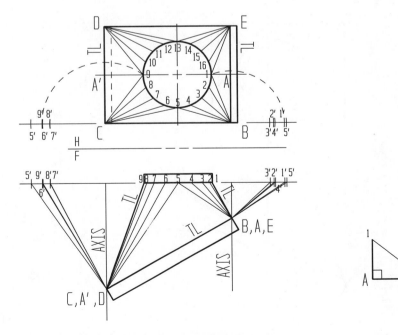

Figure 10–63 Location of first triangle.

5. Draw an arc equal to the chordal distance, distance between points 1 and 2 in the top view, across one division on the circular opening from point 1. From B, an arc equal to element B-1 is drawn (Figure 10–64). This locates point 2. This same process is repeated to locate elements B-3, B-4, and B-5 (Figure 10–65).

6. From point B, draw an arc equal to the true length of side B-C of the rectangle. Note that the true length of B-C appears in the front view. From point 5, draw an arc equal to the true length of element C-5. The intersection of these arcs establishes point C. With point C located, you can draw the elements radiating from C.

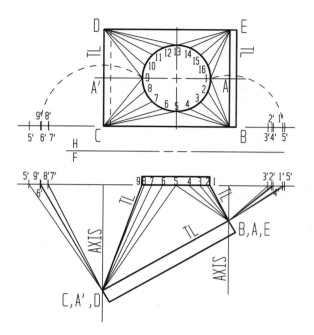

Figure 10–64 Location of line B-2.

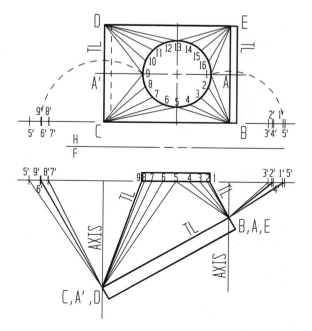
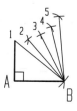

Figure 10–65 Location of lines B-3, B-4, and B-5.

7. This process continues until you have completed the development by returning to the seam at A-1. Draw a line from point 9 and the center of line C-D (A'). If line 9-A' is perpendicular to line C-D, you are proceeding correctly. (Figure 10–67)

8. Complete the development, darken in all object lines and mark bend lines. (Figure 10–68).

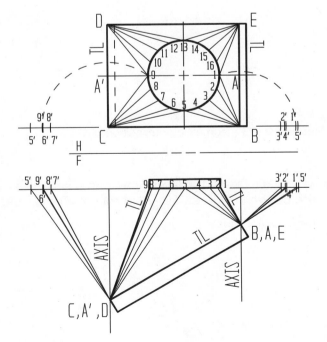

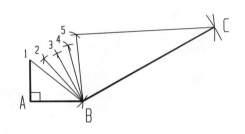

Figure 10–66 Location of line C-5.

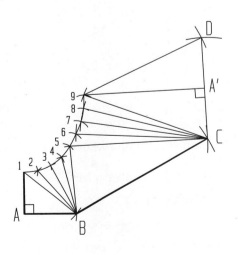

Figure 10–67 Half of development.

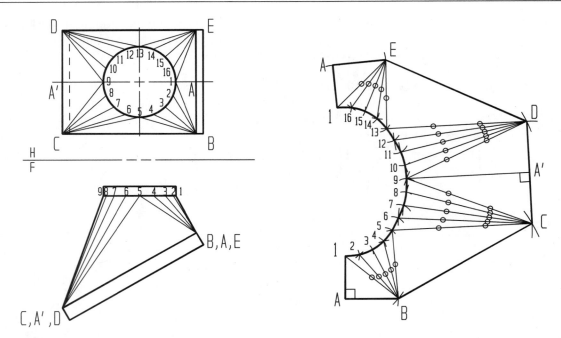

Figure 10-68 Full development for transition piece.

EXERCISE

Problem: Draw a development for the take-off transition piece shown below in Figure 10-69.

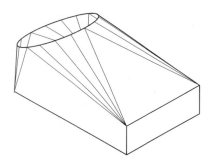

Figure 10-69 Isometric of transition piece.

Create necessary auxiliary views on the Figure 10-70. Draw the development on a separate piece of paper. The solutions for each step are shown in Figures 10-71 through 10-75.

If you are not sure where to start, answer the following questions first.

What am I **solving for**?

What is **given**?

What views are given?

Which lines are viewed true length in the given views?

What other information do I know?

What do I **need to solve** this problem? (True-length lines, stretch-out line, seam, what the development will look like, steps necessary to complete the development, etc.)

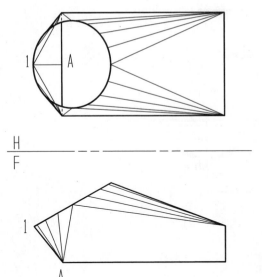

Figure 10–70   Orthographic of take-off transition piece. Line 1-A shows the location of the seam.

Answer

What am I **solving for**?

■ Development of take-off transition piece

What is **given**?

  What views are given?

  ■ Top and front

  Which lines are viewed true length in the given views?

  ■ Lines E-G, B-F, G-F, D-C, G-D, and F-C.

  What other information do I know?

  ■ The object is symmetrical; therefore, lines and curves on the front half of the transition piece are the same length as the corresponding line in the back half.

What do I **need to solve** this problem?

■ The true length of the remaining lines

■ Identify the seam location. The seam should be located at the shortest line, A-1.

■ Sketch what the development will look like (Figure 10–71).

■ Complete the following steps: (1) Segment circular view into sixteen equal segments. Draw cone elements from segment points to nearest vertexes. Project elements to the adjacent view. (2) Number in a clockwise direction, starting with the seam. (3) Create a true-length diagram, using the revolution method. (4) Locate line 1-A anywhere on the paper. Strike arcs from appropriate points the true length of appropriate lines. (4) Connect appropriate points, darken all object lines and mark bend lines. (See Figures 10–71 through 10–75.)

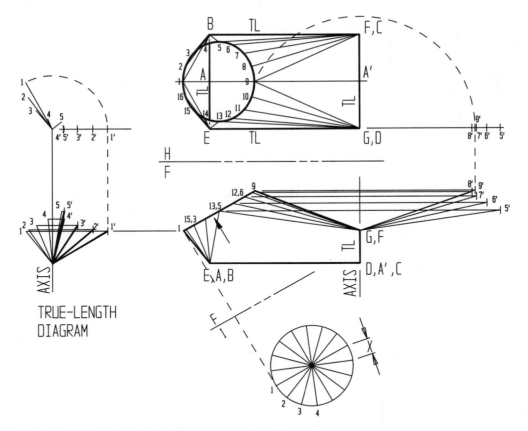

Figure 10–71 Locate the true lengths for all lines.

Figure 10–72 Create triangle 1-A-B. Point 1 may be located at any convenient location on the paper. Strike an arc the true-length distance of line A-1, from point 1. Point B is the intersection of two arcs. Strike one arc the true-length distance of line A-B, from point A; and a second arc the true-length distance of line B-1, from point 1.

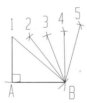

Figure 10–73 Point 2 is the intersection of two arcs. One is the true-length distance B-2, from point B and the second arc is the chordal distance from point 1 to 2 in the top view. Locate points 3, 4, and 5, using a similar process.

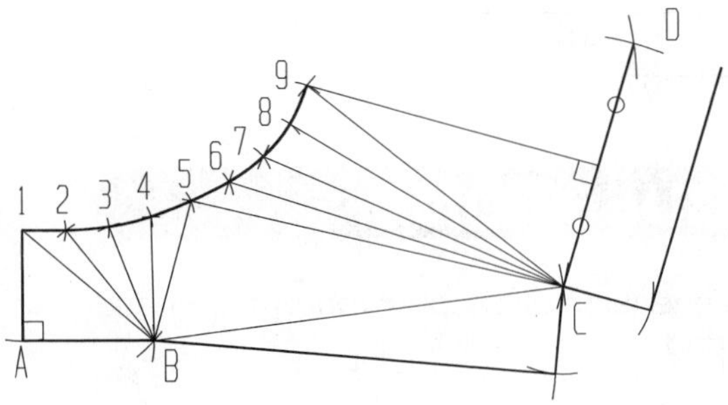

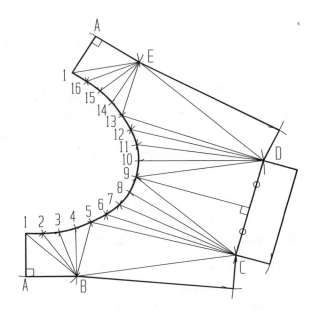

Figure 10–74 Locate line C-D. Intersection of arcs equal to true-length line C-D from point C, and 9-C from point 9.

Figure 10–75 Full development.

CHAPTER 10 STUDY QUESTIONS

The Study Questions are intended to assess your comprehension of chapter material. Please write your answers to the questions in the space provided.

1. What is meant by the development of a surface?

    _____

    _____

2. Developments are divided into three general groups. List them.

    _____

    _____

3. Compare parallel-line and radial-line developments.

    _____

    _____

4. How must all the lines to be used in a development appear before they are transferred to the development?

    _____

    _____

5. Why are developments generally joined along the shortest edge, or element?

    _____

    _____

6. Are all lines and bend marks drawn on the inside or the outside of the development?

    _____

    _____

7. When developing an oblique prism, what orthographic views would you need?

    _____

    _____

8. The development of a right cylinder has what shape?

    _____

    _____

9. How is the development of a pyramid similar to that of a cone?

    _____

    _____

10. What is a true-length diagram with regard to developments?

    _____

    _____

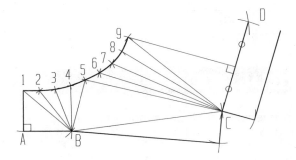

Figure 10–74 Locate line C-D. Intersection of arcs equal to true-length line C-D from point C, and 9-C from point 9.

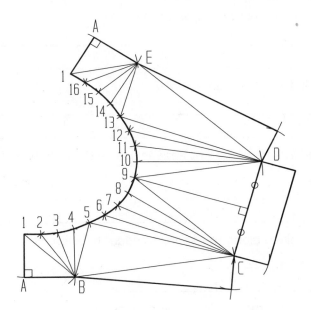

Figure 10–75 Full development.

CHAPTER 10 STUDY QUESTIONS

The Study Questions are intended to assess your comprehension of chapter material. Please write your answers to the questions in the space provided.

1.   What is meant by the development of a surface?

   _____

   _____

2.   Developments are divided into three general groups. List them.

   _____

   _____

3.   Compare parallel-line and radial-line developments.

   _____

   _____

4.   How must all the lines to be used in a development appear before they are transferred to the development?

   _____

   _____

5.   Why are developments generally joined along the shortest edge, or element?

   _____

   _____

6.   Are all lines and bend marks drawn on the inside or the outside of the development?

   _____

   _____

7.   When developing an oblique prism, what orthographic views would you need?

   _____

   _____

8.   The development of a right cylinder has what shape?

   _____

   _____

9.   How is the development of a pyramid similar to that of a cone?

   _____

   _____

10.   What is a true-length diagram with regard to developments?

   _____

   _____

11. Small parts are moved from one area of a plastics shop to another by use of a tube system. The parts are blown into the hopper shown below and then dropped into a bin. What method would be used to draw the development for the hopper shown below—parallel-line, radial-line, or triangulation development?

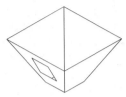

_____

_____

12. Sketch what the finished development to the hopper shown in the problem above would look like.

The problems in this chapter are based on a variety of industrial applications. In solving these problems you should try to utilize direct approaches that get to the heart of the problem. At this analysis stage of the design process, many problems use center lines and single lines to represent the problem situation. Yet, when necessary, relevant sizes are given.

The following problems may be drawn using instruments on the page provided, created using a 2D or 3D CAD system, or a combination of drawing board and CAD. Refer to Appendix A for additional dimensions needed to solve problems three-dimensionally.

Chapter 10, Problem 1: DRAW THE DEVELOPMENT OF THE RIGHT PRISM SHOWN BELOW.

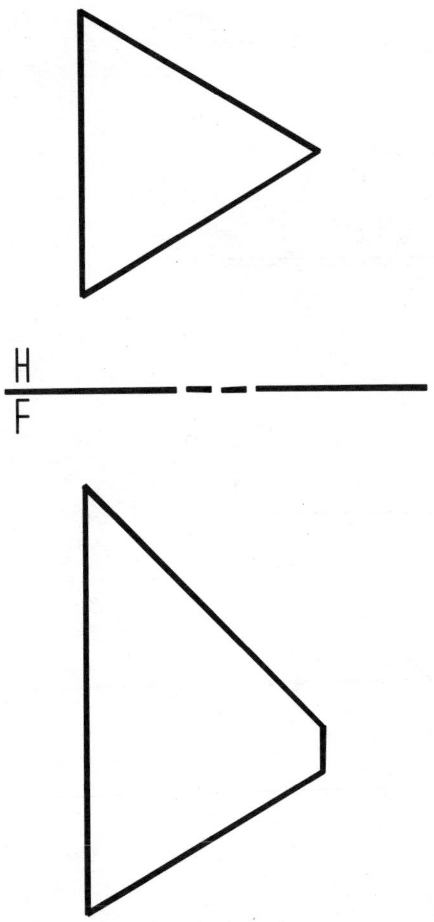

H
—
F

Chapter 10, Problem 2: A PACKAGE CHUTE IS TO EXTEND FROM THE REAR WALL THROUGH THE FLOOR AT A SLOPE ANGLE OF 22.2°
WITH THE FLOOR. COMPLETE THE FRONT VIEW. DEVELOP THE LATERAL SURFACES OF THE CHUTE.

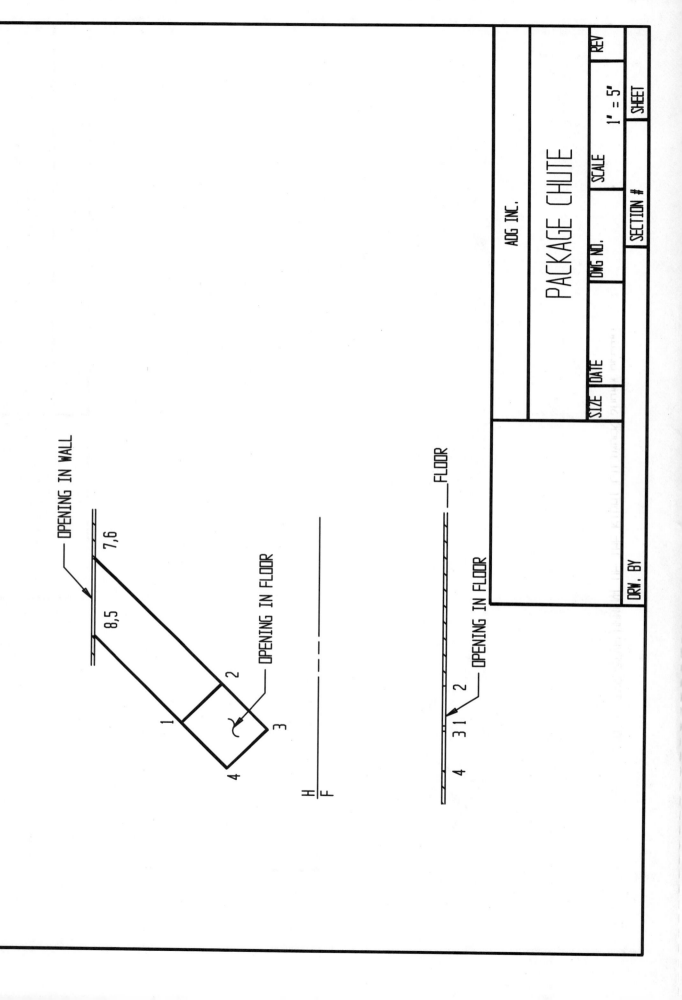

OPENING IN WALL

7,6

8,5

1

2

OPENING IN FLOOR

3

4

H
F

FLOOR

OPENING IN FLOOR

4 31 2

ADG INC.

PACKAGE CHUTE

SIZE | DATE | DWG NO. | SCALE | REV
| | | 1' = 5' |
DRW. BY | | SECTION # | SHEET |

Chapter 10, Problem 3: DRAW THE DEVELOPMENT OF THE RIGHT CYLINDER SHOWN BELOW.

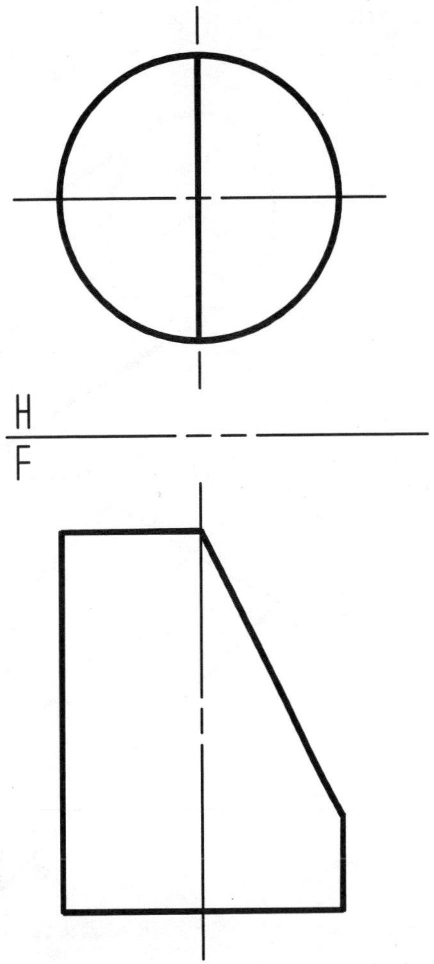

H
F

Chapter 10, Problem 4: DRAW THE DEVELOPMENT FOR THE 6-INCH DIAMETER CONNECTOR SHOWN BELOW.

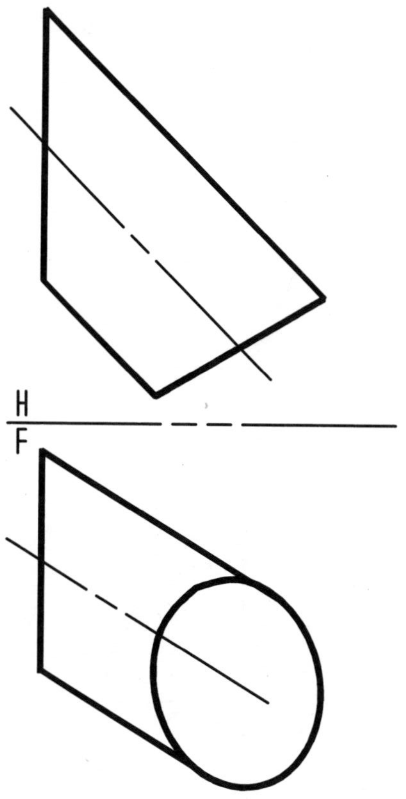

H
—
F

ADG INC.

CONNECTOR

SIZE	DATE		DWG NO.	SCALE	REV
				1" = 6"	
			SECTION #	SHEET	

DRW. BY

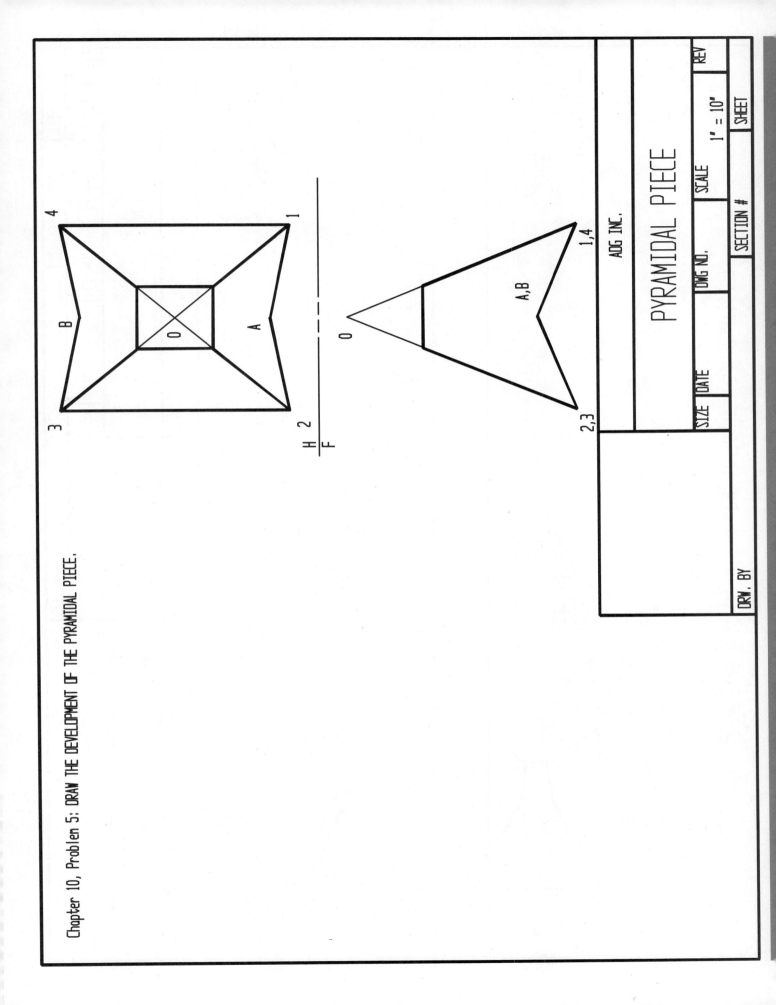

Chapter 10, Problem 5: DRAW THE DEVELOPMENT OF THE PYRAMIDAL PIECE.

ADG INC.

PYRAMIDAL PIECE

REV

SCALE   1" = 10"

SHEET

DWG NO.

SECTION #

DATE

SIZE

DRW. BY

Chapter 10, Problem 6: DRAW THE DEVELOPMENT OF THE PYRAMIDAL PORTION OF THE SAND HOPPER SHOWN BELOW. SHOW YOUR TRUE LENGTH DIAGRAM.

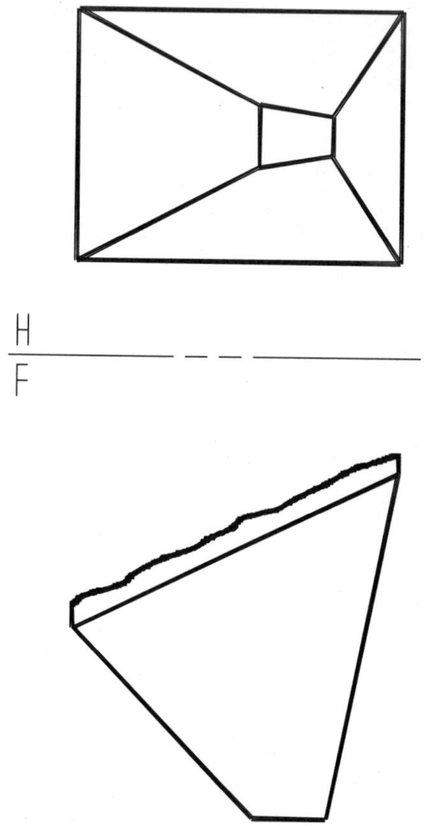

H
F

	ADG INC.		REV
	SAND HOPPER		
SIZE	DATE	DWG NO.	SCALE 1" = 3"
		SECTION #	SHEET
DRW. BY			

Chapter 10, Problem 7: COMPLETE THE TOP VIEW AND DEVELOP THE LATERAL
SURFACE OF THE RIGHT CIRCULAR CONE SHOWN BELOW.

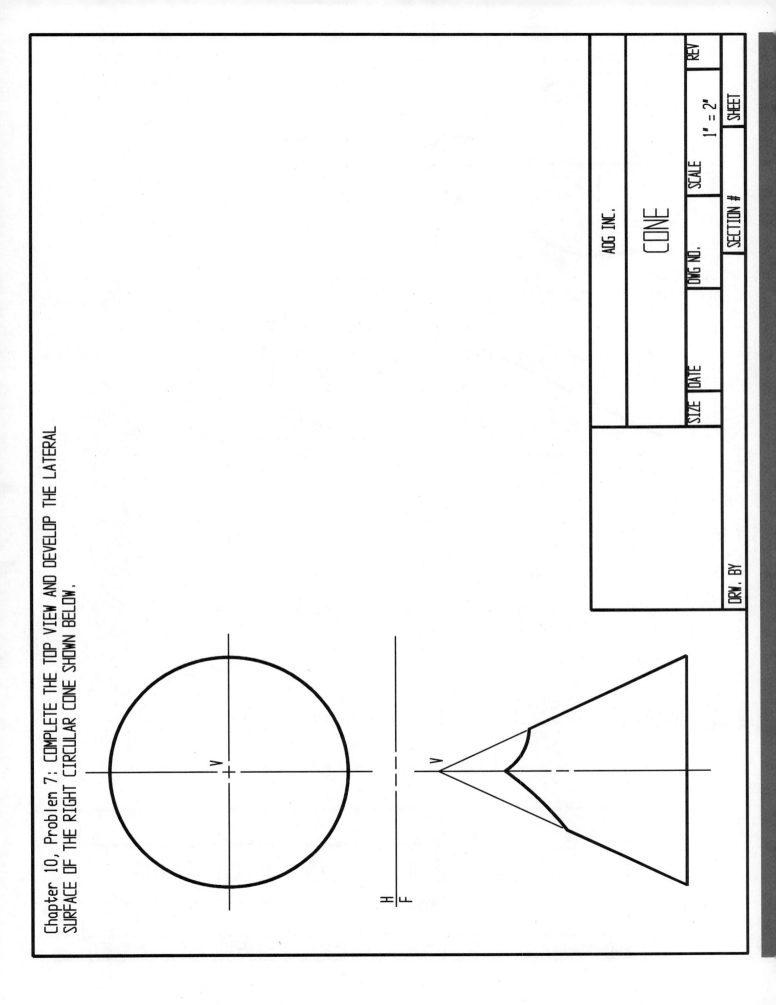

ADG INC.

CONE

SIZE | DATE
DWG NO.
SCALE   1" = 2"
SECTION #
SHEET
REV

DRW. BY

H
―
F

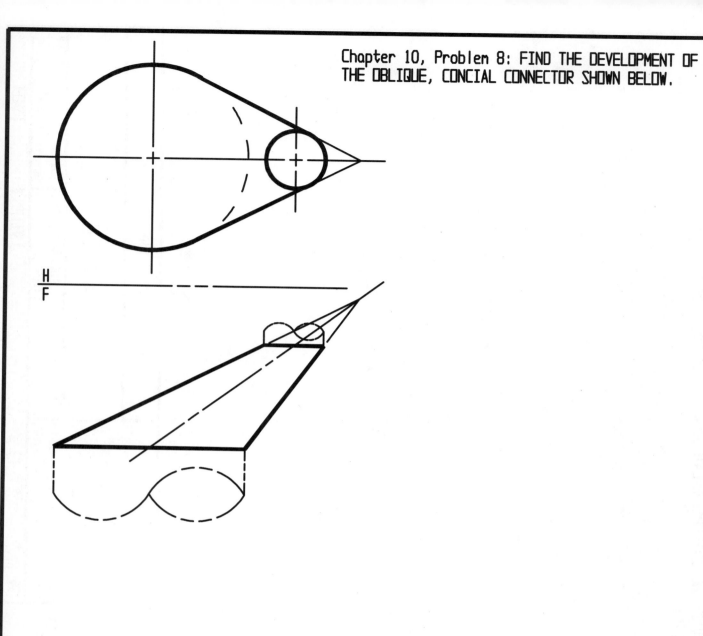

Chapter 10, Problem 8: FIND THE DEVELOPMENT OF THE OBLIQUE, CONCIAL CONNECTOR SHOWN BELOW.

H
F

	ADG INC.				
	CONICAL CONNECTOR				
	SIZE	DATE	DWG NO.	SCALE 1" = 4"	REV
DRW. BY			SECTION #		SHEET

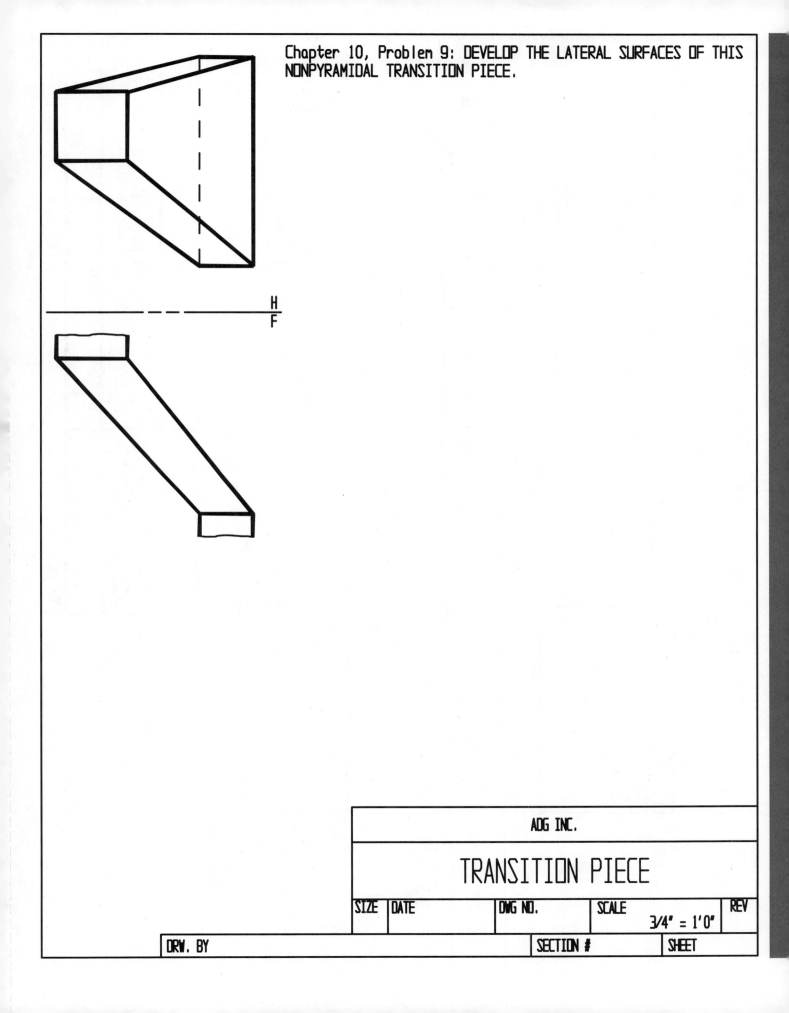

Chapter 10, Problem 9: DEVELOP THE LATERAL SURFACES OF THIS
NONPYRAMIDAL TRANSITION PIECE.

H
F

	ADG INC.				
	TRANSITION PIECE				
SIZE	DATE	DWG NO.	SCALE 3/4" = 1'0"		REV
DRW. BY			SECTION #	SHEET	

Chapter 10, Problem 10: DEVELOP THE LATERAL SURFACES OF THE TRANSITION PIECE CONNECTING THE RECTANGULAR FIREPLACE BOX TO THE CYLINDRICAL CHIMNEY FLUE.

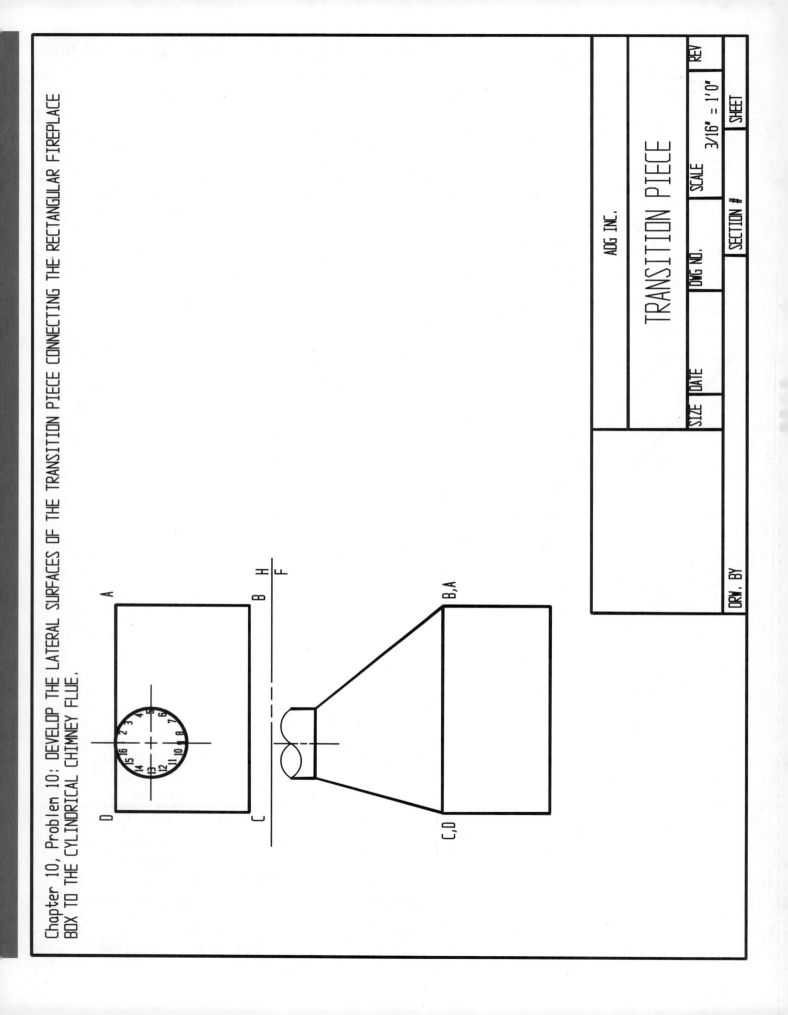

ADG INC.

TRANSITION PIECE

SIZE	DATE	DWG NO.	SCALE 3/16" = 1'0"	REV

DRW. BY

SECTION #   SHEET

Mining and Civil Engineering Applications

This photograph shows the irregularities of the earth's surface.

After completing this chapter, you will be able to:

- *Determine the strike, dip, dip direction, and thickness of a stratum of ore.*
- *Determine the probable outcrop of a stratum.*
- *Determine the cuts and fills along a level road and along a road at a specified grade.*

The principles of descriptive geometry which have been explained thus far can be particularly useful in mining and civil engineering. When solving mining or highway construction problems, you will work with **topographic maps**, representing the form of the earth's surface in a single view (top). The terrain is shown in the topographic map with **contour lines**. A contour line denotes a series of connected points at a particular elevation. A person walking on a contour line would be on a level path.

The contour line is labeled to indicate its height above sea level. For readability every fifth line is drawn darker and thicker and is called an **index contour line**. In Figure 11–1, the difference in elevation between adjacent contour lines, called the **contour interval**, is 100 feet. The small circle inside the 3,500-foot contour labeled 3,520 feet is the top of the peak. A second peak can be seen to the southeast at 3,350 feet. The two peaks side by side form a **saddle**, which is the lower area between them.

Notice that there are small dents in the contour lines. These represent **ridges** and **ravines**. If the contour lines bulge toward lower contour lines, a ridge is present. If the dents bulge toward higher contour elevations, a ravine or water course exists.

The spacing of the contour lines indicates the steepness of the ground. A steep slope is indicated by closely spaced contours, whereas greater intervals between contour lines is evidence of a gentle slope.

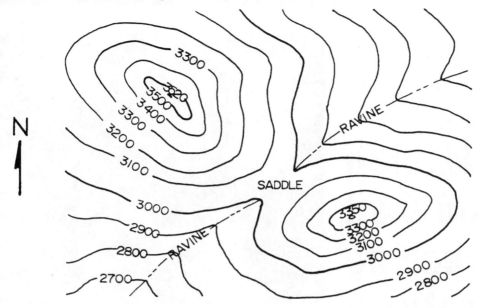

Figure 11–1 A topographical map.

Below the topsoil and loose rock that covers much of the earth's surface are a series of layers, or **strata**. Originally, the stratified rock, with a few exceptions, formed in continuous horizontal layers. In the years following their formation, geological upheaval caused these layers to become distorted. Today the originally horizontal strata are rarely level but can be considered as planes over limited areas.

Within a specific area it is reasonable to assume that a stratum (layer) of rock is uniform in thickness and that it lies between two parallel planes called **upper and lower bedding planes**. Cracks in the rock that often fill with minerals or ores are called **veins** or **lodes**. Although they vary greatly in shape, veins usually lie between parallel planes. In Figure 11–2, inclined parallel planes of sand, coal, shale, and limestone are shown.

Often these parallel strata are interrupted by fractures where one side of the bed shifts in relation to the other. This is called a **fault** and the plane of the fracture is called the **fault plane**. When the fault plane intersects the earth's surface it is called a **fault-plane outcrop**.

The term **outcrop** is also used to describe where a stratum of rock, or a vein of ore intersects the earth's surface. However, the outcrop is only visible where the soil covering the earth's surface has been removed by erosion. In Figure 11-2 a coal stratum intersects the hillside. By noting these surface conditions, from drilled test holes, and from other geological tests, a mining engineer can estimate the approximate position of an underlying stratum. With data compiled, the engineer can determine the location and depth of three or more points on upper or lower bedding planes. Since three points form a plane, the location of the whole stratum can be established, and the future working of the area can be planned.

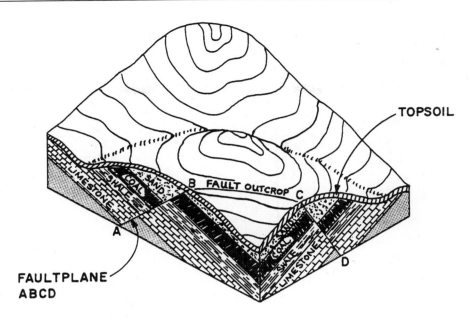

TOPSOIL

B FAULT OUTCROP C

FAULTPLANE
ABCD

Figure 11–2   A block diagram.

▼ LOCATION AND ANALYSIS OF A STRATUM

Figure 11–3 illustrates a portion of a topographical map. Points R, S, and T in the upper bedding plane are given, as well as point P on the lower bedding plane. The elevation of each point is given in parentheses. Point S, at an elevation of 1,300 feet, lies on the 1,300-foot contour lines, and thus is a point on the surface (point on the outcrop). Boreholes have been drilled to locate points R (950 feet) and T (750 feet) on the upper bedding plane, and also point P (725 feet) on the lower bedding plane.

In mining it is common practice to describe the location of a stratum using strike and dip. The **strike** is the bearing of a horizontal line in the plane of the stratum, customarily measured from north, as N65°E which is seen in Figure 11–4. The **dip** is the slope angle of the plane of the stratum, given in conjunction with the general direction of the downward slope of the plane. This **dip direction**, is always perpendicular to the strike line. The dip in Figure 11–4 is 44° downward in a southeasterly direction, or 44°SW.

Example

The steps of procedure for determining strike, dip, and dip direction are covered in detail in Chapter 6. Determining thickness of the stratum and a review of the steps necessary to locate strike, dip, and dip direction are listed below:

1. The top view, in Figure 11–4, is a copy of the map in Figure 11–3 with the contour lines deleted. The front has been drawn to show the elevation of points P, R, S, and T.

2. A horizontal line, R-X, is constructed in the front view, and projected into the top view. The strike of plane R-S-T is the bearing of this true length line, and is lettered along the line as shown (N65.4°E).

3. To find the dip angle, an edge view of the stratum must be found. The line of sight is drawn parallel to the true-length line R-X in the top view, and an edge view of the stratum is seen in view 1. Here the slope angle of 44° is the dip. The dip direc-

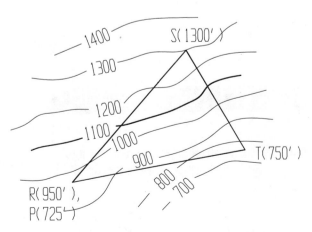

Figure 11–3   Portion of a topographical map.

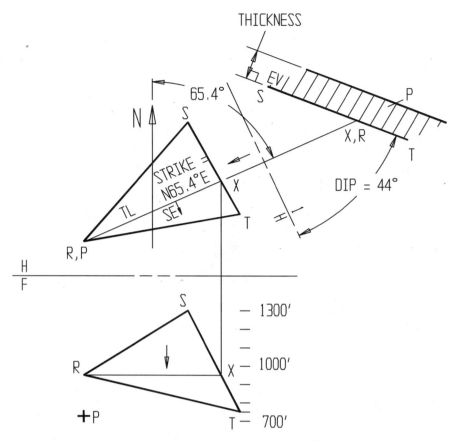

Figure 11–4 Strike, dip, dip direction, and thickness of stratum.

tion is shown in the top view as a short arrow drawn perpendicular to the strike line, and pointing toward the lower side of the plane. (Front view shows elevation.) The dip would be verbally described as 44° downward in a generally southeasterly direction, or 44°SE.

4. It is customary to assume that the upper and lower bedding planes are parallel. Thus the edge view of the lower bedding planes is drawn through P, and parallel to the edge view of R-S-T. The thickness of the stratum is measured on a perpendicular between the edge views.

EXERCISE

Problem: Points A, B, and C locate an upper bedding plane, with point P locating the lower bedding plane. Determine the strike, dip, dip direction, and thickness of the stratum (Figure 11–5).

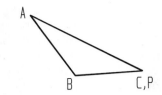

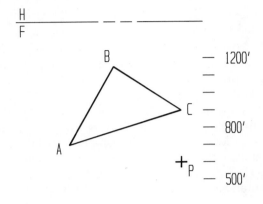

Figure 11–5 Stratum location represented by points A,B,C and P. Determine the strike, dip, dip direction, and thickness of the stratum

Answer

The stratum has a strike of N71.8°W, and a dip of 60.5° downward in a northeasterly direction, or 60.5°NE. Figure 11–6, view 1, shows where the thickness of the stratum is found.

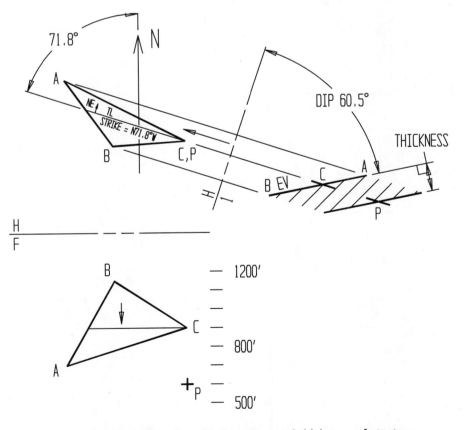

Figure 11–6 Strike, dip, dip direction, and thickness of stratum.

▼ OUTCROP OF A STRATUM OR VEIN

When a stratum or vein intersects the earth's surface, it is called an **outcrop**. When the position of the stratum has been determined, then the probable outcrop can be found, and the most economical mining of the vein can be planned. The auxiliary view showing the vein as an edge will show the contour lines as straight horizontal lines. This view shows clearly where the contour lines intersect the edges of the vein. These points of intersection can be projected to the top (map) view and connected to show the outcrop.

Example

In Figure 11–7 three points, X, Y, and Z on the upper bedding plane, and point W on the lower bedding plane, are given in both the top and front views. The following procedure is used to determine the outcrop (the cross sectioned area in Figure 11–10, top view.):

1. Since the locations of points X, Y, and Z on the upper bedding plane are known, the strike, dip, and thickness can be determined by the method discussed in the previous section. (Figure 11–7). Line A-X shows the strike to be N72°W. In view 1, the edge view is shown indicating a dip of 44.5° downward in a northeasterly direction. (The northeasterly direction is noted with an arrow in the top view.) Point W is directly below X and on the lower bedding plane. The thickness of the bedding plane is measured on a perpendicular between the edge views.

2. In view 1 (Figure 11–8) the point where the stratum intersects the 180-foot contour line is marked 1. This point of intersection is projected from view 1, where it twice crosses the 180-foot contour line as shown. Follow the same procedure to locate point of intersection 2. The point intersects the 170-foot contour line twice in the top view. Do the same for each consecutive point of intersection on the upper bedding plane.

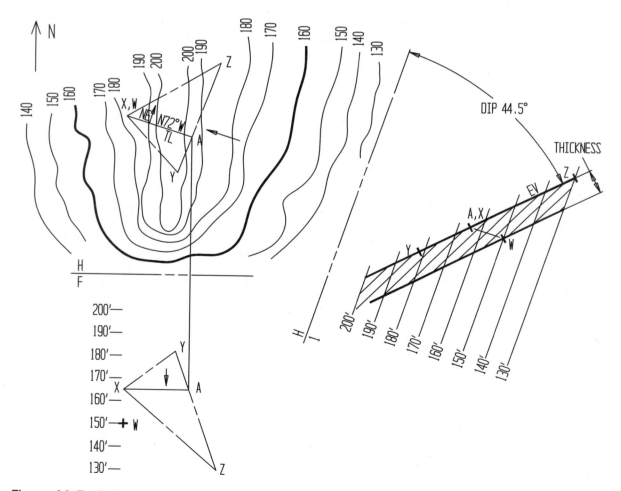

Figure 11-7 Strike, dip, dip direction of a stratum.

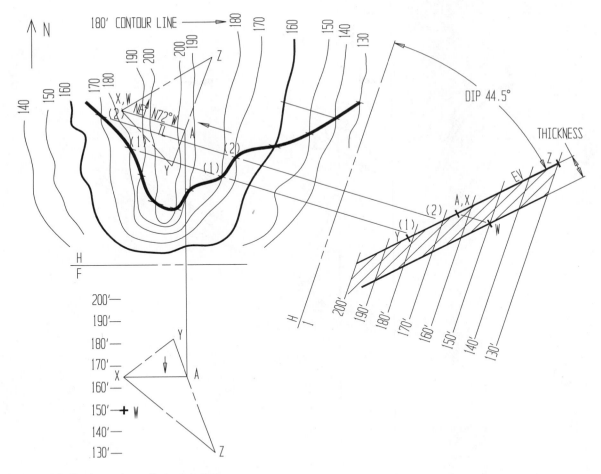

Figure 11-8 Location of upper stratum.

3. The lower outcrop line is determined in the same manner except that contour intersection points are taken on the lower bedding plane (Figure 11–9). Note that point 3 on the 180-foot contour line in the lower plane is projected to the top view, intersecting the 180-foot contour as shown. By

continuing this procedure, points of outcrop can be located on each successive contour line to establish the lower outcrop line.

4. When both outcrop lines have been determined, the area is shaded in the top view for readability (Figure 11–10).

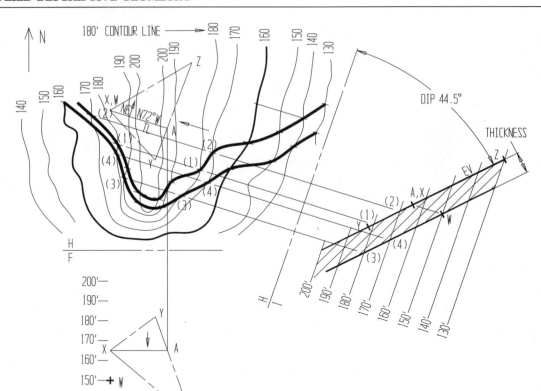

Figure 11-9 Location of lower stratum.

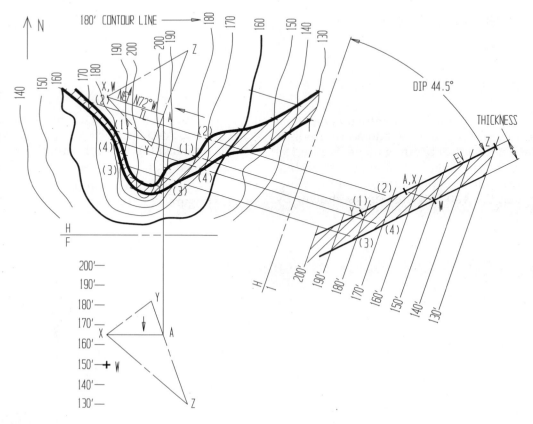

Figure 11-10 Outcrop of stratum.

EXERCISE

Problem: Determine the strike, dip, dip direction, and thickness of stratum for Figure 11–11. Show the outcrop in the top view on Figure 11–12.

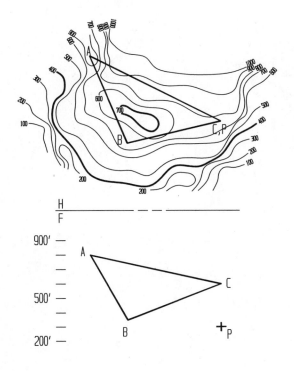

Figure 11–11 Practice problem.

Answer

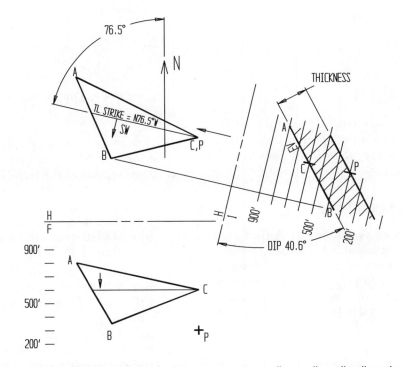

Figure 11–12 Location of where to measure dip, strike, dip direction of stratum A-B-C, and thickness between two stratums.

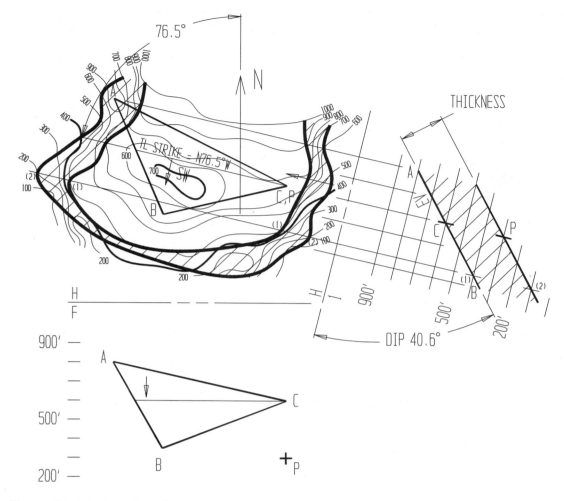

Figure 11-13 Location of outcrop.

▼ CUTS AND FILLS

Highways constructed through irregular terrain involve the principles of the intersection of contour lines and a plane (the roadways surface) to create limit lines for cut and fill.

Cuts and Fills Along a Level Road

A proposed highway, shown as center line A-B in Figure 11–14, is to be 80 feet wide at a constant elevation of 120 feet. Earth cuts (removal of earth) are to be made at a slope of 1:1 (1 unit of rise to 1 unit of run). Earth fills (addition of loose material—such as rocks and gravel) are to be made at a slope of 1:1½ (1 unit of rise to 1½ units of run).

Example
The problem is to determine the location of the limits of the required cut and fill for the example shown in

Figure 11–14. The procedure for doing so is as follows:

1. The top view shows the roadway on a given contour map. The front view shows the profile, a vertical section of the terrain along the center line of the road (A-B). The proposed highway is level; therefore, any earth above the elevation of 120 feet must be removed. Additional material must be added to where the earth's surface is below an elevation of 120 feet.

2. A cross-sectional, or edge view, (point of center line A-B) of the highway is shown in view 1 (Figure 11–15). Here the cut and fill ratios are drawn. The cut ratio of 1:1 is drawn upward, toward the top view. The fill ratio of 1:1½ is drawn downward, away from reference line H/1. In this view, the contour lines appear as straight horizontal lines (Figure 11–15).

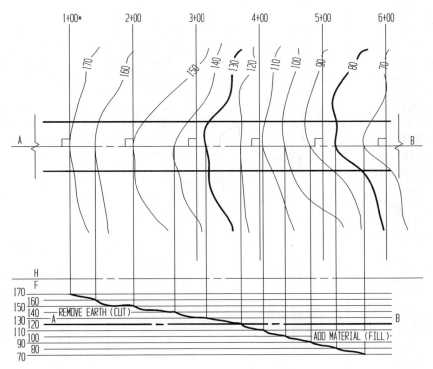

* NOTATION USED IN SURVEYING. STATION 2+00 IS 100
  FEET FROM STATION 1+00. A POINT HALF-WAY BETWEEN
  THESE TWO STATIONS WOULD BE MARKED 1+50.

Figure 11-14 Proposed level highway.

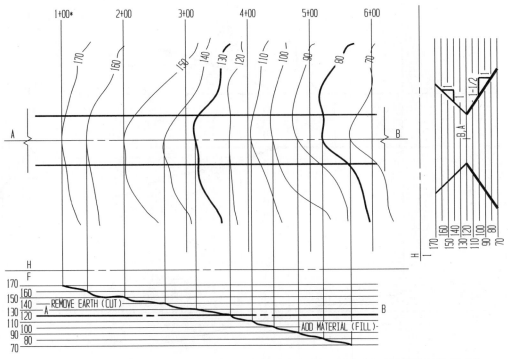

* NOTATION USED IN SURVEYING. STATION 2+00 IS 100
  FEET FROM STATION 1+00. A POINT HALF-WAY BETWEEN
  THESE TWO STATIONS WOULD BE MARKED 1+50.

Figure 11-15 Cut and fill ratios determined.

3. Begin with the fill and project the points of intersection between the fill slope and the 70-foot elevation onto the 70-foot contour line in the top view. The remaining points of intersection between the fill and the contour lines in view 1 are projected to their respective contour lines in the top view to determine the limits of the fill. The line indicating the limits of the fill is labeled "toe of fill" (Figure 11–16).

4. Again, in view 1, you will see the points where the cut slope intersects the various contour elevation lines. These points are projected onto their respective contour lines in the top view denoting the limits of the cut. The line indicating the limit of the cut is labeled "top of cut" (Figure 11–17).

Cuts and Fills Along a Grade Road

If the highway is sloping, a slightly different procedure must be followed. Line A-B, in Figure 11–18, is the center line of a proposed 80-foot wide highway having a 5% grade.

Stations are located in the front view where the highway crosses an elevation line. At station 1 + 00 the elevation of the highway is 130 feet. At station 2 + 00 the elevation of the highway is 120 feet. These points of intersection are projected to center line of the highway in the top view. The station lines are drawn in the top view perpendicular to the center line of the highway through the projected points. Because the highway has a grade of 5% in 100 feet of horizontal run, the elevation of the highway will rise 5 feet, from 120 to 125 feet. Cuts are to be made at a 1:1 slope, while fills are to be made at a 1:1½ slope.

Example
The steps required to find the required limits of cut and fill are listed below:

1. Begin at station 1 + 00 on the north edge of the road. Lay off a distance of 10 feet from the edge of the road. The elevation at this point, X, will be 140 feet. This is true because the contour lines in the vicinity of stations 1 + 00 and 2 + 00 are generally higher than the elevation of the road, thus cuts will be required. Because the cut slope is 1:1, a horizontal distance of 10 feet will mean a 10-foot increment in elevation. Lay off additional lines at intervals of 10 feet and mark their elevations.

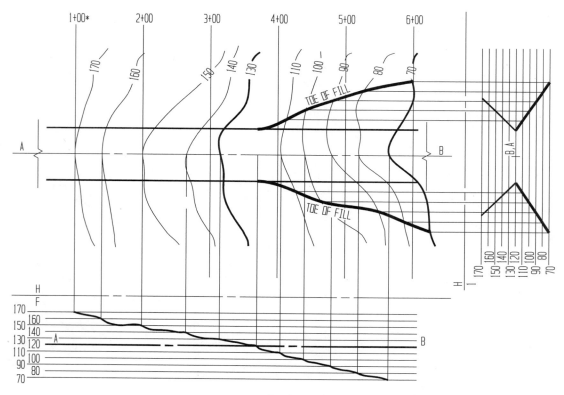

* NOTATION USED IN SURVEYING. STATION 2+00 IS 100 FEET FROM STATION 1+00. A POINT HALF-WAY BETWEEN THESE TWO STATIONS WOULD BE MARKED 1+50.

Figure 11–16 Toe of the fill.

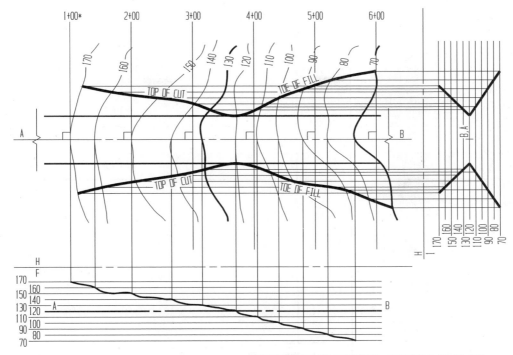

* NOTATION USED IN SURVEYING. STATION 2+00 IS 100
FEET FROM STATION 1+00. A POINT HALF-WAY BETWEEN
THESE TWO STATIONS WOULD BE MARKED 1+50.

Figure 11–17 Top of the cut.

2. At station 2 + 00 lay off a distance of 10 feet from the edge of the road. At this point the elevation will be 130 feet, because the road at this station is 120 feet in elevation. Lay out additional lines at 10-foot intervals as was done at station 1 + 00.

3. Notice that point X at station 1 + 00 is the same elevation as is the road at station 2 + 00 (point Y). The line joining these points is a contour line on the sloping 1:1 surface.

4. By connecting similar lines on the sloping surface, and by determining their intersections with the respective earth-contour lines, the top of the cut can be found. Notice the 170-foot contour line intersects with the earth's contour (elevation 170 feet) at point Z. This is a point common to the earth's surface and to the cut plane (1:1) surface. Notice that a similar process was followed on the south side of the road.

5. Approximately halfway between stations 2 + 00 and 3 + 00, observe that the contour lines are now lower than the proposed highway (Figure 11–19). This area will require fill. If we lay off a distance of 15 feet from the north edge of the road, the elevation at point P will be 100 feet. If we lay off another 15 feet from point P, the elevation will be 90 feet, and so forth. Remember that the slope of the fill is 1½:1, thus a 15-foot horizontal distance means a 10-foot drop in elevation.

6. Move to station 4 + 00 where the road elevation is 100 feet (point R). Again lay off a distance of 15 feet from R to T, the elevation of T will be 90 feet. Repeat this process to determine additional contour lines on the sloping surface, 1:1½. Connect a line from point R (elevation of 100 feet) at station 4 + 00 to 100 feet at station 3 + 00. Connect similar lines on the sloping surface, and by determining their intersections with the respective earth-contour lines, the toe of the fill can be found. This same procedure is followed for the fill on the south side of the highway.

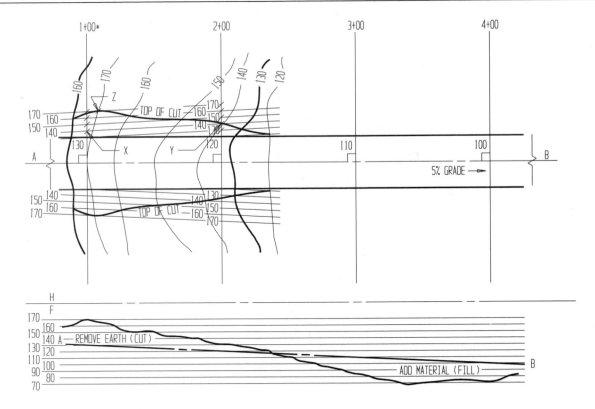

Figure 11-18 Top of cut for a highway at a 5% grade.

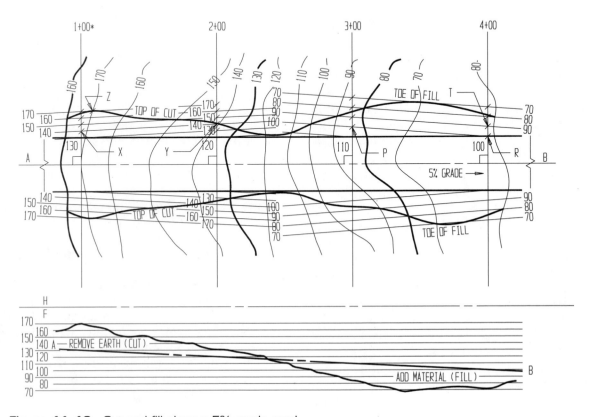

Figure 11-19 Cut and fill along a 5% grade road.

EXERCISE

Problem: A proposed level road, shown as center line A-B in Figure 1–20, is to be 30 feet wide at an elevation of 280 feet. Earth cuts (removal of earth) are to be made at a slope of 1:1. Earth fills (addition of loose material) are to be made at a slope of 1:2. Create a front view and draw the profile of the terrain along the center line of the road. Plot the lines of the cut and fill on the map. Identify the top of cut and the toe of fill. Scale 1" = 100'

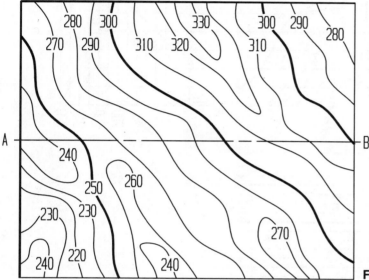

Figure 11–20 Practice problem.

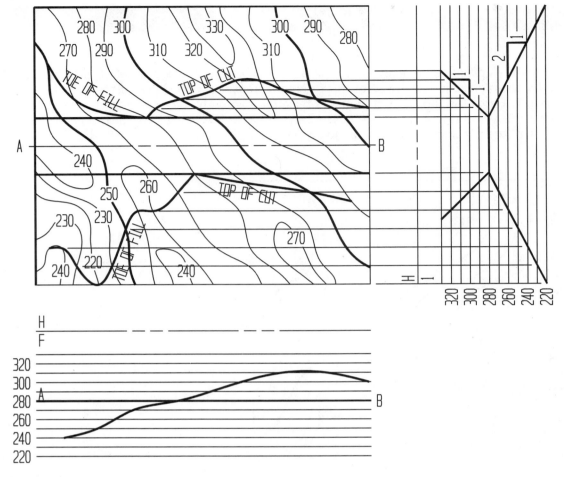

Figure 11–21 Cut and fill ratios determined.

Answer

This discussion of the application of descriptive geometry to civil engineering problems is complete for the purposes of this text. Further study would include such topics as calculating the volumes of the cuts and fills. A large amount of these kinds of calculations are currently being made using computers as are the designs and drawings for many civil engineering projects.

Name ______________________________

Date ______________________________

Course ______________________________

CHAPTER 11 STUDY QUESTIONS

The study Questions are intended to assess your comprehension of chapter material. Please write your answers to the questions in the space provided.

1. What is a topographic map?

____

____

2. What are contour lines?

____

____

3. How can you tell the difference between ridges and ravines?

____

____

4. Define the following:
   a. Stratum of rock: ______________________________
   b. Bedding planes: ______________________________
   c. Fault: ______________________________
   d. Outcrop: ______________________________
   e. Strike: ______________________________
   f. Dip: ______________________________

5. Explain how you would determine the thickness of a vein of ore.

____

____

6. In which view would you see the strike of a plane? Why?

____

____

7. In which view would you see the dip of a plane? Why?

____

____

8. The outcrop of a vein of ore represents the intersection of ______________ and____________ .

9. Explain the term **cut and fill**.

____

____

10. Explain a fill ratio of 1:2 (you may use a sketch to help illustrate your explanation).

_____

_____

11. Each point on a top-of-the-cut line or the toe-of-the-fill line is a point common to which two surfaces?

_____

_____

The problems in this chapter are based on a variety of industrial applications. In solving these problems you should try to utilize direct approaches that get to the heart of the problem. At this analysis stage of the design process, many problems use center lines and single lines to represent the problem situation. Yet, when necessary, relevant sizes are given.

The following problems may be drawn using instruments on the page provided, created using a 2D or 3D CAD system, or a combination of drawing board and CAD. Refer to Appendix A for additional dimensions needed to solve problems three-dimensionally.

Chapter 11, Problem 1: WHILE EXPLORING FOR OIL-BEARING SAND, A COMPANY DRILLS AT POINT X AND HITS OIL-BEARING SAND AT 750 FEET BELOW THE SURFACE. ANOTHER TEST WELL WAS DRILLED AT POINT Y STRIKING THE SAME SAND AT 1660 FEET BELOW THE SURFACE. A THIRD TEST WELL WAS DRILLED AT POINT Z HITTING THE STRATUM OF SAND AT 1790 FEET BELOW THE SURFACE. ALL THREE DRILLINGS INDICATE A STRATUM TRUE THICKNESS OF 40 FEET. FIND THE STRIKE, DIP, AND DIP DIRECTION OF THIS OIL-BEARING SAND LAYER. SHOW THE UPPER AND LOWER BEDDINGS PLANES IN YOUR SOLUTION.

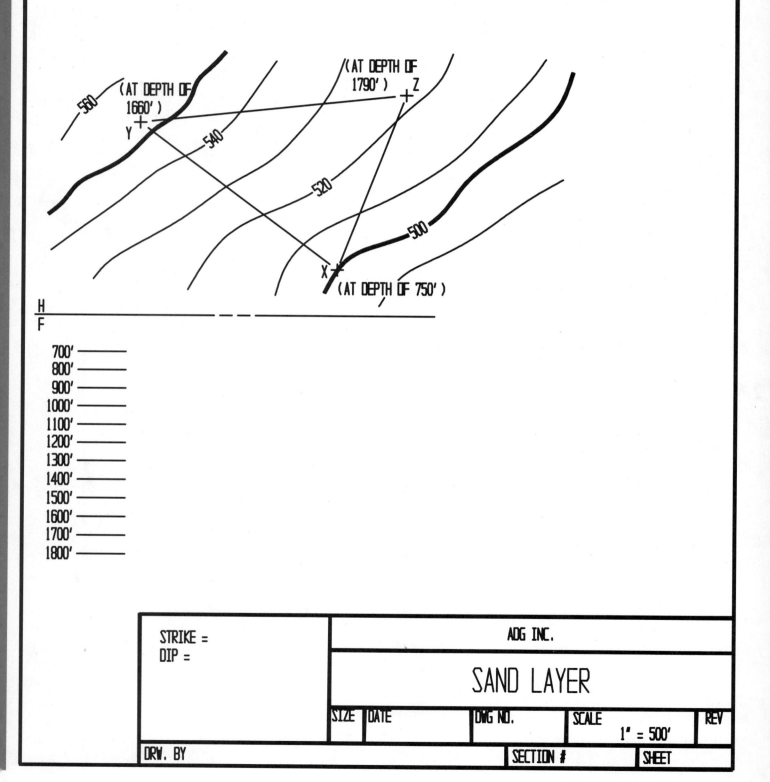

STRIKE =	ADG INC.	
DIP =	SAND LAYER	

SIZE	DATE	DWG NO.	SCALE	REV
			1" = 500'	

DRW. BY		SECTION #	SHEET

Chapter 11, Problem 2: FROM THE ORIGIN, THE LOCATIONS OF THREE VERTICAL BOREHOLES ARE SHOWN. DISTANCES ALONG THE GIVEN BEARINGS ARE MAP DISTANCES FROM THE ORIGIN. THE DRILLINGS HAVE ESTABLISHED POINTS X, Y, AND Z ON THE UPPER BEDDING PLANE OF A VEIN OF ORE. ALL THREE DRILLINGS INDICATE A LOWER BEDDING PLANE 50 FEET BELOW THE UPPER. FIND THE STRIKE, DIP, DIP DIRECTION, AND TRUE THICKNESS OF THE VEIN.

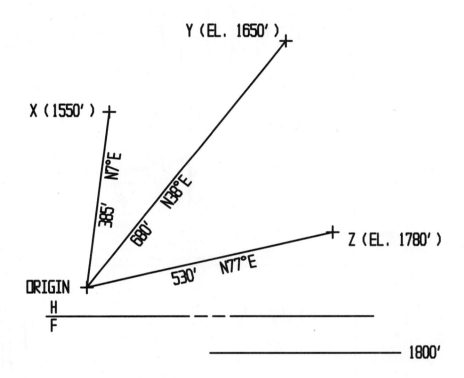

THICKNESS =	ADG INC.			
STRIKE =	VERTICAL BOREHOLES			
DIP =				
	SIZE	DATE	DWG NO.	SCALE 1" = 200'	REV
DRW. BY			SECTION #	SHEET

Chapter 11, Problem 3: POINTS A AND C ARE ON THE OUTCROP OF THE UPPER BEDDING PLANE OF A VEIN OF LOW-GRADE ORE. CONTINUED DRILLING AT A HAS INTERSECTED THE LOWER SURFACE OF THE VEIN 20 FEET BELOW A. THE STRIKE OF THE VEIN IS N75°W FROM POINT A. FIND THE DIP, THICKNESS, AND PROBABLE OUTCROP FOR THE UPPER AND LOWER PLANES. INDICATE THE OUTCROP WITH CROSSHATCHED LINES.

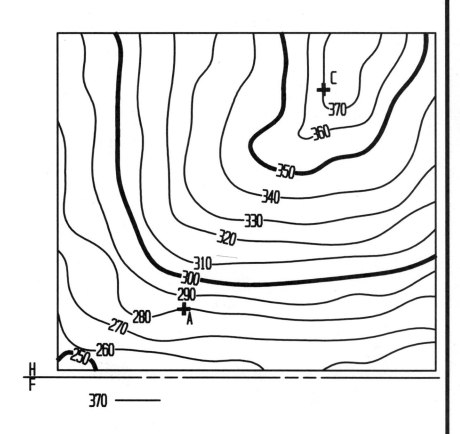

DIP =	ADG INC.			
THICKNESS =	OUTCROP			
	SIZE DATE	DWG NO.	SCALE 1" = 60'	REV
DRW. BY		SECTION #	SHEET	

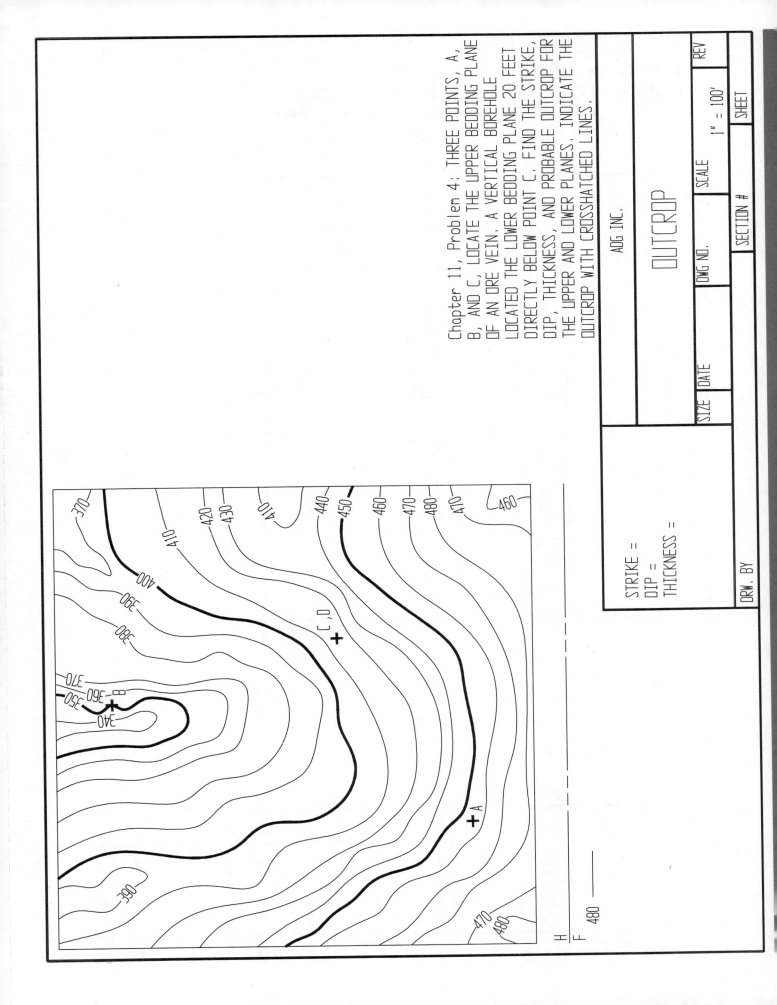

Chapter 11, Problem 4: THREE POINTS, A, B, AND C, LOCATE THE UPPER BEDDING PLANE OF AN ORE VEIN. A VERTICAL BOREHOLE LOCATED THE LOWER BEDDING PLANE 20 FEET DIRECTLY BELOW POINT C. FIND THE STRIKE, DIP, THICKNESS, AND PROBABLE OUTCROP FOR THE UPPER AND LOWER PLANES. INDICATE THE OUTCROP WITH CROSSHATCHED LINES.

ADG INC.

OUTCROP

SIZE	DATE	DWG NO.	SCALE 1" = 100'	REV
			SECTION #	SHEET

STRIKE =
DIP =
THICKNESS =

DRW. BY

Chapter 11, Problem 5: LINE XY IS THE CENTER LINE OF AN ACCESS ROAD (30 FEET WIDE) ON THE TOP OF AN EARTH DAM. THE ELEVATION OF THE ROAD IS 430 FEET. SHOW THE CUT AND FILL ON BOTH SIDES OF THE DAM. THE CUT HAD A 1:2 SLOPE, AND THE FILL A 1:3 SLOPE. DRAW A FRONT VIEW SHOWING THE PROFILE OF THE TERRAIN ALONG THE CENTER OF THE ROAD.

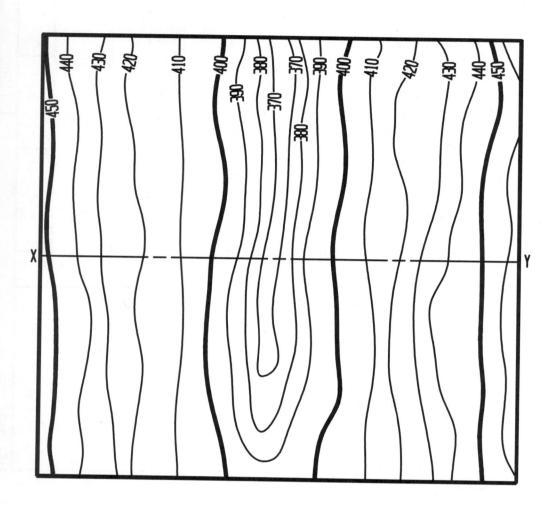

	ADG INC.			
	EARTH DAM			
SIZE	DATE	DWG NO.	SCALE 1" = 100'	REV
DRW. BY		SECTION #	SHEET	

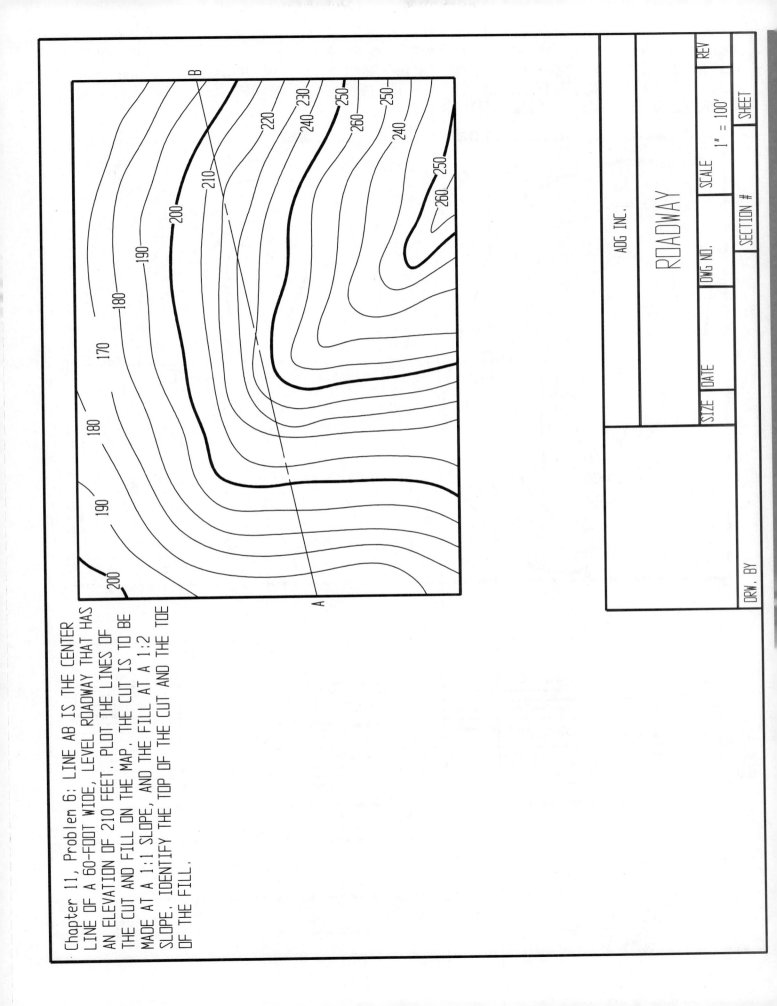

Chapter 11, Problem 6: LINE AB IS THE CENTER
LINE OF A 60-FOOT WIDE, LEVEL ROADWAY THAT HAS
AN ELEVATION OF 210 FEET. PLOT THE LINES OF
THE CUT AND FILL ON THE MAP. THE CUT IS TO BE
MADE AT A 1:1 SLOPE, AND THE FILL AT A 1:2
SLOPE. IDENTIFY THE TOP OF THE CUT AND THE TOE
OF THE FILL.

ADG INC.

ROADWAY

| SIZE | DATE | | DWG NO. | SCALE 1" = 100' | REV |

| DRW. BY | SECTION # | SHEET |

Chapter 11, Problem 7: LINE AB IS THE CENTER LINE OF A 50 FOOT WIDE ROADWAY. THE ROAD HAS A 5 PERCENT GRADE UP FROM STATION 1+00. IN THE FRONT VIEW DRAW THE PROFILE OF THE TERRAIN ALONG THE CENTER LINE OF THE ROAD. PLOT THE TOP OF THE CUT AND THE TOE OF THE FILL. THE CUT SLOPE IS 1:1, AND THE FILL SLOPE IS 1:2.

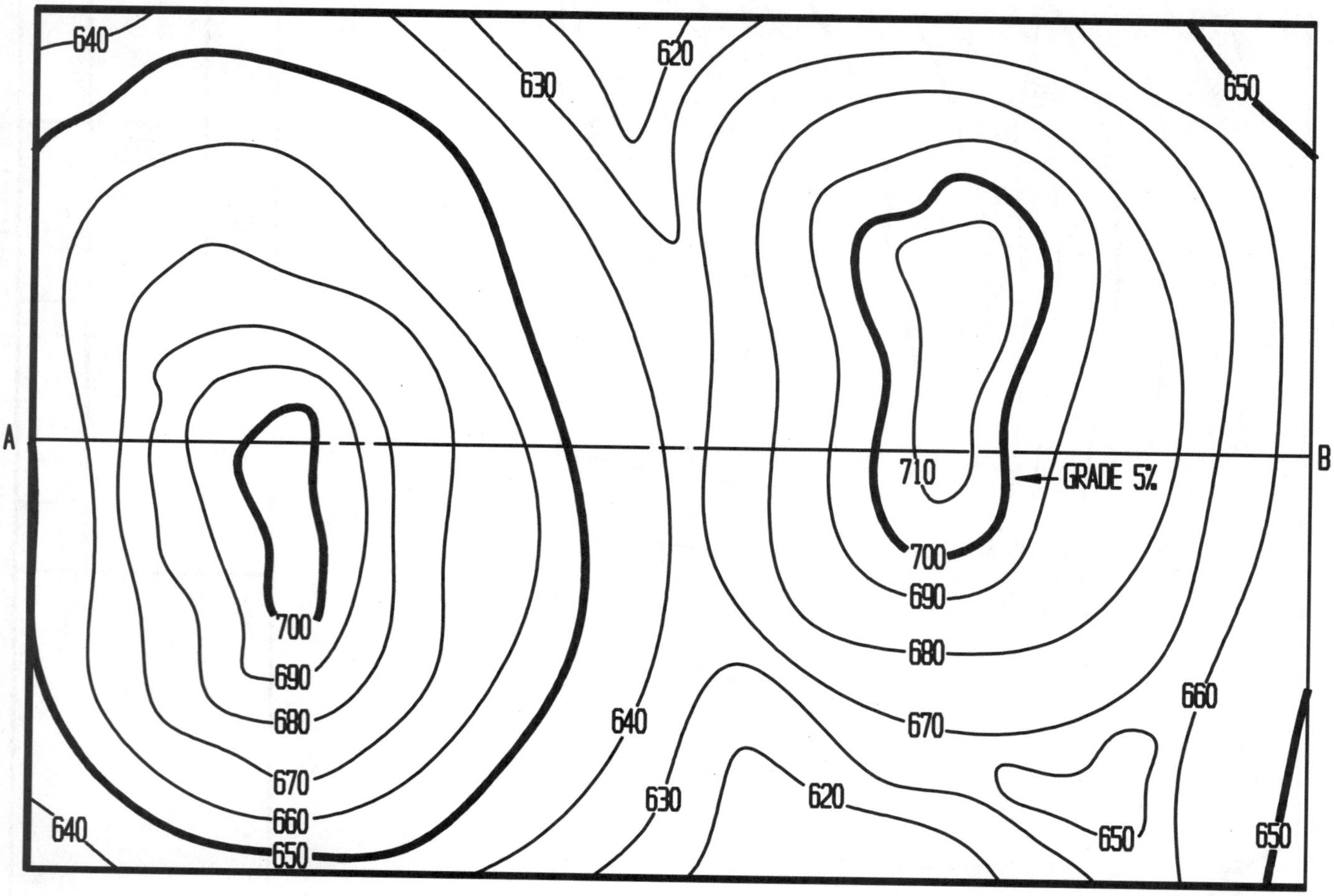

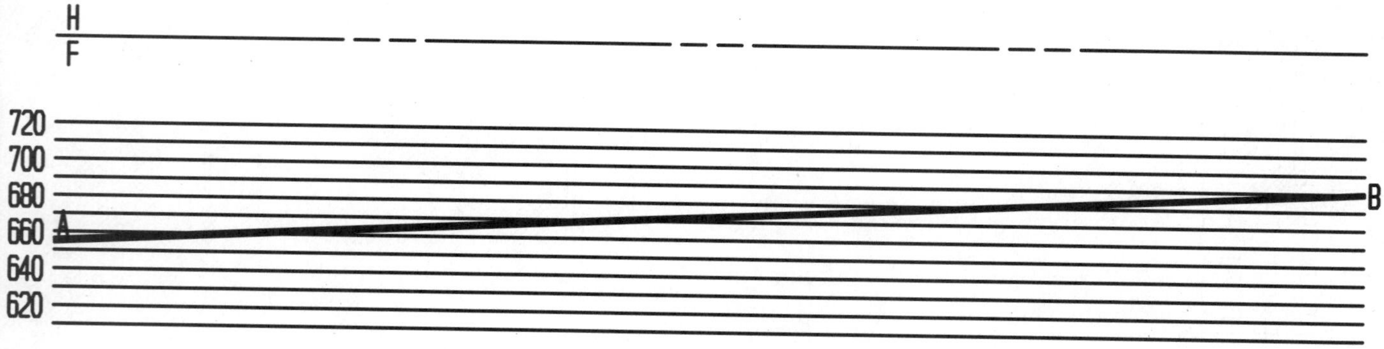

	ADG INC.			
	HIGHWAY			
SIZE	DATE	DWG NO.	SCALE 1" = 100'	REV
DRW. BY		SECTION #	SHEET	

Chapter 11, Problem 8: LOCATE THE TOP OF THE CUT AND THE TOE OF THE FILL FOR A 50-FOOT WIDE ROAD, THE CENTER LINE OF WHICH RUNS N75°W FROM POINT Y. THE ROAD HAS A 5 PERCENT GRADE UP FROM POINT Y, WHICH IS AT THE 160-FOOT ELEVATION. THE CUT SLOPE IS 1:1, AND THE FILL SLOPE IS 1:2.

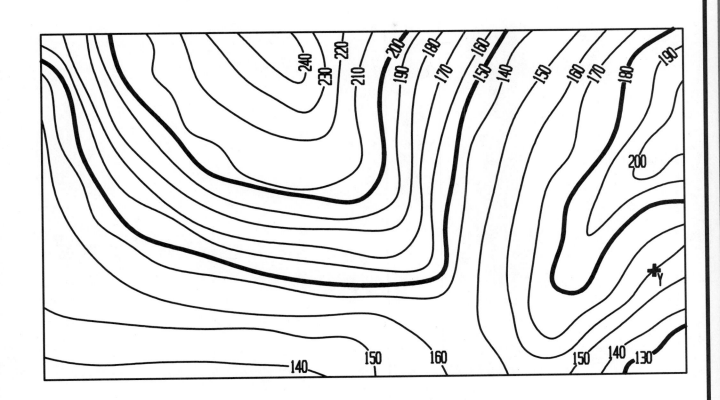

240 ——

APPENDIX A

The problems in this text are based on a variety of industrial applications. In solving these problems you should try to utilize direct approaches that get to the heart of the problem. At this analysis stage of the design process, many problems use center lines and single lines to represent the problem situation. Yet, when necessary, relevant sizes are given.

The following problems may be drawn using a 2D or 3D CAD system. Additional dimensions needed to solve problems on a CAD system are included in this section. The dimensions given are full scale, but the drawings themselves have been reduced and are not to scale.

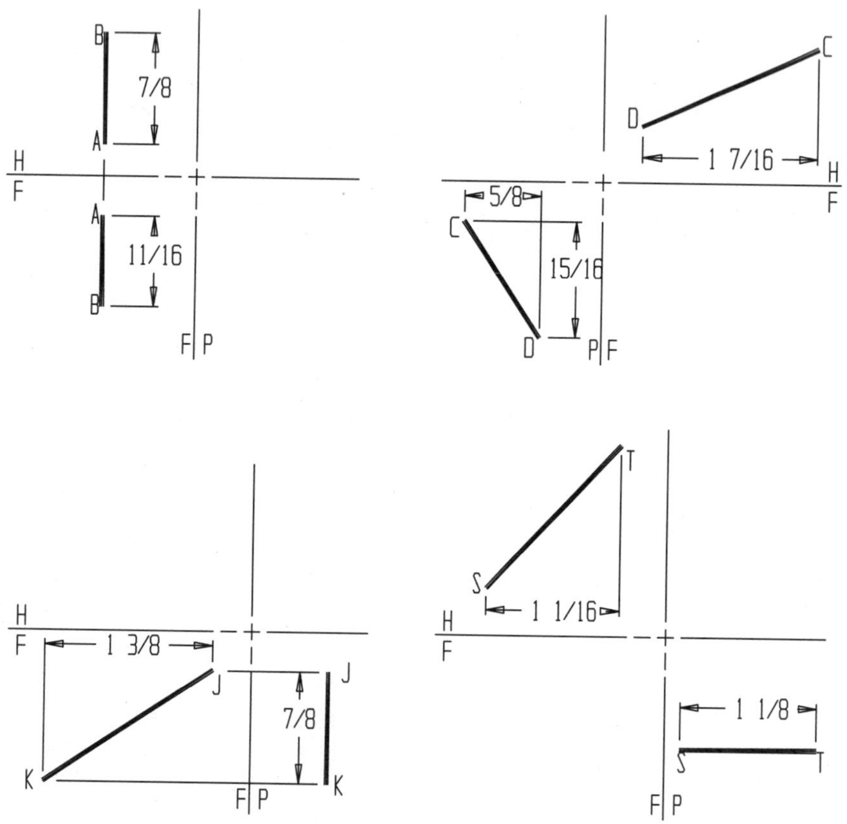

Chapter 4, Problem 2:

Chapter 4, Problem 1:

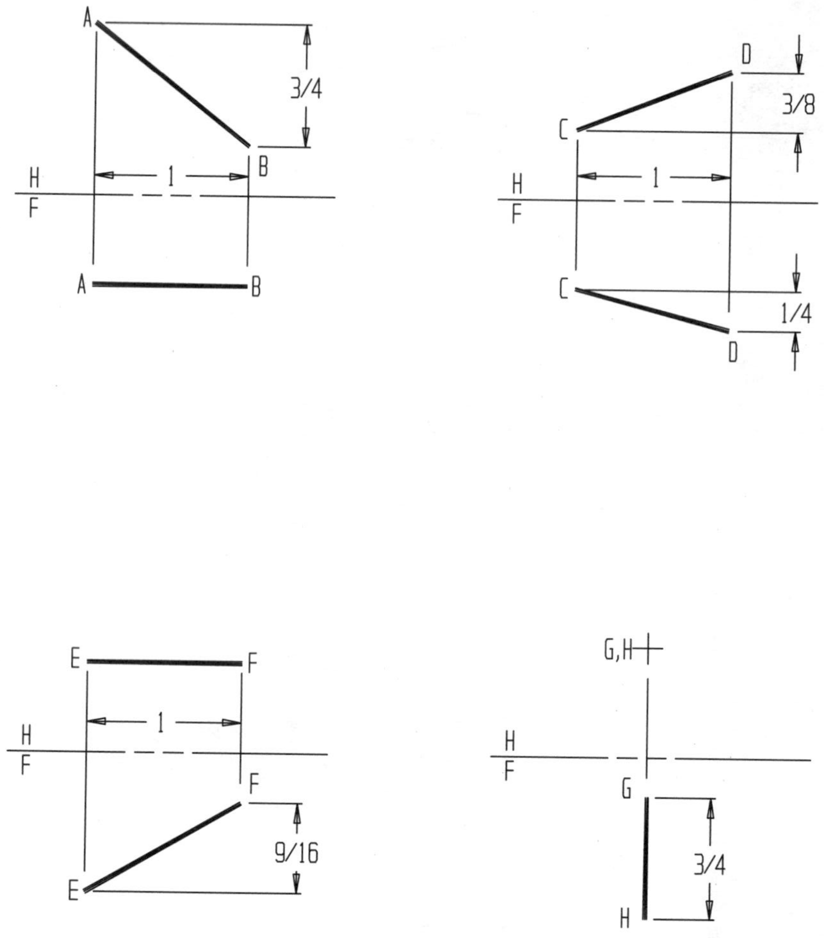

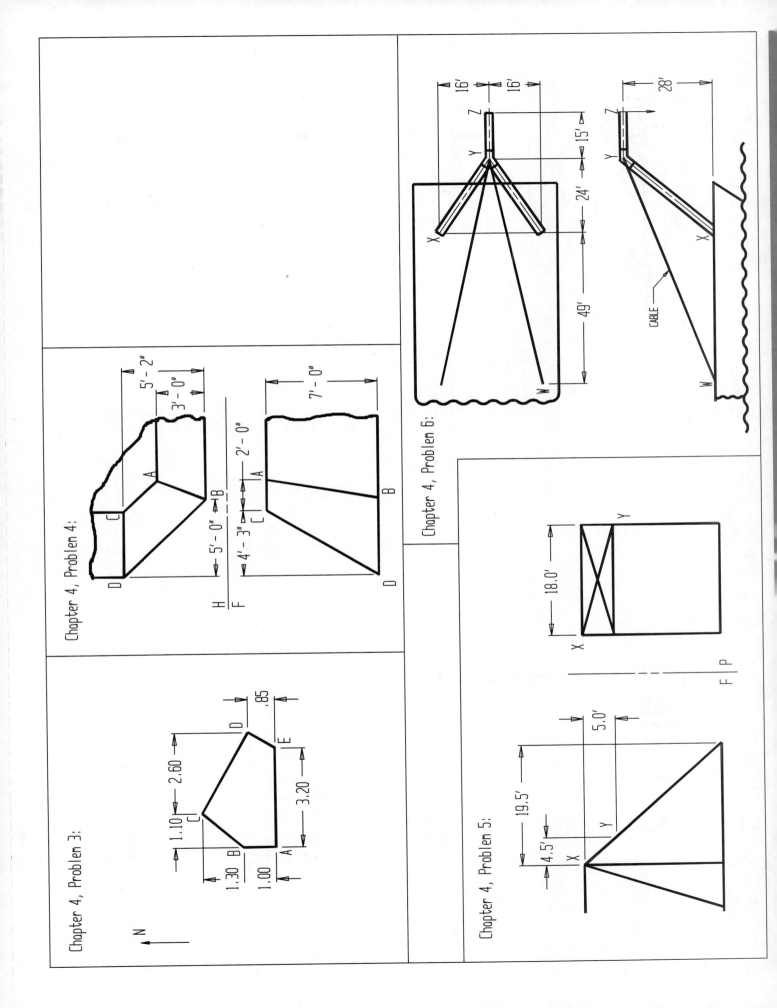

Chapter 4, Problem 3:

N

1.10
1.30   B
1.00   A
2.60
.85
D
E
3.20

Chapter 4, Problem 4:

5' – 2"
3' – 0"
A
B
C
D

H
F

5' – 0"
4' – 3"
2' – 0"
A
C
B
D
7' – 0"

Chapter 4, Problem 5:

19.5'
4.5'
5.0'
X
Y

F | P

18.0'
X
Y

Chapter 4, Problem 6:

16'
16'
Z
Y
X
W
15'
24'
49'

28'
Z
Y
X
CABLE
W

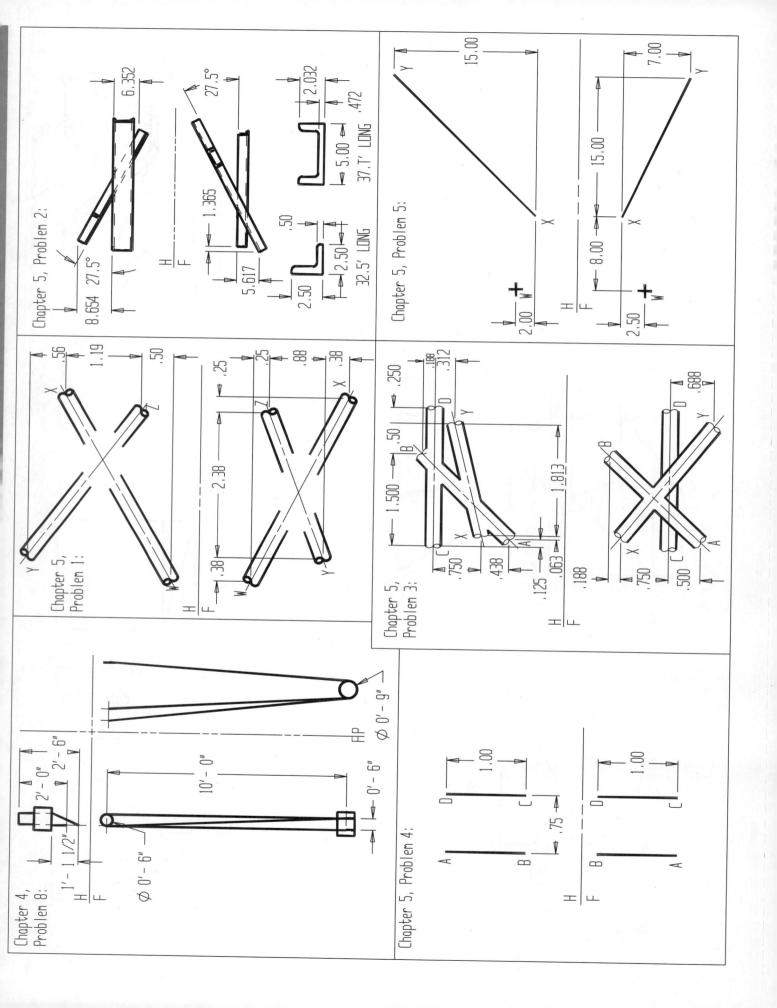

Chapter 5, Problem 2:

6.352
27.5°
8.654 27.5°
H/F
1.365
27.5°
5.617
2.032
.472
5.00
37.T' LONG
.50
2.50
2.50
32.5' LONG

Chapter 5, Problem 5:

15.00
Y
X
2.00
+W
H/F
7.00
Y
15.00
8.00
X
2.50
+W

Chapter 5, Problem 1:

X
.56
1.19
.50
Y
Z
.25
.25
.88
.38
X
2.38
.38
Z
Y
H/F
W

Chapter 5, Problem 3:

.250
.188
.312
.50
B
Y
1.500
D
.750
X
A
.438
1.813
.125
.063
H/F
.188
X
.750
C
.500
B
D
.688
Y
A

Chapter 4, Problem 8:

2'-0"
2'-6"
1'- 1 1/2"
H/F
10'-0"
Ø 0'-6"
Ø 0'-9"
HP
0'-6"

Chapter 5, Problem 4:

A
D
1.00
B
C
.75
H/F
B
D
1.00
A
C

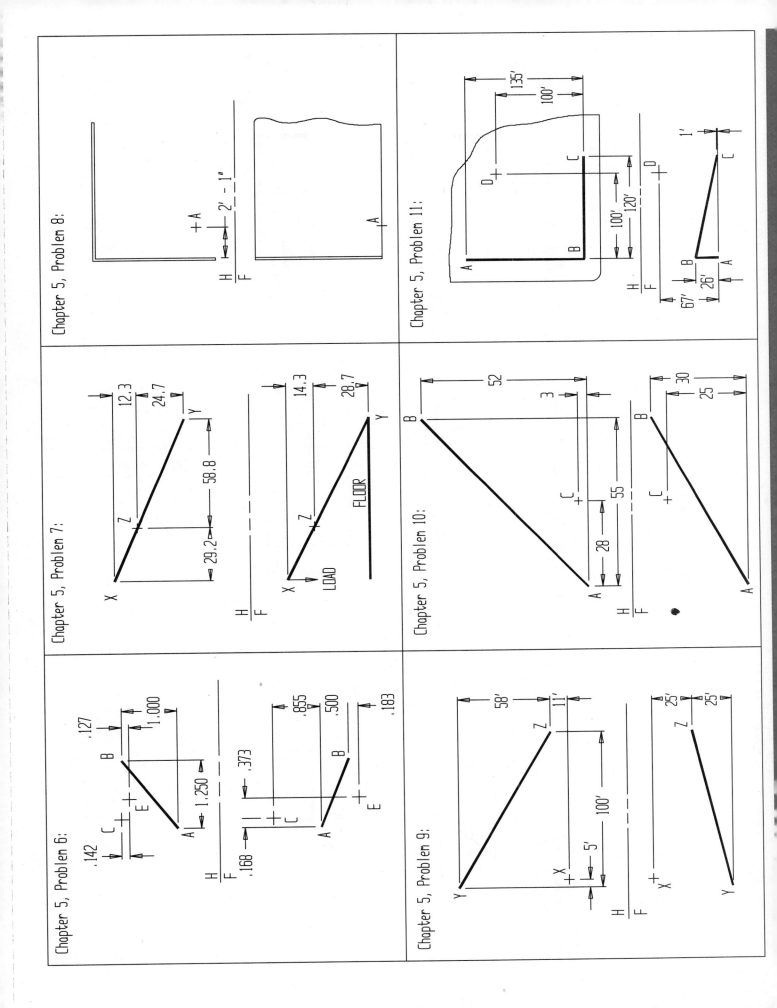

Chapter 5, Problem 6:

Chapter 5, Problem 7:

Chapter 5, Problem 8:

Chapter 5, Problem 9:

Chapter 5, Problem 10:

Chapter 5, Problem 11:

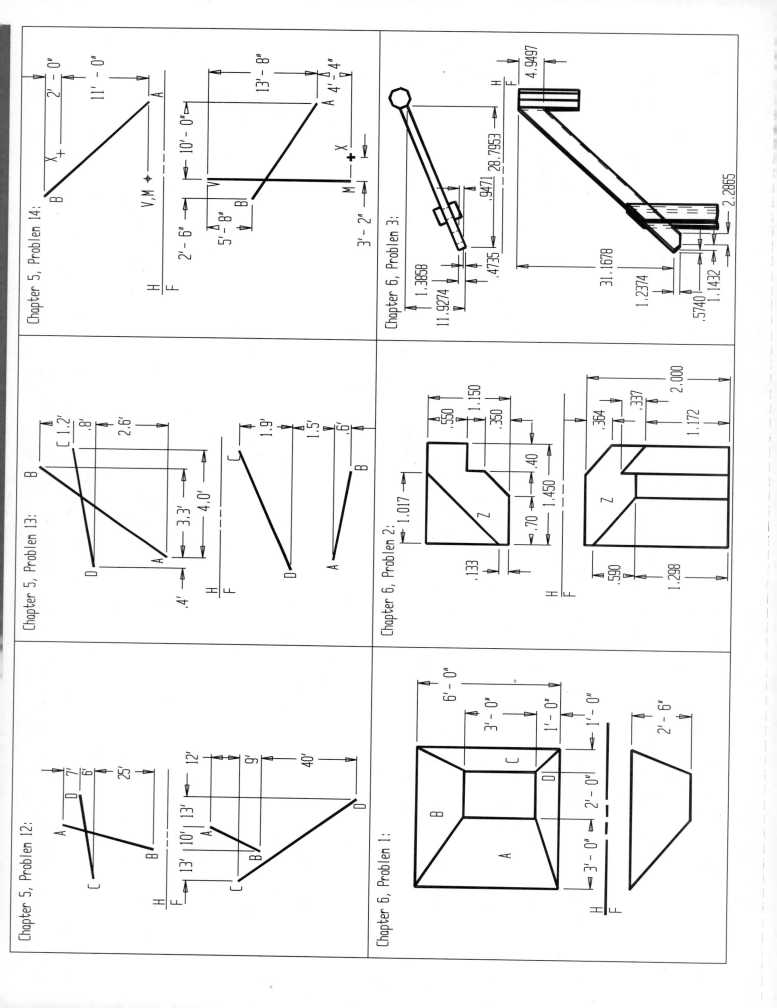

Chapter 5, Problem 14:

2' – 0"
11' – 0"
2' – 0"
A
B
X +
V,M ◆
H/F

13' – 8"
10' – 0"
A
4' – 4"
2' – 6"
5' – 8"
V
B
X +
M
3' – 2"

Chapter 5, Problem 13:

B
C 1.2'
.8'
2.6'
D
A
3.3'
4.0'
.4'
H/F

C
1.9'
1.5'
.6'
D
A
B

Chapter 5, Problem 12:

A
D
7'
6'
25'
B
C
12'
9'
40'
H/F
13' 10' 13'
A
B
C
D

Chapter 6, Problem 3:

4.9497
H/F
28.7953
.9471
.4735
1.3858
11.9274
31.1678
1.2374
.5740
1.1432
2.2865

Chapter 6, Problem 2:

1.017
.550
1.150
.350
.40
.70
1.450
.133
Z

.364
.337
2.000
1.172
Z
.590
1.298
H/F

Chapter 6, Problem 1:

6' – 0"
3' – 0"
1' – 0"
B
A
C
D
2' – 0"
1' – 0"
H/F
3' – 0"
2' – 6"

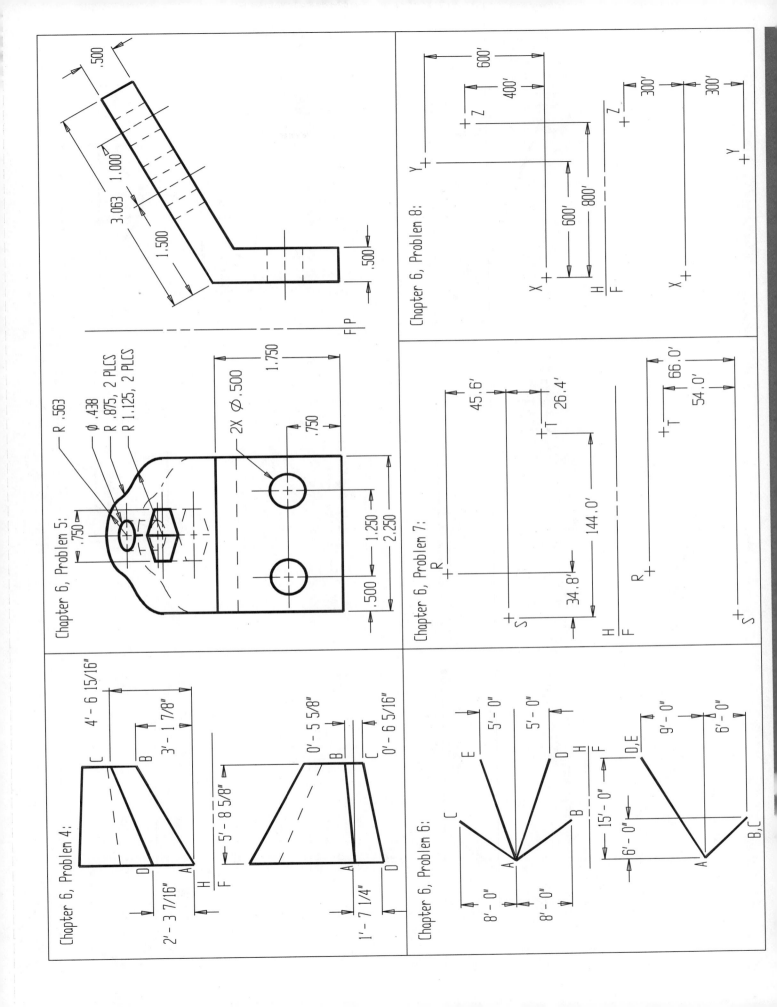

Chapter 6, Problem 4:

Chapter 6, Problem 5:

R .563
⌀ .438
R .875, 2 PLCS
R 1.125, 2 PLCS

2X ⌀ .500

Chapter 6, Problem 6:

Chapter 6, Problem 7:

Chapter 6, Problem 8:

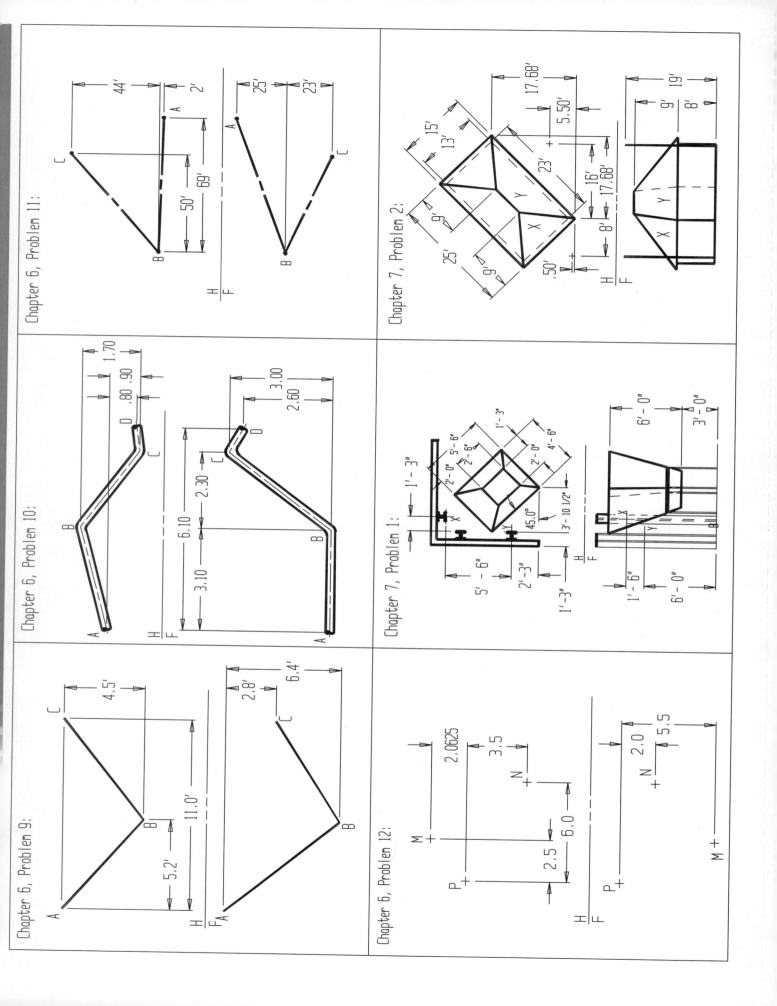

Chapter 6, Problem 9:

Chapter 6, Problem 10:

Chapter 6, Problem 11:

Chapter 6, Problem 12:

Chapter 7, Problem 1:

Chapter 7, Problem 2:

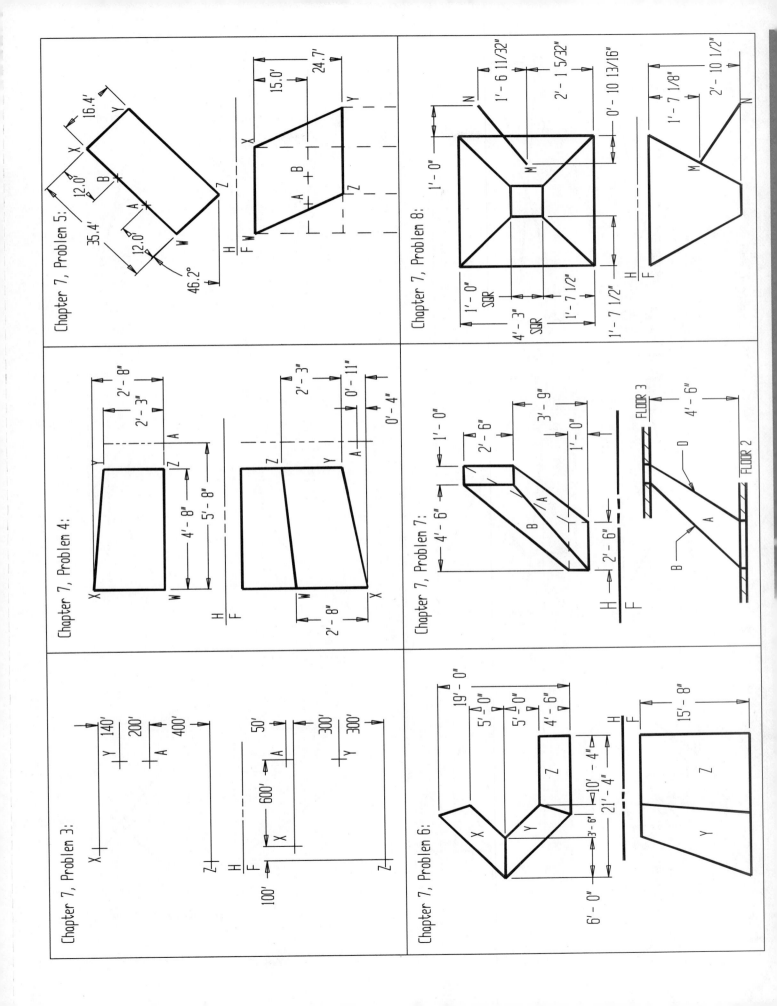

Chapter 7, Problem 3:

Chapter 7, Problem 4:

Chapter 7, Problem 5:

Chapter 7, Problem 6:

Chapter 7, Problem 7:

Chapter 7, Problem 8:

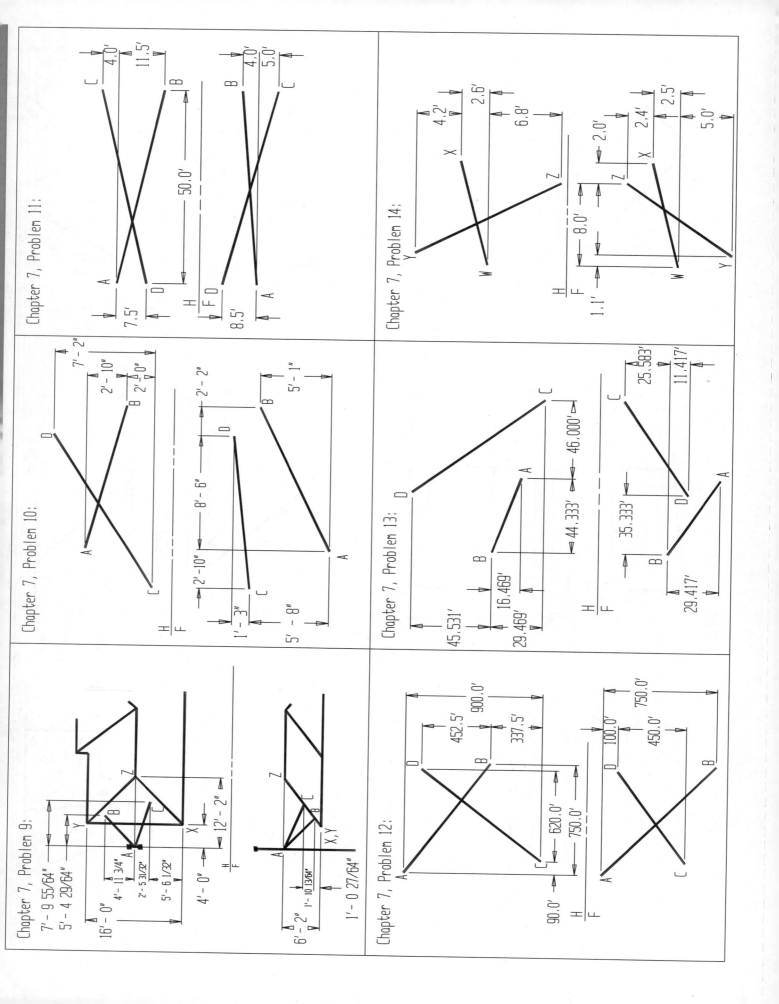

Chapter 7, Problem 11:

Chapter 7, Problem 10:

Chapter 7, Problem 9:

Chapter 7, Problem 14:

Chapter 7, Problem 13:

Chapter 7, Problem 12:

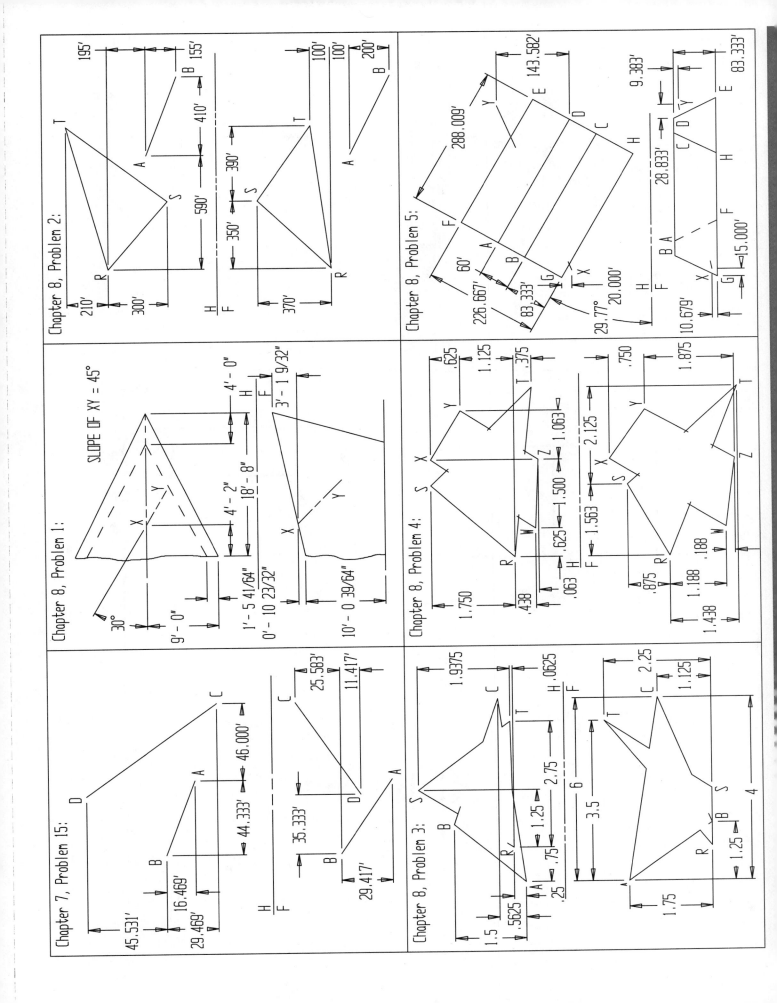

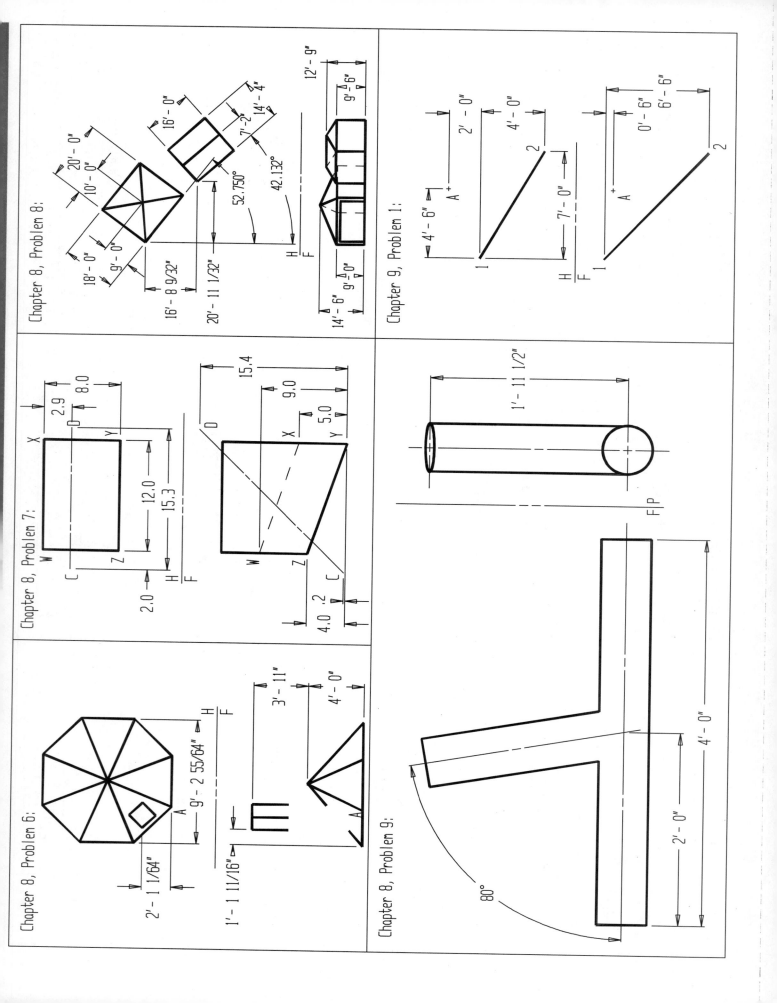

Chapter 8, Problem 8:

Chapter 9, Problem 1:

Chapter 8, Problem 7:

Chapter 8, Problem 6:

Chapter 8, Problem 9:

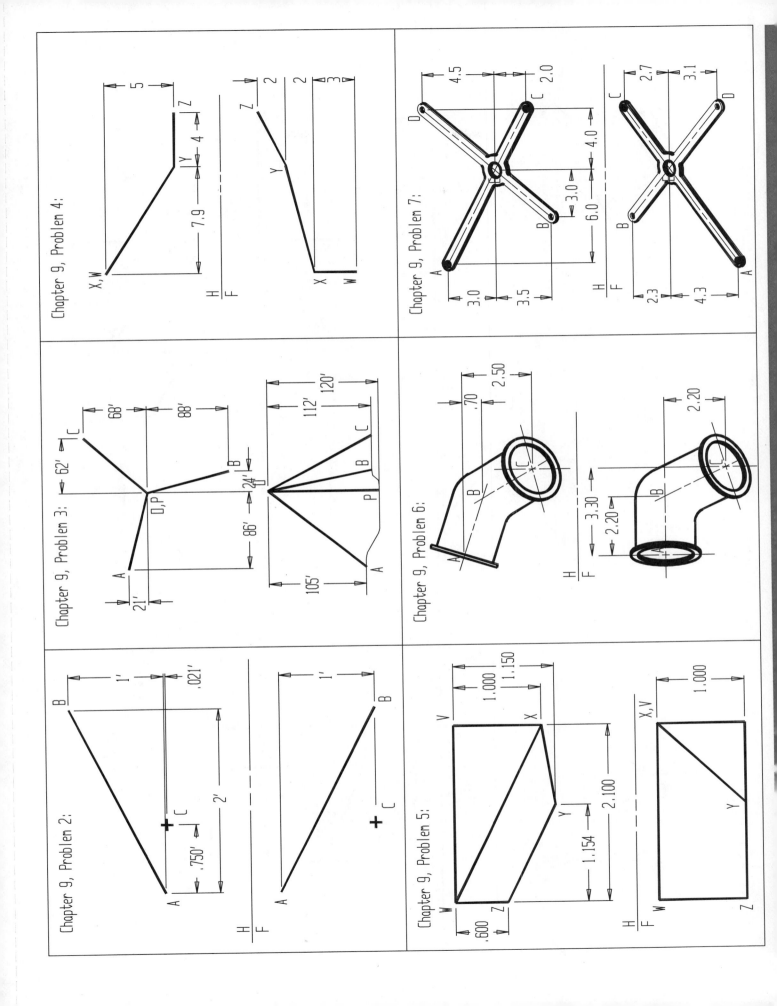

Chapter 9, Problem 4:

Chapter 9, Problem 3:

Chapter 9, Problem 2:

Chapter 9, Problem 7:

Chapter 9, Problem 6:

Chapter 9, Problem 5:

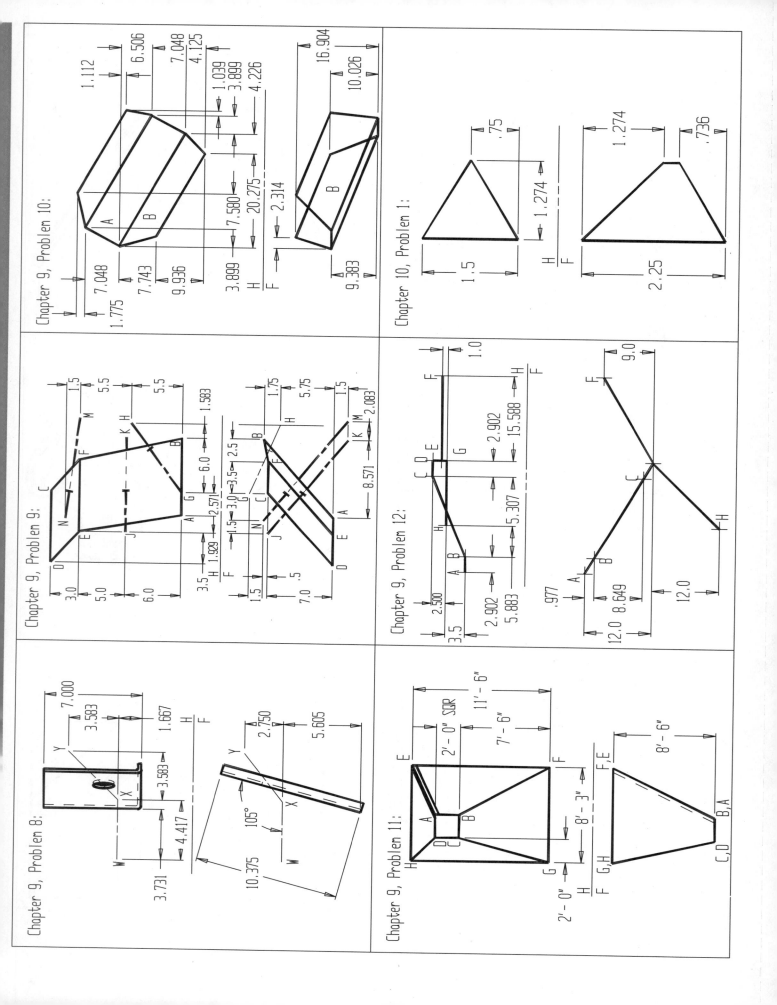

Chapter 9, Problem 10:

Chapter 10, Problem 1:

Chapter 9, Problem 9:

Chapter 9, Problem 12:

Chapter 9, Problem 8:

Chapter 9, Problem 11:

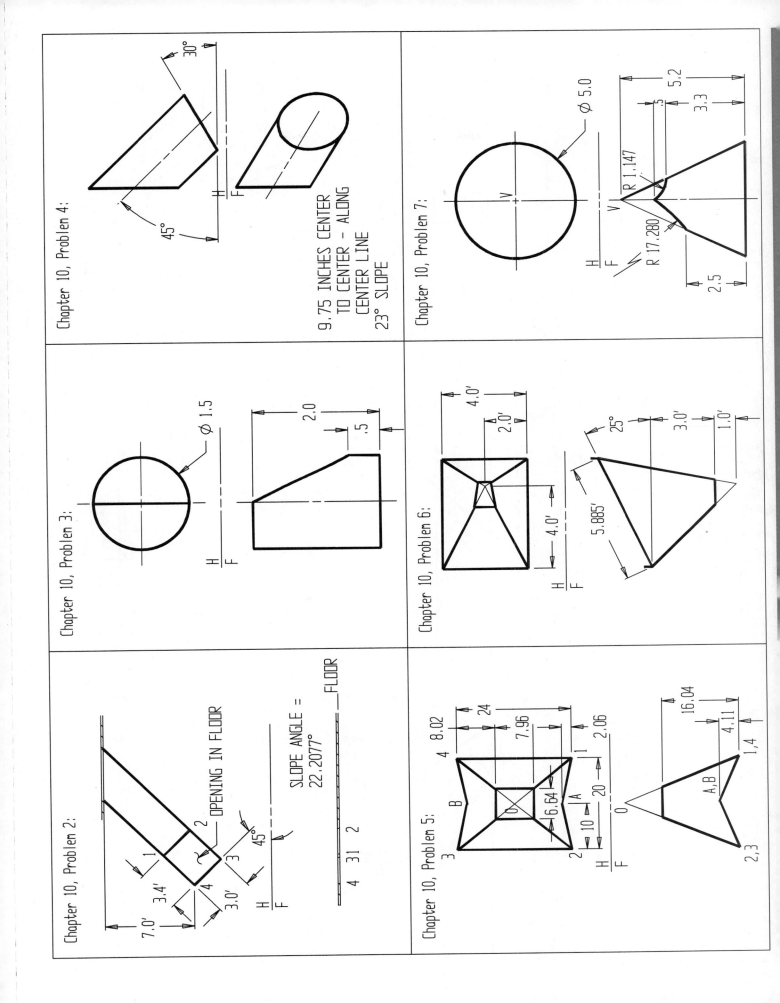

Chapter 10, Problem 4:

30°

45°

H
F

9.75 INCHES CENTER
TO CENTER - ALONG
CENTER LINE
23° SLOPE

Chapter 10, Problem 7:

Ø 5.0

V

5.2

.5

3.3

R 1.147

R 17.280

2.5

H
F

V

Chapter 10, Problem 3:

Ø 1.5

2.0

.5

H
F

Chapter 10, Problem 6:

4.0'

2.0'

4.0'

25°

3.0'

1.0'

5.885'

H
F

Chapter 10, Problem 2:

OPENING IN FLOOR

7.0'

3.4'

1

2

4

3

3.0'

45°

SLOPE ANGLE =
22.2077°

FLOOR

4 31 2

H
F

Chapter 10, Problem 5:

3

4

8.02

B

24

7.96

0

6.64

A

1

10

20

2.06

2

16.04

4.11

A,B

1,4

0

2,3

H
F

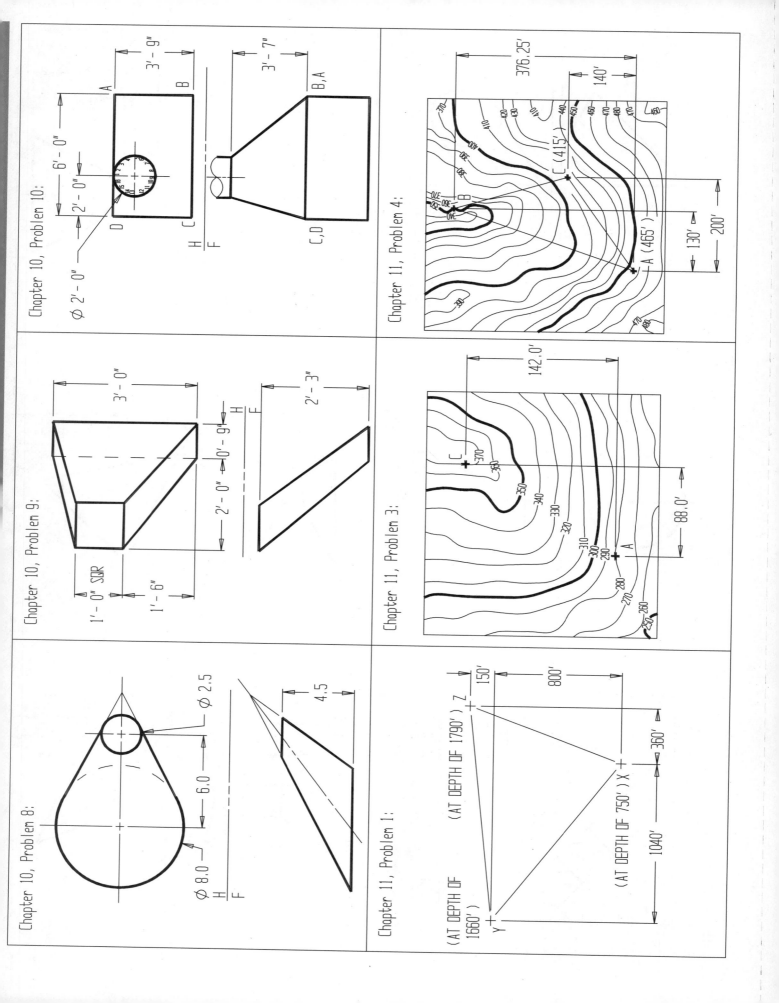

Chapter 10, Problem 10:

⌀ 2' - 0"

3' - 9"

6' - 0"

2' - 0"

A

B

D

H C
F

3' - 7"

B,A

C,D

Chapter 10, Problem 9:

3' - 0"

2' - 0"

1' - 0" SQR

1' - 6"

H
F

0' - 9"

2' - 3"

Chapter 10, Problem 8:

⌀ 2.5

⌀ 8.0

6.0

4.5

H
F

Chapter 11, Problem 4:

376.25'

140'

130'

200'

C (415')

B
A (465')

270
410
420
430
440
450
460
470
480
460
400
380
390
370
360
350
340
390
470
480

Chapter 11, Problem 3:

142.0'

88.0'

C

A

360
350
340
330
320
310
300
290
280
270
260
250

Chapter 11, Problem 1:

(AT DEPTH OF 1790')

(AT DEPTH OF 1660')

(AT DEPTH OF 750') X

150'

800'

360'

1040'

Z

Y

APPENDIX B

▼ DESCRIPTIVE GEOMETRY STEPS OF PROCEDURE REFERENCE PAGES

I. Determine **line of sight.**

    A. Locate **true-length line.**

        1. Is the true length line **in** one of the **principal planes?**

            a. If the line of sight is perpendicular to the object line in one view, it will appear true length in the adjacent view and can be measured.

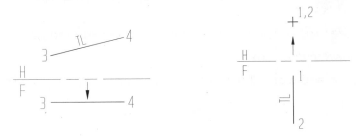

        2. If the above rule does not apply, it is an oblique line and cannot be measured as is.

            a. Draw a **primary auxiliary** view **to find the true length.**

                (1) Draw the line of sight perpendicular to the oblique line in either principal view

    B. Draw the **point view of a true-length line.**

        1. The line of sight must be drawn parallel to the true-length line.

        The line of sight locates:

        Shortest distance between a point and a line

        Shortest distance between two parallel lines

        Shortest distance, whether it be perpendicular, horizontal, or a specific slope, between two lines, whether the lines are perpendicular or skewed

    C. Draw an **edge view of a plane** (all points fall in a line).

        1. The line of sight must be drawn parallel to a true-length line that lies in the plane.

        The line of sight locates:

        Shortest distance between a point and a plane, whether the shortest distance be horizontal, perpendicular, or at a specific slope

        Angle between a line and a plane

        Angle between two planes (dihedral angle)

D. Draw the **true shape of a plane.**

    1. The line of sight must be drawn perpendicular to the edge view of a plane.

II. Draw the **reference line** perpendicular to the line of sight and label.

III. Draw **projection lines** parallel to the line of sight.

IV. Transfer **points of measurement** to the new view from the related view and label.

    A. Front-adjacent auxiliary

        1. Always projects from the front view

        2. Always utilizes the frontal reference plane for measurements

        3. Always shows true depth

    B. Top-adjacent auxiliary

        1. Always projects from the top view

        2. Always utilizes the horizontal reference plane for measurements

        3. Always shows true height

    C. Side-adjacent auxiliary

        1. Always projects from the side view

        2. Always utilizes the profile reference plane for measurements

        3. Always shows true length

V. Connect the points and **label** all true-length lines, edge views, and/or true-shape planes.

GLOSSARY

Adjacent views   Views that are aligned side by side to share a common dimension.

Auxiliary view   A view on any projection plane other than a primary or principal projection plane.

Auxiliary-elevation view   An elevation view which is not one of the principal elevation views. Auxiliary-elevation projection planes are always perpendicular to the horizontal projection plane.

Bearing of a line   The compass reading or angle of a line.

Bedding planes   Two parallel planes in which a stratum of rock in the earth's surface lies between.

Bend allowances   Allowance that accounts for the thickness of the metal when thicker than 24 gage (.025 in.).

Bend lines   Lines that a development (flat pattern) is folded along. The bend lines are often located by means of small punch marks.

Contour interval   The distance between adjacent contour lines.

Contour line   A line that denotes a series of connected points at a particular elevation representing the terrain on a topographic map.

Cut   The removal of earth from the terrain.

Descriptive geometry   The theory of orthographic projection. It can also be defined as a graphical method of solving solid (or space) analytic geometry problems.

Development   A solid shape that is unfolded or unrolled onto a flat plane.

Dihedral angle   The angle that is formed by two intersecting planes.

Dip   The geological term for slope used in conjunction with strike. The slope of a plane is the angle the edge view of the plane makes with the edge view of the horizontal plane.

Dip direction   The direction a ball will roll when placed on a plane. It will roll perpendicular to the true-length line in the top view toward the low side (found in elevation view) of the plane. It is used in conjunction with dip.

Edge view of a plane   A view in which the given plane appears as a straight line. An edge view of a plane may be found by viewing a true-length line that lies within the plane as a point.

Elevation view   A view in which the lines of sight are level. Principal elevation views are front, left side, right side, and rear.

Fault   Where parallel strata are interrupted by fractures and one side of the bed shifts in relation to the other.

Fault plane   The plane of the fracture (fault).

Fault-plane outcrop   Where the fault plane intersects the earth's surface.

Fill   The addition of loose material, such as gravel and rock, to the terrain.

Foreshortened line   A line that appears shorter than its actual length.

Frontal line   A line that is parallel to the front projection plane. Its projection will be true length in the front view.

Frontal planes (F)   Vertical planes. The lines of sight for frontal planes are horizontal (perpendicular to the frontal planes).

Grade of a line   Another way to describe the inclination of a line in relation to the horizontal plane. The percent grade is the vertical rise divided by the horizontal run multiplied by 100.

Horizontal line   A line that is parallel to the horizontal plane. Its projection will appear true length in the top view.

Horizontal, or top view, planes   Level planes. The lines of sight for horizontal planes are vertical (perpendicular to the level planes).

Inclined auxiliary view   A view in which the lines of sight of the observer are neither vertical nor horizontal.

Inclined plane A plane that appears as an edge view in one view, but is not perpendicular to a principal view line of sight.

Index contour line Every fifth contour line, drawn darker and thicker for easier readability of a topographic map.

Intersecting lines When lines are intersecting, the point of intersection is a point that lies on both lines.

Isometric A pictorial view that shows three sides of an object—usually the top, front, and right side.

Line of sight An imaginary straight line from the eye of the observer to a point on the object being observed. All lines of sight for a particular view are assumed to be parallel and are perpendicular to the projection plane involved.

Line type Each line has its own line type or name. Whatever principal plane the line is true length in is the line type—frontal, profile, or horizontal. If a line is true length in one or more principal view, it is either a horizontal, or vertical line. If a line is not true length in a principal view it is oblique.

Normal plane A plane surface that is viewed true shape and size in any one of the principal projection planes.

Oblique line A straight line that is not parallel to any of the six principal planes.

Oblique plane A plane that does not appear as an edge view in a principal view.

Origin A starting point that may be located anywhere on the object or near it.

Orthographic projection (drawing) Right-angle projection. It is a method of drawing that uses parallel lines of sight at right angles (90°) to a projection plane.

Outcrop Used to describe where a stratum of rock or view of ore intersects the earth's surface.

Plane type The plane type is determined by the principal plane in which the plane is true shape.

Parallel lines Lines that are an equal distance from each other throughout their length.

Perpendicular lines Lines with a 90° angle between them.

Piercing point A point where a line intersects a plane.

Pitch A measure of slope, expressed as a ratio of vertical rise to 12 inches of horizontal span (run).

Plane A surface that is not curved or warped. It is a surface in which any two points may be connected by a straight line, and the straight line will always be completely within the surface.

Plastic box A six-sided transparent box. The sides represent projection planes at 90° to each other. These six projection planes are known as the principal planes and the lines of sight must always be perpendicular to them. The object is set inside of the plastic box with the majority of the surfaces parallel to the box sides.

Point view of a line Seen when the line of sight is parallel to the true length of the given line.

Primary auxiliary view Any auxiliary view that is adjacent to a primary or principal plane.

Primary or principal view The view represented by the six sides of the orthographic box. They include the top, front, right-side, left-side, rear, and bottom views.

Profile line A line that is parallel to the profile projection plane. Its projection appears true length in the side view.

Profile planes (P) Vertical planes ninety degrees to frontal planes. The lines of sight for profile planes are also in a horizontal (perpendicular to the profile planes).

Projection line A line parallel to the line of sight and perpendicular to the projection plane. It transfers the two-dimensional shape from the object to the projection plane, but rarely is shown on a finished drawing.

Projection plane A flat surface that the object is projected onto, such as paper, blackboard, or plastic box. Your (the observer's) lines of sight are always perpendicular to the projection plane.

Ravine If contour lines bulge toward higher contour elevations, a ravine, or water course, exists.

Reference line (Fold line or hinged line) edges of the plastic box or the intersection of the perpendicular planes. The reference line is only drawn when needed to aid in constructing additional views.

Related views Views adjacent to the same view, sharing a common dimension which must be transferred.

Revolution An alternate method for solving descriptive geometry problems in which the observer remains stationary and the object is rotated to obtain the various views of it.

Ridge If contour lines on a topographic map bulge toward lower contour elevations, a ridge is present.

Saddle A saddle is formed by two peaks side by side.

Secondary auxiliary views An auxiliary view that is adjacent to a primary auxiliary view.

Slope of a line The angle between a true-length line and a horizontal (level) plane.

Slope of a plane Also called the dip angle, the angle (in degrees) that the edge view of a plane makes with a horizontal plane.

Stratum of rock One of the layers of a series of parallel layers of stratified rock below the topsoil and loose rock which cover the earth's surface.

Stretch-out line A line drawn the length of the perimeter of the prism or cylinder a development is being created for.

Strike The bearing (compass reading) of a level or horizontal line within a plane. Strike is a geological term used to describe direction of various layers of the earth's crust.

Surface development The representation of the surfaces of a solid unfolded, or unrolled onto a flat plane. The important characteristics of surface developments is that all lines appear true length in the development.

Surfaces The exterior of an object.

Topographic map A map that represents the irregularities (contours) of the earth's surface in a single top view.

True-length line A line that appears in its actual and true length.

True-length diagram A diagram that rotates the elements away from the views, and project the true lengths to the front view.

True shape of a plane The actual shape and size of a plane surface. A plane which appears as an edge view in one view, and is perpendicular to a principal view line of sight, such as the horizontal or frontal planes, is a true-shape plane.

Vein Cracks in the earth's strata which are often filled with minerals, or ores. Also called lodes.

Vertex The highest point of the pyramid.

INDEX